T0226257

Lecture Notes in Computer Science   10743

Commenced Publication in 1973
Founding and Former Series Editors:
Gerhard Goos, Juris Hartmanis, and Jan van Leeuwen

More information about this series at http://www.springer.com/series/7407

B. S. Panda · Partha P. Goswami (Eds.)

Algorithms and Discrete Applied Mathematics

4th International Conference, CALDAM 2018
Guwahati, India, February 15–17, 2018
Proceedings

 Springer

Editors
B. S. Panda
Indian Institute of Technology Delhi
New Delhi
India

Partha P. Goswami
University of Calcutta
Kolkata
India

ISSN 0302-9743   ISSN 1611-3349   (electronic)
Lecture Notes in Computer Science
ISBN 978-3-319-74179-6   ISBN 978-3-319-74180-2   (eBook)
https://doi.org/10.1007/978-3-319-74180-2

Library of Congress Control Number: 2017963776

LNCS Sublibrary: SL1 – Theoretical Computer Science and General Issues

Printed on acid-free paper

This Springer imprint is published by Springer Nature
The registered company is Springer International Publishing AG
The registered company address is: Gewerbestrasse 11, 6330 Cham, Switzerland

Preface

This volume contains the papers presented at CALDAM 2018: the 4th International Conference on Algorithms and Discrete Applied Mathematics held during February 15–17, 2018, in Guwahati. CALDAM 2018 was organized by the Department of Computer Science and Engineering, Indian Institute of Technology, Guwahati. The conference had papers in the areas of algorithms, graph theory, codes, polyhedral combinatorics, computational geometry, and discrete geometry. The 68 submissions had authors from 12 different countries. Each submission received at least one detailed review and nearly all were reviewed by three Program Committee members. The committee decided to accept 23 papers. The program also included four invited talks by Andreas Brandstädt, Sathish Govindarajan, J. Mark Keil, and Miklós Simonovits.

The first CALDAM was held in February 2015 at the Indian Institute of Technology, Kanpur, and had 26 papers selected from 58 submissions from ten countries. The second edition was held in February 2016 at the University of Kerala, Thiruvananthapuram (Trivandrum), India, and had 30 papers selected from 91 submissions from 13 countries. The third edition was held in February 2017 at Birla Institute of Technology and Science, Pilani (BITS Pilani), K. K. Birla Goa Campus, Goa, India, and had selected 32 papers from 103 submissions from 18 countries.

We would like to thank all the authors for contributing high-quality research papers to the conference. We express our sincere thanks to the Program Committee members and the external reviewers for reviewing the papers within a very short period of time. We thank Springer for publishing the proceedings in the *Lecture Notes in Computer Science* series. We thank the invited speakers Andreas Brandstädt, Sathish Govindarajan, J. Mark Keil, and Miklós Simonovit for accepting our invitation. We thank the Organizing Committee chaired by R. Inkulu from the Indian Institute of Technology, Guwahati, for the smooth functioning of the conference. We thank the chair of the Steering Committee, Subir Ghosh, for his active help, support, and guidance throughout. We thank our sponsors Google Inc., Microsoft Research India, and the National Board of Higher Mathematics, Department of Atomic Energy, for their financial support. We also thank Springer for its support for the two Best Paper Presentation Awards. Last but definitely most importantly, we thank the EasyChair conference management system, which was very effective in handling the entire reviewing process.

December 2017

B. S. Panda
Partha P. Goswami

Organization

Program Committee

John Ebenezer Augustine	Indian Institute of Technology, Chennai, India
Amitabha Bagchi	Indian Institute of Technology, Delhi, India
Niranjan Balachandran	Indian Institute of Technology, Mumbai, India
Partha P Bhowmick	Indian Institute of Technology, Kharagpur, India
Boštjan Brešar	University of Maribor, Slovenia
Sunil Chandran	Indian Institute of Science, Bengaluru, India
Manoj Changat	University of Kerala, India
Sandip Das	Indian Statistical Institute, Kolkata, India
Ajit A. Diwan	Indian Institute of Technology, Mumbai, India
Zachary Frigstaad	University of Alberta
Sumit Ganguly	Indian Institute of Technology, Kanpur, India
Daya Gaur	University of Lethbridge, Canada
Partha P. Goswami	Institute of Radio Physics and Electronics, University of Calcutta, Kolkata, India
Sathish Govindarajan	Indian Institute of Science, Bengaluru, India
Subrahmanyam Kalyanasundaram	Indian Institute of Technology, Hyderabad, India
Gyula O. H. Katona	Alfred Renyi Institute of Mathematics, Hungary
Sandi Klavzar	University of Ljubljana, Slovenia
Ramesh Krishnamurti	Simon Fraser University, Canada
Van Bang Le	Universität Rostock, Germany
Andrzej Lingas	Lund University, Sweden
Anil Maheshwari	Carleton University, Canada
Kazuhisa Makino	Kyoto University, Japan
Bodo Manthey	University of Twente, The Netherlands
Rogers Mathew	Indian Institute of Technology, Kharagpur, India
Bojan Mohar	Simon Fraser University, Canada
Apurva Mudgal	Indian Institute of Technology, Ropar, India
N. S. Narayanaswamy	Indian Institute of Technology, Chennai, India
Sudebkumar Prasant Pal	Indian Institute of Technology, Kharagpur, India
B. S. Panda	Indian Institute of Technology, Delhi, India
Abraham P. Punnen	Simon Fraser University, Canada
Venkatesh Raman	The Institute of Mathematical Sciences, Chennai, India
Günter Rote	Freie Universität, Berlin, Germany
Michiel Smid	Carleton University, Canada
C. R. Subramanian	The Institute of Mathematical Sciences, Chennai, India
Ambat Vijayakumar	Cochin University of Science and Technology, India
Alexander Wolff	Universität Würzburg, Germany

Organizing Committee

Santosh Biswas	Indian Institute of Technology, Guwahati
Gautam K. Das	Indian Institute of Technology, Guwahati
R. Inkulu (Chair)	Indian Institute of Technology, Guwahati
Deepanjan Kesh	Indian Institute of Technology, Guwahati
Pinaki Mitra	Indian Institute of Technology, Guwahati
S. V. Rao	Indian Institute of Technology, Guwahati

Steering Committee

Subir Kumar Ghosh (Chair)	Ramakrishna Mission Vivekananda University, India
János Pach	École Polytechnique Fédérale De Lausanne (EPFL), Lausanne, Switzerland
Nicola Santoro	School of Computer Science, Carleton University, Canada
Swami Sarvattomananda	Ramakrishna Mission Vivekananda University, India
Peter Widmayer	Institute of Theoretical Computer Science, ETH Zurich, Switzerland
Chee Yap	Courant Institute of Mathematical Sciences, New York University, USA

Additional Reviewers

Aravind, N. R.	Kern, Walter	Pradhan, D.
Basavaraju, Manu	Khodamoradi, Kamyar	Pratihar, Sanjoy
Benkoczi, Robert	Kowaluk, Miroslaw	Ramaswamy, Krithika
Bera, Sahadev	Kryven, Myroslav	Ray Chaudhury, Baskar
Biswas, Ranita	Lahiri, Abhiruk	Roy, Bodhayan
Chakraborty, Suvradip	Levcopoulos, Christos	Sarnovsky, Martin
Chaplick, Steven	Lipp, Fabian	Sen, Sagnik
Cheung, Yun Kuen	M. A., Shalu	Simon, Sunil
Das, Bireswar	Majumdar, Diptapriyo	Singh, Rishi
Dijk, Thomas C. Van	Molla, Anisur Rahaman	Soto, Mauricio
Dolžan, David	Moses Jr., William K.	Spoerhase, Joachim
Francis, Mathew	Nandakumar, Satyadev	Togni, Olivier
González Yero, Ismael	Nath, Swaprava	Tripathi, Utkarsh
Iranmanesh, Ehsan	Padinhatteeri, Sajith	Vaishali, S.
Issac, Davis	Pal, Shyamosree	Viglietta, Giovanni
Iyer, Venkitesh	Pandey, Arti	
Johansson, Thomas	Panigrahi, Pratima	

Abstracts of Invited Talks

Efficient Domination and Efficient Edge Domination: A Brief Survey

Andreas Brandstädt

Institut für Informatik, Universität Rostock, 18051 Rostock, Germany
andreas.brandstaedt@uni-rostock.de

Abstract. In a finite undirected graph $G = (V, E)$, a vertex $v \in V$ *dominates* itself and its neighbors in G. A vertex set $D \subseteq V$ is an *efficient dominating set* (*e.d.s.* for short) of G if every $v \in V$ is dominated in G by exactly one vertex of D.

The *Efficient Domination* (ED) problem, which asks for the existence of an e.d.s. in G, is known to be $\mathbb{NP}$-complete for bipartite graphs, for (very special) chordal graphs and for line graphs but solvable in polynomial time for many subclasses. For H-free graphs, a dichotomy of the complexity of ED has been reached.

An edge set $M \subseteq E$ is an *efficient edge dominating set* (*e.e.d.s.* for short) of G if every $e \in E$ is dominated in G by exactly one edge of M with respect to the line graph $L(G)$. Thus, M is an e.e.d.s. in G if and only if M is an e.d.s. in $L(G)$. An e.e.d.s. is called *dominating induced matching* in various papers.

The *Efficient Edge Domination* (EED) problem, which asks for the existence of an e.e.d.s. in G, is known to be $\mathbb{NP}$-complete even for special bipartite graphs but solvable in polynomial time for various graph classes.

The problems ED and EED are based on the $\mathbb{NP}$-complete Exact Cover problem on hypergraphs.

The Use of Dynamic Programming in Intersection Graphs

J. Mark Keil

Department of Computer Science, University of Saskatchewan, Canada

Abstract. The intersection graph of a family $\mathcal{F}$ of sets is the graph having $\mathcal{F}$ as the node set with two elements of $\mathcal{F}$ adjacent in the graph if and only if their intersection is nonempty. For example, the intersection graphs of subtrees of a tree are the chordal graphs. A graph G is a geometric intersection graph if G is the intersection graph of a set of geometric objects. If the geometric objects are intervals of the real line, then interval graphs are formed. String graphs are the intersection graphs of curves in the plane and they are among the most general geometric intersection graphs that have been studied. String graphs are a superclass of planar graphs, chordal graphs, and circle graphs. The restriction that each string touches the infinite face of the plane results in the class of outerstring graphs.

Let $G = (V, E)$ be an undirected graph with n nodes and m edges. For two specified nodes s and t in V, the k most vital nodes in G are those $k, (1 \leq k \leq n - 2)$ nodes whose removal maximizes the increase in the length of the shortest path from s to t. The problem of identifying the k most vital nodes was defined by Corley and Sha [3] in 1982 as a way to identify locations in a network that may need to be reinforced against an interdictor or a natural disaster, and shown to be NP-complete by Bar-Noy, Khuller and Schieber in 1995 [1]. In this talk I will describe polynomial time dynamic programming algorithms for the k most vital nodes problem for some classes of intersection graphs, namely interval graphs, chordal graphs, permutation graphs and interval bigraphs. This is joint work with Leizhen Cai [2].

I will also describe a dynamic programming algorithm to the maximum weight independent set problem in an outerstring graph which is polynomial in the size of the geometric input representation of the graph. This result is joint work with D. Pradhan, J. Mitchell and M. Vatshelle [4].

References

1. Bar-Noy, A., Khuller, S., Schieber, B.: The complexity of finding most vital arcs and nodes. Techinical report CS-TR-3539, University of Maryland (1995)
2. Cai, L., Keil, J.M.: Finding the most vital nodes in classes of intersection graphs (2018, submitted)
3. Corley, H.W., Sha, D.Y.: Most vital links and nodes in weighted networks. Oper. Res. Lett. **1**, 157–160 (1982)
4. Keil, J.M., Mitchell, J.S.B., Pradhan, D., Vatshelle, M.: An algorithm for the maximum weight independent set problem on outerstring graphs. Comput. Geom. Theory Appl. **60**, 19–25 (2017)

Extremal Graph Theory, Stability, and Anti-Ramsey Theorems

Miklós Simonovits

Alfréd Rényi Mathematical Institute of the Hungarian Academy of Sciences, Budapest

Extremal graph theory is one of the most developed branches of Discrete Mathematics. Stability methods introduced by the author [5] are very successful to prove sharps results in this field. We shall give some illustration of this method for graphs, hypergraphs, (among others, the Füredi-Simonovits and Füredi-Pikhurko-Simonovits theorems obtained by the Stability methods). We shall also apply stability methods to Anti-Ramsey problems. An ANTI-RAMSEY problem is where a sample graph L is fixed and we colour, e.g., the edges of a complete graph K_n without having a copy of L in which all the edges have distinct colours.

Several problems in combinatorics can be reduced to extremal graph problems. Erdős, Simonovits and Sós [4] basically reduced certain ANTI-RAMSEY problems to extremal graph problems. Some others, like the problem of $L = C_k$ were much more difficult.

In the lecture we shall also consider Dual ANTI-RAMSEY problems, coming from Theoretical Computer Science. Burr, Erdős, Graham and T. Sós [1] defined and investigated a *dual* variant of the ANTI-RAMSEY problems. Some of their results also can be found in a second paper joint with Peter Frankl [2]. As they pointed out, one of the most interesting cases they could not settle was that of C_5.

The dual Anti-Ramsey problem. Let us fix a sample graph L, and consider a (variable) graph G_n on n vertices, with

$$e = e(G_n) > \mathrm{ex}(n, L)$$

edges. Let $\chi_S(G_n, L)$ denote the *minimum* number of colours needed to colour the edges of G_n so that no $L \subseteq G_n$ has two edges of the same colour. Determine

$$\chi_S(n, e, L) := \min\{\chi_S(G_n, L) : e(G_n) = e\}.$$

Here we improve several results of [1] and [2]. We shall prove, among others, that if a graph G_n has $e = \lfloor \frac{1}{4} n^2 \rfloor + 1$ edges and we colour its edges so that every $C_5 \subset G_n$ is 5–coloured, then we have to use at least $\lfloor \frac{n}{2} \rfloor + 3$ colours, if n is sufficiently large. This result is sharp.

Theorem 1 (Erdős-Simonovits). *There exists a threshold n_0 such that if $n > n_0$, and a graph G_n has $\lfloor \frac{1}{4} n^2 \rfloor + 1$ edges and we colour its edges so that every C_5 is 5–coloured, then we have to use at least $\lfloor \frac{n}{2} \rfloor + 3$ colours.*

Theorem 2. *There exists a function $\vartheta(n) \to \infty$ such that if $0 < k = \binom{h}{2} < \vartheta(n)$, then the upper bound of Theorem 4.2/[1] is sharp for $e = \left\lfloor \frac{1}{4} n^2 \right\rfloor + k$:*

$$\chi_S(n, e, C_5) = (h+1) \left\lfloor \frac{n}{2} \right\rfloor + k.$$

Because of the monotonicity, this implies

Theorem 3. *There exists a function $\vartheta(n) \to \infty$ such that if $0 < k \leq \binom{h}{2} < \vartheta(n)$, then for $e = \left\lfloor \frac{1}{4} n^2 \right\rfloor + k$,*

$$\chi_S(n, e, C_5) = (h+1) \left\lfloor \frac{n}{2} \right\rfloor + k + O(\sqrt{k}).$$

We have several further results in this area. Altogether, mostly we restrict ourselves here to the simplest versions of our results.

This lecture is partly based on a manuscript of Erdős and Simonovits [3] from the late 1980's.

References

1. Burr, S.A., Erdős, P., Graham, R.L., Sós, V.T.: Maximal antiramsey graphs and the strong chromatic number. J. Graph Theory **13**(3), 163–182 (1989)
2. Burr, S., Erdős, P., Frankl, P., Graham, R.L., Sós, V.T.: Further results on maximal Anti–Ramsey graphs. Proc. Kalamazoo Combin. Conf. 193–206 (1989)
3. Erdős, P., Simonovits, M.: How many colours are needed to colour every pentagon of a graph in five colours? (manuscript, under publication)
4. Erdős, P., Simonovits, M., Sós, V.T.: Anti-Ramsey theorems, infinite and finite sets (Colloq., Keszthely, 1973; dedicated to P. Erdős on his 60th birthday), vol. II, pp. 633–643, Colloq. Math. Soc. János Bolyai, vol. 10, North-Holland, Amsterdam (1975)
5. Simonovits, M.: A method for solving extremal problems in graph theory. In: Erdős, P., Katona, G. (eds.): Theory of graphs (Proceedings Colloquium, Tihany (1966)), pp. 279–319. Academic Press, NY (1968)

Contents

Efficient Domination and Efficient Edge Domination: A Brief Survey

Andreas Brandstädt[(✉)]

Institut für Informatik, Universität Rostock, 18051 Rostock, Germany
andreas.brandstaedt@uni-rostock.de

Abstract. In a finite undirected graph $G = (V, E)$, a vertex $v \in V$ *dominates* itself and its neighbors in G. A vertex set $D \subseteq V$ is an *efficient dominating set* (*e.d.s.* for short) of G if every $v \in V$ is dominated in G by exactly one vertex of D.

The *Efficient Domination* (ED) problem, which asks for the existence of an e.d.s. in G, is known to be NP-complete for bipartite graphs, for (very special) chordal graphs and for line graphs but solvable in polynomial time for many subclasses. For H-free graphs, a dichotomy of the complexity of ED has been reached.

An edge set $M \subseteq E$ is an *efficient edge dominating set* (*e.e.d.s.* for short) of G if every $e \in E$ is dominated in G by exactly one edge of M with respect to the line graph $L(G)$. Thus, M is an e.e.d.s. in G if and only if M is an e.d.s. in $L(G)$. An e.e.d.s. is called *dominating induced matching* in various papers.

The *Efficient Edge Domination* (EED) problem, which asks for the existence of an e.e.d.s. in G, is known to be NP-complete even for special bipartite graphs but solvable in polynomial time for various graph classes. The problems ED and EED are based on the NP-complete Exact Cover problem on hypergraphs.

1 Exact Cover for Hypergraphs and Maximum Weight Independent Set for Graphs

A hypergraph $H = (V, \mathcal{E})$ has a finite vertex set V and for all $e \in \mathcal{E}$, $e \subseteq V$ ($\mathcal{E}$ possibly being a multiset). A *packing* (also called *matching*) in a hypergraph H is a subset of pairwise disjoint hyperedges, and a *covering* of H is a subset of edges whose union is V.

A subset of hyperedges $\mathcal{E}' \subseteq \mathcal{E}$ is an *exact cover* of H if for all $e, f \in \mathcal{E}'$ with $e \neq f$, $e \cap f = \emptyset$ and $\bigcup \mathcal{E}' = V$. In other words, $\mathcal{E}'$ is an exact cover of H if and only if it is a partition of V, i.e., $\mathcal{E}'$ is a packing and a covering of H. Clearly, not every hypergraph has an exact cover. The EXACT COVER problem asks for the existence of an exact cover in a given hypergraph H. As part of his famous list of 21 problems, Karp [56] showed that this problem is NP-complete; it is NP-complete even for 3-uniform hypergraphs, i.e., every hyperedge has size 3 (problem X3C [SP2] in [47]). For 2-uniform hypergraphs, Exact Cover corresponds to Perfect Matching for graphs.

© Springer International Publishing AG 2018
B. S. Panda and P. P. Goswami (Eds.): CALDAM 2018, LNCS 10743, pp. 1–14, 2018.
https://doi.org/10.1007/978-3-319-74180-2_1

Let $L(H) = (\mathcal{E}, F)$ denote the *line graph* of H, i.e., the graph with the hyperedges $\mathcal{E}$ of H as its nodes such that for any two hyperedges $e, e' \in \mathcal{E}$, $e \neq e'$, $ee' \in F$ if and only if $e \cap e' \neq \emptyset$. Clearly, for $\mathcal{E}' \subseteq \mathcal{E}$, $\mathcal{E}'$ is a packing in H if and only if $\mathcal{E}'$ is an independent set of nodes in $L(H)$.

The Exact Cover problem for hypergraphs can be reduced to the Maximum Weight Independent Set (MWIS) problem for graphs by the following weight function $w(e) := |e|$ of the hyperedges $e \in \mathcal{E}$. Let $\alpha_w(G)$ denote the maximum weight of an independent vertex set in G. Obviously, $\alpha_w(L(H)) \leq |V|$ and we have:

Lemma 1. *A packing $\mathcal{E}'$ in hypergraph H is an exact cover of H if and only if $\Sigma_{e \in \mathcal{E}'} w(e) = |V|$, that is, $\alpha_w(L(H)) = |V|$.*

This means that Exact Cover can be solved in polynomial time for any class of hypergraphs where MWIS can be solved in polynomial time for the line graphs of this class. By Lovász [73], a hypergraph H is *normal* if it has the Helly property and its line graph is perfect. By the famous result of Grötschel et al. [52], MWIS is solvable in polynomial time for perfect graphs. Thus, Exact Cover is polynomial for normal hypergraphs.

MWIS plays an important role as well for Efficient Domination and Efficient Edge Domination as described in the subsequent sections. Thus we collect some of these results as follows:

Theorem 1. *MWIS is solvable in polynomial time for*

(i) perfect graphs [52],
(ii) chordal graphs [44] (in linear time),
(iii) weakly chordal graphs [53,54] (in time $\mathcal{O}(n^4)$),
(iv) asteroidal-triple-free (AT-free) graphs [24,59] (in time $\mathcal{O}(n^3)$),
(v) circular-arc graphs (in time $\mathcal{O}(nm)$),
(vi) interval-filament graphs [48].

Frank's result in Theorem 1 *(ii)* is based on a perfect elimination ordering of the given chordal graph. For a circular-arc graph $G = (V, E)$, the subgraph of G induced by the non-neighborhood of each vertex $v \in V$ is an interval graph and thus chordal. This obviously leads to Theorem 1 *(v)*.

A special case of normal hypergraphs are *hypertrees* which have the Helly property and whose line graphs are chordal. Thus, by Theorem 1 *(ii)*, Exact Cover is solvable in polynomial time for hypertrees; in [16], we have shown that Exact Cover is $\mathbb{NP}$-complete for the dual variant of hypertrees, namely α-acyclic hypergraphs.

2   Efficient Domination and Efficient Edge Domination

2.1   Efficient Domination

Let $G = (V, E)$ be a finite undirected graph without loops and multiple edges; let $|V| = n$ and $|E| = m$. A vertex $v \in V$ *dominates* itself and its neighbors.

A vertex subset $D \subseteq V$ is an *efficient dominating set* (*e.d.s.* for short) of G if every vertex of G is dominated by exactly one vertex in D, that is $|D \cap N[v]| = 1$ for every $v \in V$ (where $N[v]$ denotes the closed neighborhood of x). Clearly, not every graph has an e.d.s.; the EFFICIENT DOMINATING SET (ED) problem asks for the existence of an e.d.s. in a given graph G. Clearly, ED for graph G corresponds to the Exact Cover problem for the closed neighborhood hypergraph $\mathcal{N}(G)$ of G.

The notion of efficient domination was introduced by Biggs [3] under the name *perfect code*. Efficient dominating sets are also called *independent perfect dominating sets* in various papers (see e.g. [36,85]) and *perfect dominating sets* in [42].

The ED problem is motivated by various applications, including coding theory and resource allocation in parallel computer networks; see e.g. [1–3,36,65,66,70, 76,78,84,85]. Bange et al. [1] showed that the ED problem is NP-complete (see also [2]). Moreover, using a standard reduction from the Exact Cover problem, it was shown that ED is NP-complete for $2P_3$-free (and thus, for P_7-free) chordal unipolar graphs [41,81,85] and for bipartite graphs. In [76], it was shown that ED is NP-complete for chordal bipartite graphs.

ED is NP-complete for planar bipartite graphs [76] and for planar graphs with maximum degree at most 3 [42,61], and is NP-complete for 3-regular graphs [62]. In [28], this is extended to p-regular graphs, $p > 3$. Moreover, for every $g \geq 3$, the ED problem is NP-complete for planar bipartite graphs of maximum degree 3 with girth at least g [11,17,79].

If a vertex weight function $w : V \rightarrow \mathbb{N} \cup \{\infty\}$ is given, the MINIMUM WEIGHT EFFICIENT DOMINATING SET (WED) problem asks for a minimum weight e.d.s. in G, if there is one, or for determining that G has no e.d.s. Since negative weights are allowed, the Minimum Weight Efficient Dominating Set problem is equivalent to the Maximum Weight Efficient Dominating Set problem; subsequently we restrict the problem to the minimum weight version.

The vertex weight ∞ plays a special role; vertices which are definitely not in an e.d.s. D get weight ∞, and thus, in the WED problem we are asking for an e.d.s. of finite minimum weight.

The property of having an e.d.s. is not hereditary. In [77,78], Milanič [78] characterized the graph class where every induced subgraph has an e.d.s.

2.2   Efficient Edge Domination

For a graph $G = (V, E)$, a subset $M \subseteq E$ is an *induced matching in G* [25] if the pairwise distance of edges in M is at least 2, i.e., for $e, e' \in M$ with $e \neq e'$, $e \cap e' = \emptyset$ and there is no edge between e and e'. In other words, the induced subgraph $G[V(M)]$ is the disjoint union of edges. In [45,49], an induced matching is called *strong matching of G*, and in [82], the notion was originally introduced under the name *risk-free marriage*.

The MAXIMUM INDUCED MATCHING (MIM) problem asks for an induced matching of maximum cardinality in G. In [82], the more general case of *δ-separated* *matchings* was considered; a 2-separated matching is the same as an induced matching. While it is well known that the maximum matching problem is solvable in polynomial time (based on Jack Edmonds' results [40]), Stockmeyer and Vazirani [82] showed that for every $\delta \geq 2$, Maximum δ-Separated Matching (including the MIM problem) is NP-complete for bipartite graphs of maximum degree 4.

A subset $M \subseteq E$ of edges is an *efficient edge dominating set* (*e.e.d.s.* for short) of $G = (V, E)$ if with respect to the line graph $L(G)$, every $e \in E$ is dominated in $L(G)$ by exactly one edge of M. Thus, we have:

$$M \text{ is an e.e.d.s. in } G \Longleftrightarrow M \text{ is an e.d.s. in } L(G). \tag{1}$$

In various papers such as [13,29,30,55], an e.e.d.s. is called *dominating induced matching* (*d.i.m.*); the main reason for that is the following obvious fact:

Observation 1. *M is an e.e.d.s. in G if and only if M is an induced matching in G such that $V \setminus V(M)$ is an independent vertex set in G.*

Clearly, not every graph (in fact, not every tree) has an e.e.d.s.; the EFFICIENT EDGE DOMINATING SET (EED) problem asks for the existence of an e.e.d.s. in a given graph G. By Observation 1, the EED problem is also called DIM. Grinstead et al. [51] showed that EED is NP-complete. This implies that the ED problem is NP-complete for line graphs. Actually, ED is NP-complete even for line graphs of bipartite graphs [75].

See [13,18,29,74,75] for various other NP-completeness results for EED. In [79], it is shown that EED is NP-complete for planar bipartite graphs with maximum degree at most 3 and girth at least g for every fixed g. Moreover, for each $p \geq 3$, the EED problem is NP-complete for p-regular graphs [28].

3 Complexity of Efficient Domination

3.1 A Dichotomy for H-free Graphs

For a subset $U \subseteq V$, let $G[U]$ denote the *induced subgraph* of G with vertex set U. For a graph H, a graph G is *H-free* if G does not contain any induced subgraph isomorphic to H. Let P_k denote a chordless path with k vertices. $H + H'$ denotes the disjoint union of graphs H and H'; for example, $2P_3$ denotes $P_3 + P_3$. H is a *linear forest* if H is claw-free and C_k-free for every $k \geq 3$, that is, H is the disjoint union of chordless paths. Recall that ED is NP-complete for chordal graphs, bipartite graphs, and claw-free graphs. Thus, whenever H contains C_k, $k \geq 3$, or claw then ED is NP-complete for H-free graphs.

For linear forests H, recall that ED is NP-complete for $2P_3$-free graphs. Thus, if two of the components of H contain P_3 or one of its components contains $2P_3$ (such as P_7) then ED is NP-complete for H-free graphs. Moreover, ED is solvable in linear time for $2P_2$-free graphs [17], and for P_5-free graphs [4,20], and if ED is polynomial for H-free graphs then it is polynomial for $(H + kP_2)$-free graphs for every fixed k [10].

The complexity of ED for P_6-free graphs was the last open question for H-free graphs [10]; it was the main open question in [17]. As partial results, based on [46], ED was solved in polynomial time for P_6-free chordal graphs [7], and in [14,57], for some further subclasses of P_6-free graphs.

Recently, it has been shown by Lokshtanov, Pilipczuk and van Leeuwen [71] that ED is solvable in polynomial time for P_6-free graphs (the time bound is more than $\mathcal{O}(n^{500})$). Their result for ED is based on their quasi-polynomial algorithm for the MWIS problem on P_6-free graphs. Independently, in [19,20] we found a polynomial time solution for ED on P_6-free graphs using a direct approach.

Theorem 2 ([19,20,71]). *For P_6-free graphs, the WED problem is solvable in polynomial time.*

Thus Theorem 2 finally lead to a dichotomy for the ED problem on H-free graphs. Our approach in [19,20] is simpler than the one in [71] and leads to the better time bound $\mathcal{O}(n^5 m)$.

In [22], the complexity of ED for H-free chordal graphs is analyzed (without reaching a dichotomy).

3.2 Further Polynomial Time Results for ED

In many papers, polynomial time or even linear time algorithms for the weighted ED (and consequently the ED) problem on special graph classes were found:

- trees [84],
- co-comparability graphs [31,36],
- split graphs [34] (linear time),
- interval graphs [35,36] and on their superclasses AT-free graphs [9], dually chordal graphs [9] (linear time), and circular-arc graphs [35],
- permutation graphs [65],
- trapezoid graphs [65,66],
- bipartite permutation graphs [76] (linear time),
- convex bipartite graphs [16] and on their superclass interval bigraphs [16],
- hereditary efficiently dominatable graphs [78],
- block graphs [85], distance-hereditary graphs [76] (linear time), and, more generally, graphs of bounded clique-width [38].

Some of these results can be reached by using the G^2 approach described in the next section.

3.3 Solving Efficient Domination for G via MWIS for G^2

The *square* of a graph $G = (V, E)$ is the graph $G^2 = (V, E^2)$ such that $uv \in E^2$ if and only if $d_G(u, v) \in \{1, 2\}$. It is easy to see that the dual $\mathcal{N}(G)^*$ of the closed neighborhood hypergraph $\mathcal{N}(G)$ is $\mathcal{N}(G)$ itself, and for any graph G, we have:

$$G^2 \text{ is isomorphic to the line graph } L(\mathcal{N}(G)). \qquad (2)$$

In [16,64,77,78], the following relationship between the ED problem on a graph G, the minimum weight dominating set problem on G and the MWIS problem on G^2 is used. For this, we need the following notions: For $G = (V, E)$ and $v \in V$, let $w(v) := |N_G[v]| = deg_G(v) + 1$, and for $D \subseteq V$, let $w(D) := \Sigma_{d \in D} w(d)$. Obviously, the following holds:

Proposition 1. *Let $G = (V, E)$ be a graph and $D \subseteq V$.*

(i) If D is a dominating vertex set in G then $w(D) \geq |V|$.
(ii) If D is an independent vertex set in G^2 then $w(D) \leq |V|$.

Recall Lemma 1 for e.d.s. D as the exact cover of $\mathcal{N}(G)$.

Lemma 2. *Let $G = (V, E)$ be a graph and $w(v) := |N[v]|$ a vertex weight function for G. Then the following are equivalent for any subset $D \subseteq V$:*

(i) D is an efficient dominating set in G.
(ii) D is a minimum weight dominating set in G with $w(D) = |V|$.
(iii) D is a maximum weight independent set in G^2 with $w(D) = |V|$.

Thus, the ED problem on a graph class $\mathcal{C}$ can be reduced to the MWIS problem on the squares of graphs in $\mathcal{C}$. Let ω denote the matrix multiplication exponent; by [83], $\omega < 2.2737$.

Theorem 3 ([16,78]). *Let $\mathcal{C}$ be a graph class for which the MWIS problem is solvable in time $T(|G|)$ on squares of graphs from $\mathcal{C}$. Then the ED problem is solvable on graphs in $\mathcal{C}$ in time $\mathcal{O}(\min(n^\omega, nm + n) + T(|G^2|))$.*

In [9], Theorem 3 is extended to the weighted version of ED (with the same time bound).

Squares of circular-arc graphs are circular-arc graphs [80], and correspondingly for trapezoid graphs [43]. Since trapezoid graphs are AT-free, by Theorem 1 (iv) and (v), WED is solvable in polynomial time for these graph classes.

An important example is the class of dually chordal graphs: In [76], the complexity of ED for strongly chordal graphs was mentioned as an open problem. Actually, a graph G is strongly chordal if and only if every induced subgraph of G is dually chordal, that is, strongly chordal graphs are the hereditarily dually chordal graphs [6].

Theorem 4 ([5,6,39]). *Let G be a graph and H be a hypergraph.*

(i) G is dually chordal if and only if $\mathcal{N}(G)$ is a hypertree.
(ii) G is dually chordal if and only if G^2 is chordal and $\mathcal{N}(G)$ has the Helly property.
(iii) If H is α-acyclic then its line graph $L(H)$ is dually chordal.
(iv) If H is a hypertree then its 2-section graph $2sec(H)$ is dually chordal.

Recall Theorem 1 (ii). Thus, ED is solvable in polynomial time for every graph class $\mathcal{C}$ such that for $G \in \mathcal{C}$, G^2 is chordal. In particular, for dually chordal graphs, ED is solvable in polynomial time. In [16], we refined this approach by avoiding the explicit construction of G^2: For dually chordal graphs, and thus, also for strongly chordal graphs, the ED problem is solvable in linear time. In [9], this is extended to linear time for WED on the same graph classes.

Another important example is the class of AT-free graphs: In [33], it is shown that for AT-free graphs G, G^2 is a co-comparability graph (which is AT-free).

Recall Theorem 1 (iv). Thus for AT-free graphs, the WED problem is solvable in polynomial time. This extends the result of [36] showing that WED is polynomial for co-comparability graphs.

Finally we mention an interesting result about P_6-free graphs with e.d.s. by Friese [46] (see [7,8]): If G is P_6-free and has an e.d.s. then G^2 is hole-free (i.e., C_k-free for every $k \geq 5$). Friese's conjecture is that in this case, G^2 is also odd-antihole-free which, by the Strong Perfect Graph Theorem [37] would imply that G^2 is perfect (which would be another approach for solving ED in polynomial time for P_6-free graphs).

4 Complexity of Efficient Edge Domination

4.1 Direct Approaches

Recall that by Observation 1, EED and DIM are equivalent; in this section, we use the notions of d.i.m. and the DIM problem. In [29,74], the complexity of the DIM problem for weakly chordal graphs was mentioned as an open problem. Using a direct approach, we showed in [13] that DIM is solvable in polynomial time for weakly chordal graphs.

Theorem 5 ([13]). *For weakly chordal graphs, the DIM problem can be solved in polynomial time.*

Details of the direct approach are the following facts: An edge $e \in E$ is *forced* if it is contained in every d.i.m. of $G = (V, E)$.

Observation 2 ([13]). *Let M be a d.i.m. in G.*

(i) M contains at least one edge of every odd cycle C_{2k+1} in G, $k \geq 1$, and exactly one edge of every odd cycle C_3, C_5, C_7 of G.
(ii) No edge of any C_4 can be in M.
(iii) For each C_6 either exactly two or none of its edges are in M.

As a consequence, the clique K_4 of size 4 (and any graph containing K_4) has no d.i.m., the mid-edge of diamond and the two peripheral edges of butterfly are forced edges, and gem has no d.i.m. Thus graph G can be reduced such that it is $(K_4, \text{diamond, butterfly})$-free. In particular, the treewidth of chordal graphs having a d.i.m. is bounded which implies a linear time algorithm for DIM on chordal graphs.

Moreover, long anti-holes (i.e., $\overline{C_k}$, $k \geq 6$) have no d.i.m., i.e., a hole-free graph having an d.i.m. is weakly chordal. Thus, the polynomial-time solution for weakly chordal graphs implies that for hole-free graphs, the DIM problem is solvable in polynomial time (in [13], a direct approach for this was described):

Corollary 1. *DIM can be solved in polynomial time for hole-free graphs.*

An improved time bound $\mathcal{O}(nm)$ for Minimum Weight DIM on hole-free graphs is given in [79].

For indices $i, j, k \geq 0$, let $S_{i,j,k}$ denote the graph with vertices $u, x_1, \ldots, x_i$, $y_1, \ldots, y_j, z_1, \ldots, z_k$ such that the subgraph induced by $u, x_1, \ldots, x_i$ forms a P_{i+1} $(u, x_1, \ldots, x_i)$, the subgraph induced by $u, y_1, \ldots, y_j$ forms a P_{j+1} $(u, y_1, \ldots, y_j)$, and the subgraph induced by $u, z_1, \ldots, z_k$ forms a P_{k+1} $(u, z_1, \ldots, z_k)$, and there are no other edges in $S_{i,j,k}$. Thus, *claw* is $S_{1,1,1}$, and P_k is isomorphic to e.g. $S_{0,0,k-1}$.

Recall Observation 1: G has a d.i.m. if and only if V has a partition into an independent vertex set I and the vertex set of an induced matching.

Thus every graph having a d.i.m. is monopolar. In [29], this partition is called *black-white* partition, i.e., $V(G)$ is partitioned into W (the *white* vertices) and B (the *black* vertices) such that W is an independent set and $G[B]$ is a 1-regular subgraph. Based on this partition, a polynomial-time algorithm for DIM on claw-free graphs is found in [29].

For various graph classes defined by forbidden $S_{i,j,k}$, using a direct approach, DIM is solvable in polynomial time:

Theorem 6. *DIM is solvable in polynomial time for H-free graphs if H is:*

(i) $S_{1,1,1}$ [29]
(ii) $S_{1,2,3}$ [60]
(iii) $S_{2,2,2}$ [55]
(iv) $S_{1,2,4}$ [23]
(v) P_7 [18] *(even in linear time)*
(iv) P_8 [21]

Note that P_7 is isomorphic to $S_{0,3,3}$, and $S_{1,2,4}$ contains P_7 as an induced subgraph. The results *(i)*–*(iii)* are done via the black-white approach (the proof for result *(iii)* is very long and technical) while in *(iv)*, this black-white approach is combined with the direct approach done in [18,21]. In [67,68], DIM is solved in time $\mathcal{O}(n)$ for $S_{1,1,1}$-free graphs.

In [55], it is conjectured that for every fixed i, j, k, DIM is solvable in polynomial time for $S_{i,j,k}$-free graphs; this also includes P_k-free graphs for $k \geq 9$.

4.2   Solving Efficient Edge Domination for G via MWIS for $L(G)^2$

Recall Lemma 2 and the fact that M is an e.e.d.s. of G if and only if M is an e.d.s. of $L(G)$. Since the EED problem for graph G corresponds to the ED problem for $L(G)$, EED for G can be reduced to MWIS for $L(G)^2$. There are various examples of graph classes $\mathcal{C}$ where for any $G \in \mathcal{C}$ we have $L(G)^2 \in \mathcal{C}$ (most of these results were motivated by the Maximum Induced Matching (MIM) problem). The class of chordal graphs is the first example:

Theorem 7 ([25]). *If G is*

(i) *chordal then so is $L(G)^2$ [25].*
(ii) *HHD-free (HHDA-free, respectively) then so is $L(G)^2$ [63].*
(iii) *weakly chordal then so is $L(G)^2$ [27].*
(iv) *a circular-arc graph then so is $L(G)^2$ [49].*
(v) *a co-comparability graph then so is $L(G)^2$ [50].*
(vi) *an interval-filament graph then so is $L(G)^2$ [26].*
(vii) *AT-free then so is $L(G)^2$ [26,32].*
(viii) *P_5-free then so is $L(G)^2$ [58].*

Recall Theorem 1 (*iii*). Thus, by Theorem 7 (*iii*), EED can be solved in polynomial time for weakly chordal graphs (and the "open problem" mentioned in [29,74] was solved already by this approach).

For circular-arc graphs, recall Theorem 1 (*v*). Thus, by Theorem 7 (*iv*), EED is polynomial for circular-arc graphs.

Since interval graphs and polygon-circle graphs [26] are subclasses of interval-filament graphs, Gavril's result in Theorem 1 (*vi*) implies by Theorem 7 (*vi*) that EED is polynomial for interval-filament graphs.

Co-comparability graphs, permutation graphs and trapezoid graphs are AT-free (see e.g. [15]). Thus, by Theorem 1 (*iv*) and Theorem 7 (*vii*), EED is polynomial for AT-free graphs.

By the result of [72], (*viii*) implies that MIM can be solved in polynomial time for P_5-free graphs; for DIM, the direct approach however is much simpler.

For line graphs, the complexity of DIM and MIM is different: While Kobler and Rotics [58] showed that MIM is NP-complete for line graphs, Cardoso et al. [29] showed that DIM is polynomial even for claw-free graphs and more general cases - see Theorem 6.

In [69], a linear-time algorithm for EED on circular-arc graphs is given.

MIM is solvable in linear time for chordal graphs [12]. In contrast to this, MIM is NP-complete for dually chordal graphs [16].

By Theorem 7 (*i*), EED is solvable in polynomial time for chordal graph but recall that EED is solvable in linear time for chordal graphs [74] (and by bounded treewidth).

By Observation 2, a graph having an e.e.d.s. is gem-free and W_4-free (W_4 is a C_4 plus a universal vertex). This allows us to solve the EED problem for dually chordal graphs using the following lemma:

Lemma 3 ([16]). *If G is a graph with an e.e.d.s. then G is chordal if and only if G is dually chordal.*

Corollary 2 ([16]). *EED is solvable in linear time for dually chordal graphs.*

In [67,68], DIM is solved in time $\mathcal{O}(n)$ for dually chordal graphs. The subsequent scheme summarizes some of our results in [16]; NP-c. means $\mathbb{NP}$-complete, pol. (linear) means polynomial-time (linear-time) solvable, and XC means the Exact Cover problem.

	Chordal gr.	Dually chordal gr.	α-acyclic hypergr.	Hypertrees
ED	NP-c. [85]	linear	NP-c.	pol.
EED	linear [74]	linear	pol.	NP-c.
MIM	pol. [25]	NP-c.	pol.	NP-c.
XC			NP-c.	pol.

5   Conclusion

The results presented in this survey do not explain all the other aspects which are important for Efficient Domination and Efficient Edge Domination. Some of them are approximation, exact algorithms, fixed-parameter tractability and the influence of eigenvalues and spectra. Moreover, due to the space limitation, the survey cannot present typical and important proofs. Efficient Domination and Efficient Edge Domination is still an attractive task for many researchers, and for many open problems, it is impossible to solve them immediately.

Acknowledgment. The author thanks all of his coauthors for working on these topics.

References

1. Bange, D.W., Barkauskas, A.E., Slater, P.J.: Efficient dominating sets in graphs. In: Ringeisen, R.D., Roberts, F.S. (eds.) Applications of Discrete Mathematics, pp. 189–199. SIAM, Philadelphia (1988)
2. Bange, D.W., Barkauskas, A.E., Host, L.H., Slater, P.J.: Generalized domination and efficient domination in graphs. Discrete Math. **159**, 1–11 (1996)
3. Biggs, N.: Perfect codes in graphs. J. Comb. Theory Ser. B **15**, 289–296 (1973)
4. Brandstädt, A.: Weighted efficient domination for P_5-free graphs in linear time. CoRR arXiv:1507.06765v1 (2015)
5. Brandstädt, A., Chepoi, V.D., Dragan, F.F.: The algorithmic use of hypertree structure and maximum neighbourhood orderings. Discrete Appl. Math. **82**, 43–77 (1998)
6. Brandstädt, A., Dragan, F.F., Chepoi, V.D., Voloshin, V.I.: Dually chordal graphs. SIAM J. Discrete Math. **11**, 437–455 (1998)
7. Brandstädt, A., Eschen, E.M., Friese, E.: Efficient domination for some subclasses of P_6-free graphs in polynomial time. CoRR arXiv:1503.00091 (2015). Extended abstract. In: Mayr, E.W. (ed.) Proceedings of WG 2015. LNCS, vol. 9224, pp. 78–89 (2015)

8. Brandstädt, A., Eschen, E.M., Friese, E., Karthick, T.: Efficient domination for classes of P_6-free graphs. Discrete Appl. Math. **223**, 15–27 (2017)
9. Brandstädt, A., Fičur, P., Leitert, A., Milanič, M.: Polynomial-time algorithms for weighted efficient domination problems in AT-free graphs and dually chordal graphs. Inf. Process. Lett. **115**, 256–262 (2015)
10. Brandstädt, A., Giakoumakis, V.: Weighted efficient domination for $(P_5 + kP_2)$-free graphs in polynomial time. CoRR arXiv:1407.4593v1 (2014)
11. Brandstädt, A., Giakoumakis, V., Milanič, M., Nevries, R.: Weighted efficient domination for F-free graphs. Manuscript (2014, submitted)
12. Brandstädt, A., Hoàng, C.T.: Maximum induced matchings for chordal graphs in linear time. Algorithmica **52**, 440–447 (2008)
13. Brandstädt, A., Hundt, C., Nevries, R.: Efficient edge domination on hole-free graphs in polynomial time. In: López-Ortiz, A. (ed.) LATIN 2010. LNCS, vol. 6034, pp. 650–661. Springer, Heidelberg (2010). https://doi.org/10.1007/978-3-642-12200-2_56
14. Brandstädt, A., Karthick, T.: Weighted efficient domination in two subclasses of P_6-free graphs. Discrete Appl. Math. **201**, 38–46 (2016)
15. Brandstädt, A., Le, V.B., Spinrad, J.P.: Graph Classes: A Survey. SIAM Monographs on Discrete Mathematics and Applications, vol. 3. SIAM, Philadelphia (1999)
16. Brandstädt, A., Leitert, A., Rautenbach, D.: Efficient dominating and edge dominating sets for graphs and hypergraphs. In: Chao, K.-M., Hsu, T., Lee, D.-T. (eds.) ISAAC 2012. LNCS, vol. 7676, pp. 267–277. Springer, Heidelberg (2012). https://doi.org/10.1007/978-3-642-35261-4_30. Full version: CoRR arXiv:1207.0953v2, [cs.DM] (2012)
17. Brandstädt, A., Milanič, M., Nevries, R.: New polynomial cases of the weighted efficient domination problem. In: Chatterjee, K., Sgall, J. (eds.) MFCS 2013. LNCS, vol. 8087, pp. 195–206. Springer, Heidelberg (2013). https://doi.org/10.1007/978-3-642-40313-2_19. Full version: CoRR arXiv:1304.6255v1
18. Brandstädt, A., Mosca, R.: Dominating induced matchings for P_7-free graphs in linear time. In: Asano, T., Nakano, S., Okamoto, Y., Watanabe, O. (eds.) ISAAC 2011. LNCS, vol. 7074, pp. 100–109. Springer, Heidelberg (2011). https://doi.org/10.1007/978-3-642-25591-5_12. Full version: Algorithmica **68**, 998–1018 (2014)
19. Brandstädt, A., Mosca, R.: Weighted efficient domination for P_6-free graphs in polynomial time. CoRR arXiv:1508.07733 (2015). Based on a manuscript by R. Mosca, Weighted Efficient Domination for P_6-Free Graphs, July 2015
20. Brandstädt, A., Mosca, R.: Weighted efficient domination for P_6-free and for P_5-free graphs. In: Heggernes, P. (ed.) WG 2016. LNCS, vol. 9941, pp. 38–49. Springer, Heidelberg (2016). https://doi.org/10.1007/978-3-662-53536-3_4. Full version: SIAM J. Discrete Math. **30**(4) (2016)
21. Brandstädt, A., Mosca, R.: Dominating induced matchings for P_8-free graphs in polynomial time. Algorithmica **77**, 1283–1302 (2017)
22. Brandstädt, A., Mosca, R.: On efficient domination for some classes of H-free chordal graphs. CoRR arXiv:1701.03414 (2017). Extended abstract in Conference Proceedings of LAGOS 2017, Marseille, Electron. Notes Discrete Math. **62**, 57–62 (2017)
23. Brandstädt, A., Mosca, R.: Dominating induced matchings in $S_{1,2,4}$-free graphs. CoRR arXiv:1706.09301 (2017)
24. Broersma, H.J., Kloks, T., Kratsch, D., Müller, H.: Independent sets in asteroidal-triple-free graphs. SIAM J. Discrete Math. **12**, 276–287 (1999)

25. Cameron, K.: Induced matchings. Discrete Appl. Math. **24**, 97–102 (1989)
26. Cameron, K.: Induced matchings in intersection graphs. Discrete Math. **278**, 1–9 (2004)
27. Cameron, K., Sritharan, R., Tang, Y.: Finding a maximum induced matching in weakly chordal graphs. Discrete Math. **266**, 133–142 (2003)
28. Cardoso, D.M., Cerdeira, J.O., Delorme, C., Silva, P.C.: Efficient edge domination in regular graphs. Discrete Appl. Math. **156**, 3060–3065 (2008)
29. Cardoso, D.M., Korpelainen, N., Lozin, V.V.: On the complexity of the dominating induced matching problem in hereditary classes of graphs. Discrete Appl. Math. **159**, 521–531 (2011)
30. Cardoso, D.M., Lozin, V.V.: Dominating induced matchings. In: Lipshteyn, M., Levit, V.E., McConnell, R.M. (eds.) Graph Theory, Computational Intelligence and Thought. LNCS, vol. 5420, pp. 77–86. Springer, Heidelberg (2009). https://doi.org/10.1007/978-3-642-02029-2_8
31. Chang, M.-S.: Weighted domination of co-comparability graphs. Discrete Appl. Math. **80**, 135–148 (1997)
32. Chang, J.-M.: Induced matchings in asteroidal-triple-free graphs. Discrete Appl. Math. **132**, 67–78 (2004)
33. Chang, J.-M., Ho, C.-W., Ko, M.-T.: Powers of asteroidal-triple-free graphs with applications. Ars Comb. **67**, 161–173 (2003)
34. Chang, M.-S., Liu, Y.C.: Polynomial algorithms for the weighted perfect domination problems on chordal graphs and split graphs. Inf. Process. Lett. **48**, 205–210 (1993)
35. Chang, M.-S., Liu, Y.C.: Polynomial algorithms for the weighted perfect domination problems on interval and circular-arc graphs. J. Inf. Sci. Eng. **11**, 549–568 (1994)
36. Chang, G.J., Pandu Rangan, C., Coorg, S.R.: Weighted independent perfect domination on co-comparability graphs. Discrete Appl. Math. **63**, 215–222 (1995)
37. Chudnovsky, M., Robertson, N., Seymour, P., Thomas, R.: The strong perfect graph theorem. Ann. Math. **164**, 51–229 (2006)
38. Courcelle, B., Makowsky, J.A., Rotics, U.: Linear time solvable optimization problems on graphs of bounded clique-width. Theory Comput. Syst. **33**, 125–150 (2000)
39. Dragan, F.F., Prisacaru, C.F., Chepoi, V.D.: Location problems in graphs and the Helly property. Discrete Math. **4**, 67–73 (1992). Moscow, (in Russian), the full version appeared as preprint: Dragan, F.F., Prisacaru, C.F., Chepoi, V.D.: r-domination and p-center problems on graphs: special solution methods and graphs for which this method is usable, Kishinev State University, preprint Mold-NIINTI, N. 948-M88 (1987), (in Russian)
40. Edmonds, J.: Paths, trees, and flowers. Canad. J. Math. **17**, 449–467 (1965)
41. Eschen, E., Wang, X.: Algorithms for unipolar and generalized split graphs. Discrete Appl. Math. **162**, 195–201 (2014)
42. Fellows, M.R., Hoover, M.N.: Perfect domination. Australas. J. Comb. **3**, 141–150 (1991)
43. Flotow, C.: On powers of m-trapezoid graphs. Discrete Appl. Math. **63**, 187–192 (1995)
44. Frank, A.: Some polynomial algorithms for certain graphs and hypergraphs. In: Proceedings of the 5th British Combinatorial Conference (Aberdeen 1975), Congressus Numerantium, No. XV, pp. 211–226 (1976)
45. Fricke, G., Laskar, R.: Strong matchings on trees. Congr. Numer. **89**, 239–243 (1992)

46. Friese, E.: Das Efficient-Domination-Problem auf P_6-freien Graphen. Master thesis. University of Rostock, Germany (2013). (in German)
47. Garey, M.R., Johnson, D.S.: Computers and Intractability – A Guide to the Theory of NP-completeness. Freeman, San Francisco (1979)
48. Gavril, F.: Maximum weight independent sets and cliques in intersection graphs of filaments. Inf. Process. Lett. **73**, 181–188 (2000)
49. Golumbic, M.C., Laskar, R.: Irredundancy in circular arc graphs. Discrete Appl. Math. **44**, 79–89 (1993)
50. Golumbic, M.C., Lewenstein, M.: New results on induced matchings. Discrete Appl. Math. **101**, 157–165 (2000)
51. Grinstead, D.L., Slater, P.L., Sherwani, N.A., Holmes, N.D.: Efficient edge domination problems in graphs. Inf. Process. Lett. **48**, 221–228 (1993)
52. Grötschel, M., Lovász, L., Schrijver, A.: The ellipsoid method and its consequences in combinatorial optimization. Combinatorica **1**, 169–197 (1981). Corrigendum. Combinatorica **4**, 291–295 (1984)
53. Hayward, R.B., Spinrad, J.P., Sritharan, R.: Weakly chordal graph algorithms via handles. In: Proceedings of the 11th Symposium on Discrete Algorithms (SODA), pp. 42–49 (2000)
54. Hayward, R.B., Spinrad, J.P., Sritharan, R.: Improved algorithms for weakly chordal graphs. ACM Trans. Algorithms **3**, Article No. 14 (2007)
55. Hertz, A., Lozin, V.V., Ries, B., Zamaraev, V., de Werra, D.: Dominating induced matchings in graphs containing no long claw. CoRR arXiv:1505.02558 (2015)
56. Karp, R.M.: Reducibility among combinatorial problems. In: Miller, R.E., Thatcher, J.W., Bohlinger, J.D. (eds.) Complexity of Computer Computations, pp. 85–103. Plenum Press, New York (1972)
57. Karthick, T.: New polynomial case for efficient domination in P_6-free graphs. In: Ganguly, S., Krishnamurti, R. (eds.) CALDAM 2015. LNCS, vol. 8959, pp. 81–88. Springer, Cham (2015). https://doi.org/10.1007/978-3-319-14974-5_8
58. Kobler, D., Rotics, U.: Finding maximum induced matchings in subclasses of claw-free and P_5-free graphs, and in graphs with matching and induced matching of equal maximum size. Algorithmica **37**, 327–346 (2003)
59. Köhler, E.: Graphs without asteroidal triples. Ph.D. thesis, Technical University of Berlin (1999)
60. Korpelainen, N., Lozin, V.V., Purcell, C.: Dominating induced matchings in graphs without a skew star. J. Discrete Algorithms **26**, 45–55 (2014)
61. Kratochvíl, J.: Perfect codes in general graphs, Rozpravy Československé Akad. Věd Řada Mat. Přírod Vě 7. Akademia, Praha (1991)
62. Kratochvíl, J.: Regular codes in regular graphs are difficult. Discrete Math. **133**, 191–205 (1994)
63. Krishnamurthy, C.M., Sritharan, R.: Maximum induced matching problem on HHD-free graphs. Discrete Appl. Math. **160**, 224–230 (2012)
64. Leitert, A.: Das dominating induced matching problem für azyklische Hypergraphen. Diploma thesis, University of Rostock, Germany (2012)
65. Liang, Y.D., Lu, C.L., Tang, C.Y.: Efficient domination on permutation graphs and trapezoid graphs. In: Jiang, T., Lee, D.T. (eds.) COCOON 1997. LNCS, vol. 1276, pp. 232–241. Springer, Heidelberg (1997). https://doi.org/10.1007/BFb0045090
66. Lin, Y.-L.: Fast algorithms for independent domination and efficient domination in trapezoid graphs. In: Chwa, K.-Y., Ibarra, O.H. (eds.) ISAAC 1998. LNCS, vol. 1533, pp. 267–275. Springer, Heidelberg (1998). https://doi.org/10.1007/3-540-49381-6_29

67. Lin, M.C., Mizrahi, M.J., Szwarcfiter, J.L.: Fast algorithms for some dominating induced matching problems. Inf. Process. Lett. **114**(10), 524–528 (2014)
68. Lin, M.C., Mizrahi, M.J., Szwarcfiter, J.L.: $O(n)$ time algorithms for dominating induced matching problems. In: Pardo, A., Viola, A. (eds.) LATIN 2014. LNCS, vol. 8392, pp. 399–408. Springer, Heidelberg (2014). https://doi.org/10.1007/978-3-642-54423-1_35
69. Lin, M.C., Mizrahi, M.J., Szwarcfiter, J.L.: Efficient and perfect domination on circular-arc graphs. CoRR arXiv:1502.01523v1 (2015). Electron. Notes Discrete Math. **50**, 307–312 (2015)
70. Livingston, M., Stout, Q.: Distributing resources in hypercube computers. In: Proceedings of Third Conference on Hypercube Concurrent Computers and Applications, pp. 222–231 (1988)
71. Lokshtanov, D., Pilipczuk, M., van Leeuwen, E.J.: Independence and efficient domination on P_6-free graphs. CoRR arXiv:1507.02163v2 (2015). Conference Proceedings of SODA 2016, pp. 1784–1803
72. Lokshtanov, D., Vatshelle, M., Villanger, Y.: Independent set in P_5-free graphs in polynomial time. In: Proceedings of the Twenty-Fifth Annual ACM-SIAM Symposium on Discrete Algorithms, pp. 570–581 (2014)
73. Lovász, L.: Normal hypergraphs and the perfect graph conjecture. Discrete Math. **2**, 253–267 (1972)
74. Lu, C.L., Ko, M.-T., Tang, C.Y.: Perfect edge domination and efficient edge domination in graphs. Discrete Appl. Math. **119**, 227–250 (2002)
75. Lu, C.L., Tang, C.Y.: Solving the weighted efficient edge domination problem on bipartite permutation graphs. Discrete Appl. Math. **87**, 203–211 (1998)
76. Lu, C.L., Tang, C.Y.: Weighted efficient domination problem on some perfect graphs. Discrete Appl. Math. **117**, 163–182 (2002)
77. Milanič, M.: A hereditary view on efficient domination. In: Proceedings of the 10th Cologne-Twente Workshop. Extended Abstract, pp. 203–206 (2011)
78. Milanič, M.: Hereditary efficiently dominatable graphs. J. Graph Theory **73**, 400–424 (2013)
79. Nevries, R.: Efficient domination and polarity. Ph.D. thesis, University of Rostock (2014)
80. Raychaudhuri, A.: On powers of strongly chordal and circular-arc graphs. Ars Combin. **34**, 147–160 (1992)
81. Smart, C.B., Slater, P.J.: Complexity results for closed neighborhood order parameters. Congr. Numer. **112**, 83–96 (1995)
82. Stockmeyer, L.J., Vazirani, V.V.: NP-completeness of some generalizations of the maximum matching problem. Inf. Process. Lett. **15**, 14–19 (1982)
83. Williams, V.V.: Multiplying matrices faster than Coppersmith-Winograd. In: Proceedings of the 44th Symposium on Theory of Computing, STOC 2012, pp. 887–898. ACM, New York (2012)
84. Yen, C.-C.: Algorithmic aspects of perfect domination. Ph.D. thesis, Institute of Information Science, National Tsing Hua University, Taiwan (1992)
85. Yen, C.-C., Lee, R.C.T.: The weighted perfect domination problem and its variants. Discrete Appl. Math. **66**, 147–160 (1996)

Mixed Unit Interval Bigraphs

Ashok Kumar Das[(✉)] and Rajkamal Sahu

Department of Pure Mathematics, University of Calcutta, Kolkata, India
ashokdas.cu@gmail.com, rajkamalmath@gmail.com

Abstract. The class of intersection bigraphs of unit intervals of the real line whose ends may be open or closed is called mixed unit interval bigraphs. This class of bigraphs is a strict superclass of the class of unit interval bigraphs. We provide several infinite families of forbidden induced subgraphs of mixed unit interval bigraphs. We also pose a conjecture concerning characterization of mixed unit interval bigraphs and verify parts of it.

Keywords: Interval bigraphs · Unit interval bigraphs
Mixed unit interval bigraphs

1 Introduction

Interval graphs are the intersection graphs of intervals of the real line. *Unit interval graphs* are interval graphs where all the intervals are of unit length. *Proper interval graphs* are interval graphs where no interval is properly contained in another. Interval graphs and their subclasses like unit/proper interval graphs have been extensively studied by several researchers from structural [7,10], algorithmic [2,3] and application [9] view point.

However, most of the researchers do not specify which type of interval is used, that is, whether the ends of the intervals are open, closed or semi-closed. This is acceptable because the class of graphs does not actually depend on this. Frankl and Meahara [8] observed that using only open intervals or only closed intervals leads to the same class of graphs. In [6] it was shown that this is even true when we allow all possible types of intervals in the intersection representation. This is no longer true for the class of unit interval graphs. Rautenbach and Szwarcfiter [15] showed that the class of intersection graphs of unit intervals of open and closed intervals is a strict superclass of the class of unit interval graphs. They also characterized this class of graphs, by a finite list of forbidden induced subgraphs. Dourado et al. [6] generalized the result of [15] to mixed unit interval graphs allowing all four distinct types of unit intervals. Felix Joos [12] gave a complete characterization of mixed unit interval graphs in terms of infinite families of forbidden induced subgraphs.

A bipartite graph (in short, bigraph) $B = (X, Y, E)$ is an *interval bigraph* if there exist a one-to-one correspondence between the vertex set $X \cup Y$ of B and a collection of intervals $\{I(v) : v \in X \cup Y\}$ on the real line such that two vertices

© Springer International Publishing AG 2018
B. S. Panda and P. P. Goswami (Eds.): CALDAM 2018, LNCS 10743, pp. 15–29, 2018.
https://doi.org/10.1007/978-3-319-74180-2_2

are adjacent if and only if their corresponding intervals intersect and they belong to different partite sets. The collection of intervals $\{I(v) : v \in X \cup Y\}$ is called an interval representation of B. We simply denote the interval representation of B by I (which is a function from the vertex set $X \cup Y$ to a collection of intervals).

An interval bigraph is a *unit interval bigraph* if all the intervals in the interval representation are of unit length. An interval bigraph $B = (X, Y, E)$ is a *proper interval bigraph* if in the interval representation no interval is properly contained in another. Hell and Huang [11] proved that an interval bigraph is a unit interval bigraph if and only if it does not contain the bipartite claw (H_1), the bipartite net (H_2) or the bipartite tent (H_3) as an induced subgraph (see Fig. 1). In [4] we observe that the bigraphs H_1, H_2 and H_3 have intersection representation with unit open and closed intervals. In the same paper we give a characterization of the class of finite intersection bigraphs of unit open and closed intervals in terms of forbidden induced bigraphs.

In the present paper we generalize the results of [4] to the mixed unit interval bigraphs where we allow all four types of unit intervals namely closed, open, left closed-right open and right closed-left open unit interval in the interval representation. Here we show that the list of forbidden induced subgraphs for mixed unit interaval bigraphs is infinite.

In Sect. 2 we introduce basic definitions, terminology, and results related to our work. In Sect. 3 we give some forbidden induced subgraphs of mixed unit interval bigraphs. In Sect. 4 we pose a conjecture concerning characterization of mixed unit interval bigraphs and verify parts of it.

2 Preliminary Results

We consider only simple, finite and connected bigraphs. For a bigraph $B = (X, Y, E)$ the neighbourhood of a vertex $u \in X \cup Y$ is denoted by $N_B(u)$. Two distinct vertices u and v of B are copies if $N_B(u) = N_B(v)$. If no two vertices of B are copies then B is copy-free. If $\mathcal{F}$ is a set of graphs and a graph G does not contain a graph in $\mathcal{F}$ as an induced subgraph then G is $\mathcal{F}$-free.

Let $\mathcal{M}$ be a family of sets. An $\mathcal{M}$-intersection representation of a bigraph is a function $f : X \cup Y \to \mathcal{M}$ such that for any two distinct vertices u and v of a bigraph B, we have $uv \in E$ if and only if $f(u) \cap f(v) \neq \emptyset$. A bigraph is an $\mathcal{M}$-bigraph if it has an $\mathcal{M}$-intersection representation.

For two real numbers a and b, we denote the open interval $\{x \in \mathbb{R} | a < x < b\}$ by (a, b), the closed interval $\{x \in \mathbb{R} | a \leq x \leq b\}$ by $[a, b]$, the open-closed interval $\{x \in \mathbb{R} | a < x \leq b\}$ by $(a, b]$ and the closed-open interval $\{x \in \mathbb{R} | a \leq x < b\}$ by $[a, b)$. For an interval I, let $l(I) = \inf(I)$ and $r(I) = \sup(I)$. We suppose $\mathcal{I}^{++}$ is the set of closed intervals, $\mathcal{I}^{--}$ is the set of open intervals, $\mathcal{I}^{+-}$ is the set of closed-open intervals and $\mathcal{I}^{-+}$ is the set of open-closed intervals. Also suppose $\mathcal{U}^{++}$ is the set of unit closed intervals, $\mathcal{U}^{--}$ is the set of unit open intervals, $\mathcal{U}^{+-}$ is the set of unit closed-open intervals and $\mathcal{U}^{-+}$ is the set of unit open-closed intervals. In addition, let $\mathcal{I}^{\pm} = \mathcal{I}^{++} \cup \mathcal{I}^{--}$, $\mathcal{U}^{\pm} = \mathcal{U}^{++} \cup \mathcal{U}^{--}$, $\mathcal{I} = \mathcal{I}^{++} \cup \mathcal{I}^{--} \cup \mathcal{I}^{+-} \cup \mathcal{I}^{-+}$, and $\mathcal{U} = \mathcal{U}^{++} \cup \mathcal{U}^{--} \cup \mathcal{U}^{+-} \cup \mathcal{U}^{-+}$.

Our first result shows that as in the case of interval graphs, the class of interval bigraphs does not depend on the type interval used in the intersection representation.

Proposition 1. *The classes of* $\mathcal{I}^{++}$*-bigraphs,* $\mathcal{I}^{--}$*-bigraphs,* $\mathcal{I}^{\pm}$*-bigraphs,* $\mathcal{I}^{+-}$*-bigraphs,* $\mathcal{I}^{-+}$*-bigraphs and* $\mathcal{I}$*-bigraphs are the same.*

The following proposition extends the result of Proposition 2 of [6] which showed that a bigraph is a $\mathcal{U}^{++}$-bigraph if and only if it is a $\mathcal{U}^{--}$-bigraph.

Proposition 2. *The classes of* $\mathcal{U}^{++}$*-bigraphs,* $\mathcal{U}^{--}$*-bigraphs,* $\mathcal{U}^{+-}$*-bigraphs,* $\mathcal{U}^{-+}$*-bigraphs and* $\mathcal{U}^{+-} \cup \mathcal{U}^{-+}$*-bigraphs are the same.*

The proofs of the above propositions are similar to the proof of Dourado et al. [6] and so omitted.

Fig. 1. The bipartite claw (H_1), net (H_2) and tent (H_3)

Interval bigraphs coincide with $\mathcal{I}^{++}$-bigraphs and unit interval bigraphs coincide with $\mathcal{U}^{++}$-bigraphs. Following theorem relates the class of interval bigraphs, unit interval bigraphs and proper interval bigraphs.

Theorem 3 ([11,13,16]). *An interval bigraph is a unit interval bigraph if and only if it is a proper interval bigraph if and only if it does not contain H_1, H_2 or H_3 as an induced subgraph.*

As mentioned in the introduction, the bigraphs H_1, H_2 and H_3 have $\mathcal{U}$-intersection representation; see Figs. 2, 3 and 4.

Fig. 2. The bipartite claw H_1 and its $\mathcal{U}$-intersection representation.

As observed in [4] each intersection representation of H_1, H_2 and H_3 is unique upto trivial modifications (these trivial modifications include suitable interval shifts that preserve intersections and relative positions between intervals, changes in the types (open, closed or half closed) of some intervals, reflection of the

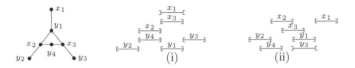

Fig. 3. The bipartite Net H_2 and its two $\mathcal{U}$-intersection representation.

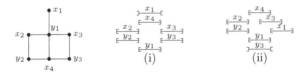

Fig. 4. The bipartite tent H_3 and its two $\mathcal{U}$-intersection representation.

entire model about a point on the real line, translation of the entire model, and relabeling of some intervals).

Therefore the class of $\mathcal{U}^{\pm}$-bigraphs is a strict superclass of the class of unit interval bigraphs. We have characterized these class of bigraphs in [4] (Fig. 5).

For an $\mathcal{I}^{++}$-bigraph if two vertices u and v are copies then they belong to the same partite set. And in the $\mathcal{I}^{++}$-interval representation we can take same interval for these two vertices. Thus we consider that our bigraphs are copy free.

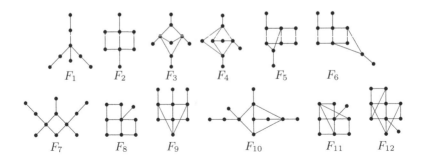

Fig. 5. Forbidden induced subgraphs of $\mathcal{U}^{\pm}$-bigraphs.

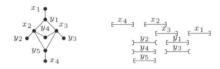

Fig. 6. The bigraph F_3 and its $\mathcal{U}$-intersection representation.

Theorem 4 ([4]). *For a copy-free bipartite graph B, the following statements are equivalent.*

(i) B *is a* $\{F_1, F_2, F_3, F_4, F_5, F_6, F_7, F_8, F_9, F_{10}, F_{11}, F_{12}\}$-*free interval bigraph.*
(ii) B *is an almost proper interval bigraph.*
(iii) B *is a* $\mathcal{U}^{++} \cup \mathcal{U}^{--}$-*bigraph.*

3 Forbidden Induced Subgraphs of Mixed Unit Interval Bigraphs

It can be observed that the bigraphs F_2, F_4, F_5, F_8, F_9, F_{10}, F_{11} and F_{12} have no $\mathcal{U}$-intersection representation. In this section we shall give some other forbidden induced subgraphs of mixed unit interval bigraphs (Figs. 7, 8, 9, 10, 11 and 12).

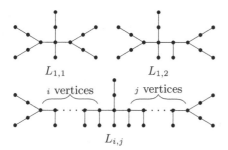

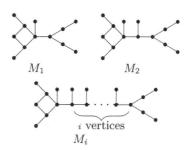

Fig. 7. The class $\mathcal{L}$

Fig. 8. The class $\mathcal{M}$.

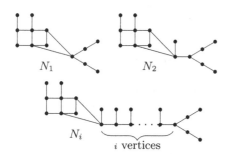

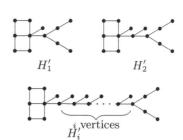

Fig. 9. The class $\mathcal{N}$.

Fig. 10. The class $\mathcal{H}'$.

Fig. 11. The bigraph B_1 **Fig. 12.** The bigraph B_2

Lemma 5. *The bigraph F_1 has unique $\mathcal{U}$-intersection upto trivial modifications.*

Fig. 13. The graph F_1 and its $\mathcal{U}$-representation.

Proof. The proof follows from the Proposition 5 of [4]. The bigraph $F_1 - y_3$ is the bipartite claw (H_1). Thus from that Proposition, H_1 has a unique $\mathcal{U}$-intersection as shown in Fig. 2. Since the vertex y_3 is adjacent to x_1 only, so we can take $I(y_3)$ as in Fig. 13 to get the $\mathcal{U}$-intersection representation of F_1. Again $I(y_3)$ can be taken as closed-open copy of $I(y_2)$ and make some trivial modifications to get the same representation of F_1 as earlier. This completes the proof. $\qquad\square$

From the above Lemma we have the following corollary.

Corollary. *The bigraph B_1 is the minimal forbidden induced subgraph of the class of $\mathcal{U}$-intersection bigraphs (Fig. 14).*

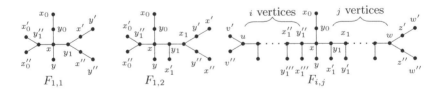

Fig. 14. The class $\mathcal{F}'$ of bigraphs

In the bigraph $F_{i,j}$, if i is even then $u \in X$ and $v', v'' \in Y$; and if i is odd then $u \in Y$ and $v', v'' \in X$. Similarly if j is even then $w \in X$ and if j is odd then $w \in Y$ and w, z are vertices of different partite sets. The two vertices v' and v'' are the *special vertices* of $F_{i,j}$. In the next lemma, we show that the bigraph $F_{i,j}$ has a unique $\mathcal{U}$-intersection representation upto trivial modifications.

Lemma 6. *Let $i, j \in \mathbb{N}$*

(a) A $\mathcal{U}$-intersection representation $I : V(F_{i,j}) \to \mathcal{U}$ of $F_{i,j}$, where $I(V(F_{i,j}))$ consists of the following intervals

- $I(x) = I(y_0) = [0,1]$, $I(x_0) = (0,1)$, $I(y) = (-1,0]$
- $I(y_k) = [2(k-1)+1, 2(k-1)+2]$, $I(x_k) = [2(k-1)+2, 2(k-1)+3]$, $(k \geq 1)$
- $I(x_k') = (2(k-1)+1, 2(k-1)+2)$, $I(y_k') = (2(k-1)+2, 2(k-1)+3)$, $(k \geq 1)$
- $I(y_k'') = [-2(k-1)-1, -2(k-1)]$, $I(x_k'') = [-2(k-1)-2, -2(k-1)-1]$, $(k \geq 1)$
- $I(x_k''') = (-2(k-1)-2, -2(k-1)-1]$, $I(y_k''') = (-2k-1, -2k]$, $(k \geq 1)$
- $I(w) = [j, j+1]$, $I(z'') = I(w'') = (j, j+1)$, $I(z') = [j+1, j+2]$, $I(w') = [j+2, j+3]$ or $[j+2, j+3)$ *and*
- $I(u) = [-i, -i+1]$, $I(v') = I(v'') = [-i-1, -i]$ or $(-i-1, -i]$

is unique upto trivial modifications.

(b) $L_{i,j}$ is a minimal forbidden induced subgraph for the class of $\mathcal{U}$-bigraphs.

Proof. (a) It can be easily observed that every $F_{i,j}$ contains F_1 as an induced subgraph. Consider $F_{1,1}$, it contains F_1 as an induced subgraph, where $V(F_1) = \{x_0, y_0, x, y, y_1, y_1'', x_0', x'\}$. Without loss of generality we consider the $\mathcal{U}$-intersection of F_1 for the vertices $x_0, y_0, x, y, y_1, y_1'', x_0', x'$ as follows: $I(x) = I(y_0) = [0,1]$, $I(x_0) = (0,1)$, $I(y_1) = [1,2]$, $I(x') = [2,3]$, $I(y'') = [-1,0]$, $I(y) = (-1,0]$, $I(x_0') = [-2,-1]$ or $(-2,-1]$. Next we take $I(x'') = I(y'') = (1,2)$, $I(y') = [3,4]$ or $[3,4)$ and $I(x_0'') = I(x_0')$. Now consider $F_{i,j}$ and let $j = 2n$. Then $w = x_n$, and the path $y_1, x_1, y_2, x_2, \ldots, y_k, x_k, \ldots, y_n, x_n$ has the representation $I(y_1) = [1,2]$, $I(x_1) = [2,3]$, $I(y_2) = [3,4]$, $I(x_2) = [4,5]$, and by induction $I(y_k) = [2(k-1)+1, 2(k-1)+2]$ and $I(x_k) = [2(k-1)+2, 2(k-1)+3]$. Then we have $I(x_n) = [2(n-1)+2, 2(n-1)+3] = [2n, 2n+1]$. In the other case, if $j = 2n+1$, then $w = y_{n+1}$ and we have $I(y_{n+1}) = [2(n+1-1)+1, 2n+2] = [2n+1, 2n+2]$. Thus $I(w) = [j, j+1]$, $I(z') = [j+1, j+2]$, $I(w') = [j+2, j+3]$ or $[j+2, j+3)$. As z'' is adjacent to w only in this path and w'' is adjacent to z'' so we take $I(z'') = I(w'') = (j, j+1)$. As the vertices x_0, y_0, x, y belong to $F_{i,j}$, we take the interval representation of these vertices as before (i.e. as in the case of $F_{1,1}$).

Next x_k' is adjacent to y_k and y_k' is adjacent to x_k only so we take $I(x_k')$ as the open copy of $I(y_k)$ and $I(y_k')$ as the open copy of $I(x_k)$. Again consider the path $x, y_1'', x_1'', \ldots, y_k'', x_k'', \ldots, u$. As $I(x) = [0,1]$ we take the interval representation of $I(y_1'') = [-1,0]$, $I(x_1'') = [-2,-1]$, $I(y_2'') = [-3,-2]$, $I(x_2'') = [-4,-3]$, and by induction $I(y_k'') = [-2(k-1)-1, -2(k-1)]$ and $I(x_k'') = [-2(k-1)-2, -2(k-1)-1]$. If $i = 2m$, then $u = x_m$; so $I(u) = I(x_m) = [-2(m-1)-2, -2(m-1)-1] = [-2m, -2m+1]$. And if $i = 2m+1$, then $u = y_{m+1}$, so $I(u) = [-2(m+1-1)-1, -2(m+1-1)] = [-2m-1, -2m]$. Thus $I(u) = [-i, -i+1]$ and $I(v_0')$ and $I(v_0'') = [-i-1, -i]$ or $(-i-1, -i]$. Finally, x_k''' is adjacent to y_k'' only and y_k''' is adjacent to x_k'' only. So we take $I(x_k''')$ as the open-closed copy of $I(x_k'')$ and $I(y_k''')$ as the open-closed copy of $I(y_{k+1}'')$. Which completes the proof of (a).

(b) $L_{i,j}$ is obtained from $F_{i,j}$ by adjoining two distinct vertices u' and u'' with v' and v'', where u' is adjacent to v' and u'' is adjacent to v''. Thus from $\mathcal{U}$-intersection representation of $F_{i,j}$, it follows that $L_{i,j}$ is the minimal forbidden induced subgraph of $\mathcal{U}$-bigraphs. □

As before the two vertices v' and v'' are the special vertices of M_i'. In the following lemma we show that the bigraph M_i' has a unique $\mathcal{U}$-intersection representation upto trivial modifications.

Lemma 7. *Let* $i \in \mathbb{N}$

(a) *A $\mathcal{U}$-intersection representation $I : V(M_i') \to \mathcal{U}$ of M_i', where $I(V(M_i'))$ consists of the following intervals*
  - $I(x_2'') = I(y_4'') = [0,1]$, $I(x_3'') = I(y_1) = [1,2]$, $I(y_2'') = [-1,0]$ *or* $(-1,0]$, $I(y_3'') = (1,2)$
  - $I(x_k) = [2k, 2k+1]$, $I(x_k') = [2k, 2k+1)$, $(k \geq 1)$
  - $I(y_k) = [2k-1, 2k]$, $I(y_{k-1}') = [2k-1, 2k)$, $(k \geq 2)$ *and*
  - $I(u) = [i+1, i+2]$, $I(v') = I(v'') = [i+2, i+3]$ *or* $[i+2, i+3)$

  is unique upto trivial modifications.

(b) *M_i is a minimal forbidden induced subgraph for the class of $\mathcal{U}$-bigraphs (Fig. 15).*

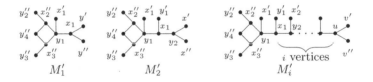

Fig. 15. The class $\mathcal{M}'$ of bigraphs.

Proof. (a) It can be easily observed that the bigraph M_i' contains H_2 as an induced subgraph, where $V(H_2) = \{x_1, y_1, x_2'', y_2'', x_3'', y_3'', y_4''\}$. From Proposition 6 of [4] we take the intervals corresponding to these vertices as $I(x_2'') = I(y_4'') = [0,1]$, $I(y_2'') = [-1,0]$ or $(-1,0]$, $I(x_3'') = I(y_1) = [1,2]$, $I(y_3'') = (1,2)$, $I(x_1) = [2,3]$. As x_1' is adjacent to y_1 only we must take $I(x_1') = [2,3)$. In M_i' consider the path $P : y_1, x_1, y_2, x_2, \ldots, y_k, x_k, \ldots, u$. If $i = 2n$, then $u = y_{n+1}$ and if $i = 2n+1$, then $u = x_{n+1}$. Now intervals corresponding to the vertices y_2, x_2, y_3, x_3 are $I(y_2) = [3,4]$, $I(x_2) = [4,5]$, $I(y_3) = [5,6]$, $I(x_3) = [6,7]$. Thus by induction, $I(y_k) = [2k-1, 2k]$ and $I(x_k) = [2k, 2k+1]$. Hence $I(y_{n+1}) = [2n+1, 2n+2]$ and $I(x_{n+1}) = [2n+2, 2n+3]$.

Thus $I(u) = [i+1, i+2]$ and $I(v)$ and $I(v'') = [i+2, i+3]$ or $[i+2, i+3)$. Now x_k' is adjacent to y_k only so we take $I(x_k')$ as the closed-open copy of $I(x_k)$. Again y_{k-1}' is adjacent to x_{k-1} only, and as y_k is adjacent to x_{k-1} in P. So we take $I(y_{k-1}')$ as the closed-open copy of $I(y_k)$, $(k \geq 2)$. This completes the proof of (a).

(b) From the above representation of M_i' it follows that M_i has no $\mathcal{U}$-intersection representation and hence M_i is a minimal forbidden induced subgraph for the class of $\mathcal{U}$-bigraphs.   □

The vertices v' and v'' are the special vertices of N_i'. In the next lemma we show that the bigraph N_i' has a unique $\mathcal{U}$-intersection representation upto trivial modifications.

Lemma 8. *Let $i \in \mathbb{N}$.*

(a) A $\mathcal{U}$-intersection representation $I : V(N_i') \to \mathcal{U}$ of N_i', where $I(V(N_i'))$ consists of the following intervals
- $I(x_2'') = I(y_2'') = [-2,-1]$, $I(x_4'') = I(y_1'') = [-1,0]$, $I(x_1'') = (-1,0)$
- $I(x_3'') = [0,1]$, $I(y_3'') = [0,1)$
- $I(y_4'') = [-3,-2]$ or $(-3,2]$
- $I(y_k) = [2k-2, 2k-1]$, $I(x_k) = [2k-1, 2k]$, $(k \geq 1)$
- $I(x_k') = [2k-1, 2k)$, $(k \geq 1)$ and $I(y_k') = [2k-2, 2k-1)$, $(k \geq 2)$ and
- $I(u) = [i-1, i]$, $I(v')$ and $I(v'') = [i, i+1]$ or $[i, i+1)$

is unique upto trivial modifications.

(b) N_i is a minimal forbidden induced subgraph for the class of $\mathcal{U}$-bigraphs (Fig. 16).

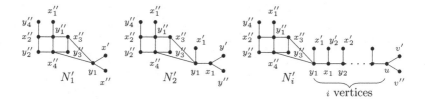

Fig. 16. The class $\mathcal{N}'$ of bigraphs.

Proof. (a) It is easy to obeserve that each of the bigraph N_i' contains the graph $H_3 + y_4''$ as an induced subgraph, where $V(H_3 + y_4'') = \{x_1'', y_1'', x_2'', y_2'', x_3'', y_3'', x_4'', y_4''\}$. Without loss of generality we take the following representation of $H_3 + y_4''$. Where $I(y_2'') = I(x_2'') = [-2,-1]$, $I(x_4'') = I(y_1'') = [-1,0]$, $I(x_1'') = (-1,0)$, $I(x_3'') = [0,1]$ and $I(y_3'') = [0,1)$. As y_1 is adjacent to x_3'' and x_4'' and x_1 is adjacent to y_1 so we take $I(y_1) = [0,1]$ and $I(x_1) = [1,2]$. Again as y_2 is adjacent to x_1, x_2 is adjacent to y_2 and so on; we can take the interval representation of the path $y_1, x_1, y_2, x_2, \ldots, y_k, x_k, \ldots, u$ where $I(y_1) = [0,1]$, $I(x_1) = [1,2]$, $I(y_2) = [2,3]$, $I(x_2) = [3,4]$. And by induction, we have $I(y_k) = [2k-2, 2k-1]$ and $I(x_k) = [2k-1, 2k]$. Now if $i = 2m$, then $u = x_m$. Then $I(x_m) = [2m-1, 2m] = [i-1, i]$. In the other case, if $i = 2m-1$, then $u = y_m$. And $I(y_m) = [2m-2, 2m-1] = [i-1, i]$. As v and v' are adjacent to u, we take $I(v) = I(v') = [i, i+1]$ or $[i, i+1)$. Again as x_k' is adjacent to y_k only also y_k is adjacent to x_k we take $I(x_k') = [2k-1, 2k)$, $(k \geq 1)$. As y_k' is adjacent to only x_{k-1} also x_{k-1} is adjacent to y_k, we take $I(y'_k) = [2k-2, 2k-1)$, $(k \geq 2)$. This completes the proof of (a).

(b) From the representation of N'_i it follows that N_i is the minimal forbidden induced subgraphs for $\mathcal{U}$-bigraphs.   □

The two vertices v' and v'' are the special vertices of H''_i. In the next lemma we show that the bigraph H''_i has a unique $\mathcal{U}$-intersection representation upto trivial modifications.

Lemma 9. *Let $i \in \mathbb{N}$.*

(a) *A $\mathcal{U}$-intersection representation $I : V(H''_i) \to \mathcal{U}$ of H''_i, where $I(V(H''_i))$ consists of the following intervals*
  - $I(x''_2) = [-1, 0]$ *or* $(-1, 0]$, $I(y''_2) = [-1, 0]$ *or* $(-1, 0]$
  - $I(x''_4) = I(y''_1) = [0, 1]$, $I(y''_3) = (0, 1)$, $I(x''_3) = [\frac{1}{2}, \frac{3}{2}]$
  - $I(x_1) = [1, 2]$, $I(x''_1) = [1, 2)$
  - $I(x_k) = [2k - 1, 2k]$, $I(y_k) = [2k, 2k + 1]$, $(k \geq 1)$
  - $I(y'_k) = [2k, 2k + 1)$, $I(x'_k) = [2k + 1, 2k + 2)$, $(k \geq 1)$
  - $I(u) = [i, i + 1]$, $I(v')$ *and* $I(v'') = [i + 1, i + 2]$ *or* $[i + 1, i + 2)$
  is unique upto trivial modifications.
(b) *H'_i is a minimal forbidden induced subgraph for $\mathcal{U}$-bigraph (Fig. 17).*

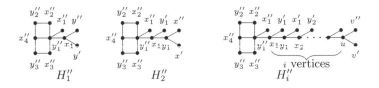

Fig. 17. The class $\mathcal{H}''$ of bigraphs.

Proof. (a) Every H''_i contains H_3 as an induced subgraph, where $V(H_3)$ is $\{x''_2, y''_2, x''_3, y''_3, x''_4, y''_1, x_1\}$. Consider the following representation of H_3, where $I(x''_2) = I(y''_2) = [-1, 0]$ or $(-1, 0]$, $I(x''_4) = I(y''_1) = [0, 1]$, $I(y''_3) = (0, 1)$, $I(x''_3) = [\frac{1}{2}, \frac{3}{2}]$, $I(x_1) = [1, 2]$. Next, consider the interval representation of the path $y''_1, x_1, y_1, x_2, y_2, \ldots, x_k, y_k, \ldots, u$, where $I(x_1) = [1, 2]$, $I(y_1) = [2, 3]$, $I(x_2) = [3, 4]$, $I(y_2) = [4, 5]$. And by induction $I(x_k) = [2k - 1, 2k]$, $I(y_k) = [2k, 2k + 1]$, $(k \geq 1)$. For the ith vertex u, if $i = 2m$ then $u \in Y$ and $u = y_m$, and $I(u) = [2m, 2m + 1] = [i, i + 1]$. Again if $i = 2m + 1$, $u \in X$ and $u = x_{m+1}$, then $I(u) = [2(m + 1) - 1, 2(m + 1)] = [2m + 1, 2m + 1 + 1] = [i, i + 1]$. Consequently $I(v')$ and $I(v'')$ are $[i + 1, i + 2]$ or $[i + 1, I + 2)$. Again x''_1 is adjacent to y''_1 only so $I(x''_1) = [1, 2)$. Similarly y'_k is adjacent to x_k only so $I(y'_k) = [2k, 2k + 1)$ as y_k is also adjacent to x_k. Next x'_k is adjacent to y_k and x_{k+1} is also adjacent to y_k, so $I(x'_k) = [2k + 1, 2k + 2)$, $(k \geq 1)$, this completes the proof of (a).

(b) From the $\mathcal{U}$-representation of H''_i it follows that H'_i is minimal induced forbidden subgraph for $\mathcal{U}$-bigraphs.   □

Lemma 10. *The bigraph B_2 is minimal forbidden induced subgraph for $\mathcal{U}$-bigra-phs.*

Proof. B_2 contains H_1 as an induced subgraph, where $V(H_1) = \{y, x, y_1, x_2, y_2, x_2', y_2'\}$. As H_1 has a unique $\mathcal{U}^{\pm}$-representation upto trivial modifications, we take the following representation of it, $I(x) = I(y_1) = [1, 2]$, $I(y) = (1, 2)$, $I(x_2) = [0, 1]$, $I(x_2') = [2, 3]$ and $I(y_2) = [-1, 0]$ or $(-1, 0]$, $I(y_2') = [3, 4]$ or $[3, 4)$. As x_3 is adjacent to y_1 but not to y_2 we take $I(x_3) = (0, 1]$. Similarly $I(x_3') = [2, 3)$. Again as y_4 is adjacent to x_2 and x_3 only so we may take $I(y_4) = (0, 1)$. Similarly $I(y_4') = (2, 3)$. Now it is not possible to give an interval representation for the vertex x_1.

Also it may be noted that B_2 contains $H_2 - y_3$ as an induced subgraph, where $V(H_2 - y_3) = \{x_1, y_1, x_2, y_2, x_3, y_4\}$. Consider the representation of $H_2 - y_3$, where $I(x_3) = [1, 2]$, $I(x_1) = (1, 2)$ or $(1, 2]$, $I(y_1) = [1, 2]$, $I(x_2) = [0, 1]$, $I(y_4) = [0, 1]$, $I(y_2) = [-1, 0]$ or $(-1, 0]$. Now x_2' and x_3' are adjacent to y_1 and y_2' is adjacent to x_2'. So we take $I(x_2') = [2, 3]$, $I(x_3') = [2, 3)$, $I(y_2') = [3, 4]$ or $[3, 4)$. Again y_4' is adjacent to x_2' and x_3' only so $I(y_4') = (2, 3)$. As x is adjacent to y_1 only so we take $I(x) = (1, 2)$. But now, it is not possible to give an interval representation for the vertex y. Also it can be verified that for other representation of $H_2 - y_3$, it is not possible to give an interval representation of B_2. This completes the proof of the lemma. $\qquad\square$

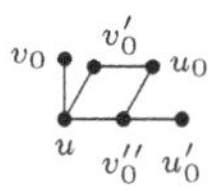

Fig. 18. The bigraph B_0

Lemma 11. *In the bigraphs $F_{i,j}$, M_i', N_i', H_i'' if we have the bigraph B_0 containing u as an induced subgraph (the vertices v' and v'' are absent) then the resulting bigraphs are still minimal forbidden induced subgraphs for $\mathcal{U}$-bigraphs (u and v are vertices of different partite sets) (Fig. 18).*

Proof. In the $\mathcal{U}$-intersection representation of any of the bigraphs $F_{i,j}$, M_i', N_i' or H_i'', let the interval corresponding to u is $I(u) = [a, a + 1]$. As v_0' and v_0'' are adjacent to u, u_0 is adjacent to v_0', v_0'' and u_0' is adjacent to v_0'' only, we take intervals corresponding to these vertices as follows: $I(v_0'') = [a + 1, a + 2]$, $I(v_0') = [a + 1, a + 2)$, $I(u_0) = (a + 1, a + 2)$ and $I(u_0') = [a + 2, a + 3]$ or $[a + 2, a + 3)$. Now the interval representation of v_0 is not possible as there exists an interval $I(u') = [a, a + 1]$ in the interval representation of each of the bigraphs $F_{i,j}$, M_i', N_i', H_i''. This completes the proof of the lemma. $\qquad\square$

4   A Conjecture for Mixed Unit Interval Bigraphs

In the previous Section we have seen that the bigraphs $F_2, F_4, F_5, F_8, F_9, F_{10}, F_{11}$, F_{12}, B_1 and B_2 are minimal forbidden induced subgraphs for $\mathcal{U}$-bigraphs. Also several infinite families of bigraphs, namely, $\mathcal{L}, \mathcal{M}, \mathcal{N}, \mathcal{H}'$ that are the forbidden families of $\mathcal{U}$-bigraphs. Next, we observe that in the bigraph $F_{i,j}$, the vertex u is adjacent to two special vertices v' and v''. Now if there exist two distinct vertices u' and u'' such that u' is adjacent to v' and u'' is adjacent to v'' we have the bigraph $L_{i,j}$. Similar observation can be made for the bigraphs M_i, N_i and H_i''. Also in the Lemma 11, we proved that for the bigraphs $F_{i,j}, M_i', N_i', H_i''$ where vertices v' and v'' are deleted, if we have the bigraph B_0 as an induced subgraph containing the vertex u, then the resulting graph is also a forbidden induced subgraph for $\mathcal{U}$-bigraphs. These results inspire us to pose a conjecture. But before that we introduce a new definition. For notational convenience we write $l(I(v)) = l(v)$ and $r(I(v)) = r(v)$

A bigraph $B = (X, Y, E)$ is a *mixed proper interval bigraph* if it has an $\mathcal{I}$-intersection representation $I : V(B) \to \mathcal{I}$ such that

(i) for two distinct vertices u and v of B with $I(u), I(v) \in \mathcal{I}^{++}$, $I(u) \not\subset I(v)$ and $I(v) \not\subset I(u)$, and

(ii) for every vertex u of B with $I(u) \notin \mathcal{I}^{++}$, there is a vertex v of B with $I(v) \in \mathcal{I}^{++}$, $l(u) = l(v)$ and $r(u) = r(v)$, that is no closed interval is properly contained in another closed interval and for any non closed interval, there is a closed interval with same end point.

Let $\mathcal{B}'$ be the class of bigraphs, where $\mathcal{B}' = \mathcal{F}' \cup \mathcal{M}' \cup \mathcal{N}' \cup \mathcal{H}''$. We are now in a position to phrase our conjecture.

Conjecture 12. *For a bigraph B, the following statements are equivalent.*

(a)
  - B is $\{B_1, B_2, F_2, F_4, F_5, F_8, F_9, F_{10}, F_{11}, F_{12}\} \cup \mathcal{L} \cup \mathcal{M} \cup \mathcal{N} \cup \mathcal{H}'$-free interval bigraph and
  - for every induced subgraph H of B that is isomorphic to one of the bigraphs of the class $\mathcal{B}'$ and any vertex $u^* \in V(B) \setminus V(H)$ is such that u^* is adjacent to exactly one of the special vertices of H,
  - if $H' = H \setminus \{v', v''\}$ then $H' \cup B_0$ is not an induced subgraph of B.

(b) *B is a mixed proper interval bigraph.*

(c) *B is a mixed unit interval bigraph.*

In the last two results we verify the conjecture partly and we leave open the problem of finding the complete list of forbidden bigraphs of mixed unit interval bigraphs.

Proposition 13. *The implication $(c) \Rightarrow (a)$ of Conjecture 12 is true.*

Proof. Let B be a $\mathcal{U}$-bigraph, and let I be a $\mathcal{U}$-intersection representation of B. Then obviously B is $F_2, F_4, F_5, F_8, F_9, F_{10}, F_{11}, F_{12}$-free interval bigraphs. Also corollary of Lemmas 5 and 10 imply that B is B_1 and B_2-free. And from Lemmata 6, 7, 8 and 9, B is $\mathcal{L} \cup \mathcal{M} \cup \mathcal{N} \cup \mathcal{H}'$-free interval bigraph

Now the H be an induced subgraph of B that is isomorphic to any bigraph of the class $\mathcal{B}'$. Let the vertices in H be denoted as in the definition of the bigraphs in the class $\mathcal{B}'$. Then the two pendant vertices v' and v'' are special vertices which are adjacent to u. And $v', v'', u \in V(H)$. Let $u^* \in V(B) \setminus V(H)$ be such that u^* is adjacent to v' but not to v''. By Lemmata 6, 7, 8 and 9 we may assume that $I(v') = [a, a+1]$ and $I(v'') = [a, a+1)$, where $a \in \mathbb{R}$. $r(I(v)) \leq a$ for any $v \in V(H) \setminus \{v', v''\}$. Thus $I(u^*)$ can be taken as any of intervals $[a+1, a+2]$ or $[a+1, a+2)$ and this implies that u^* is adjacent to v' only. From Lemma 11, it follows that $H' \cup B_0$ is forbidden induced subgraph of B. This completes the proof. $\square$

Theorem 14. *A bigraph is a mixed proper interval bigraph if and only if it is a $\mathcal{U}$-bigraph; that is; statements (b) and (c) of Conjecture 12 are equivalent.*

Proof. The 'only if' part of the proof is similar to the proof of Theorem 8 in [6]. For the sake of completeness, we give here details.

Let B be a mixed proper interval bigraph and I be a mixed proper interval representation of B. Let V_1 denote the set of vertices u of B such that $I(u) \in \mathcal{U}^{++}$. By the definition of mixed proper interval bigraphs, the subgraph $B[V_1]$ induced by the vertex set V_1 is a proper interval bigraph. And the interval representation of $B[V_1]$ is given by the corresponding intervals of I. Bogart and West [1], gave a constructive method how a proper interval representation I_1 produces to a unit interval representation I_2 gradually converting the intervals into unit intervals by means of successive contraction, dilations and translation. In this procedure it may be noted that two intervals intersect at a single point in I_1 if and only if the corresponding intervals intersect at a single point in I_2. And two intervals intersect more than one point in I_1 if and only if corresponding intervals intersects more than one point in I_2. Also two intervals do not intersect in I_1 if and only if they do not intersect in I_2. This implies that reinserting the mixed intervals corresponding to the vertices in $V(B) \setminus V_1$ as mixed copies of the corresponding closed intervals results in a $\mathcal{U}$-intersection representation of B, and this completes the 'only if' part.

For the 'if' part let B be a $\mathcal{U}$-bigraph and $I : V(B) \to \mathcal{U}$-intersection representation of B. Let $I(u)$ be an open interval of I. As $I(u)$ is forced to be open there must exist v_1 and v_2, such that $I(v_1)$ and $I(v_2)$ are closed and $r(v_1) = l(u)$, $l(v_2) = r(u)$. Now $I(v_1)$ and $I(v_2)$ must not be moved to the left and the right respectively as then $I(u)$ can be made to a closed interval, so there exist u' such that $I(u')$ is closed and $l(u) = l(u')$, $r(u) = r(u')$, u and v vertices of different partite sets.

Next, let $I(u)$ be an open-closed interval (i.e. $l(u)$ is open and $r(u)$ is closed). By the similar reason there exists a closed interval $I(v)$ such that $l(u) = r(v)$. Since B is connected we have $I(u')$ intersecting $I(v)$. Now $I(u') \in \mathcal{U}^{++}$ and $l(u') = l(u)$ and $r(u') = r(u)$, otherwise $I(v)$ can be moved to the left and may makes $l(u)$ closed. Thus for every $I(z) \notin \mathcal{U}^{++}$, $z \in V(B)$, we have closed interval with the same end points as of $I(z)$. This completes the proof. $\square$

5   Conclusion

In this paper we provide some forbidden subgraphs and four infinite families of forbidden subgraphs of mixed unit interval bigraphs. We also put forward a conjecture and hope that this will motivate to give a complete characterization of the class of mixed unit interval bigraphs in terms of forbidden induced subgraphs. In an earlier paper [4] we give the forbidden subgraph characterization of unit interval bigraphs of open and closed intervals, but the forbidden subgraph characterization of interval bigraphs is an interesting open problem. In [5] Das et al. have made considerable progress to solve it. The complexity of the only known recognition of interval bigraphs given by Müller [14] is very high. The problem of finding a recognition algorithm for interval graphs (or bigraphs) of open and closed intervals is still open. However, in a very recent paper [17] Talon and Kratochvíl have given a quadratic-time algorithm to recognize the class of mixed unit interval graphs.

References

1. Bogart, K.P., West, D.B.: A short proof that "proper=unit". Discret. Math. **201**, 21–23 (1999)
2. Corneil, D.G.: A simple 3-sweep LBFS algorithm for recognition of unit interval graphs. Discret. Appl. Math. **138**, 371–379 (2004)
3. Corneil, D.G., Olariu, S., Stewart, L.: The ultimate interval graph recognition algorithm? (Extended abstract). In: Proceedings of the 9th Annual ACMSIAM Symposium on Discrete Algorithms (SODA), pp. 175–180 (1998)
4. Das, A.K., Sahu, R.: A characterization of unit interval bigraphs of open and closed intervals. J. Graph Theory (under revision)
5. Das, A.K., Das, S., Sen, M.: Forbidden substructure for interval digraphs/bigraphs. Discret. Math. **339**, 1028–1051 (2016)
6. Dourado, M.C., Le, V.B., Protti, F., Rautenbach, D., Szwarcfiter, J.L.: Mixed unit interval graphs. Discret. Math. **312**, 3357–3363 (2012)
7. Fishburn, P.C.: Interval Orders Interval Graphs. A Study of Partially Ordered Sets. Wiley, New York (1985)
8. Frankl, P., Maehara, H.: Open-interval graphs versus closed-interval graphs. Discret. Math. **63**, 97–100 (1987)
9. Goldberg, P.W., Golumbic, M.C., Kaplan, H., Shamir, R.: Four strikes against physical mapping of DNA. J. Comput. Biol. **2**, 139–152 (1995)
10. Golumbic, M.C.: Algorithmic Graph Theory and Perfect Graphs. Annals of Discrete Mathematics, vol. 57. North-Holland Publishing Co., Amsterdam (2004)
11. Hell, P., Huang, J.: Interval bigraphs and Circular arc graphs. J. Graph Theory. **46**, 313–327 (2004)
12. Joos, F.: A characterization of mixed unit interval graphs. J. Graph Theory. **79**, 267–281 (2015)
13. Lin, I.J., West, D.B.: Interval digraphs that are indifference digraphs. In: Graph Theory, Combinatorics and Algorithms (Proceedings of the Quadrennial Conference Kalamazo, MI, 1992), pp. 751–765. Wiley-Interscience (1995)
14. Müller, H.: Recognizing interval digraphs and interval bigraphs in polynomial time. Discret. Appl. Math. **78**, 189–205 (1997)

15. Rautenbach, D., Szwarcfiter, J.L.: Unit interval graphs of open and closed intervals. J. Graph Theory. **72**, 418–429 (2013)
16. Sen, M., Sanyal, B.K.: Indifference digraphs: a generalization of indifference graphs and semi orders. SIAM J. Disc. Math. **7**, 157–165 (1994)
17. Talon, A., Kratochvíl, J.: Completion of the mixed unit interval graphs hierarchy. J. Graph Theory, 1–15 (2017). https://doi.org/10.1002/jgt.22159

Hamiltonian Path in $K_{1,t}$-free Split Graphs- A Dichotomy

Pazhaniappan Renjith[(✉)] and Narasimhan Sadagopan

Indian Institute of Information Technology, Design and Manufacturing,
Kancheepuram, India
{coe14d002,sadagopan}@iiitdm.ac.in

Abstract. In this paper, we investigate Hamiltonian path problem in the context of split graphs and produce a dichotomy result on the complexity of the problem. That is, unless P = NP, Hamiltonian path problem has no polynomial-time solution in $K_{1,5}$-free split graphs and polynomial-time solvable in $K_{1,4}$-free split graphs.

1 Introduction

Hamiltonian path problem is a well studied problem of finding a spanning path in a connected graph. This problem has been studied in various perspectives. In the initial stages of study, researchers explored the problem on structural perspective. That is, necessary conditions and sufficient conditions for the existence of Hamiltonian paths in connected graphs. Further, special graphs with bounded graph parameters such as degree, toughness, connectivity, independence number, etc., have been explored for obtaining Hamiltonian paths [1]. Another interesting view on the Hamiltonian problems have been obtained on graphs with forbidden sub graph structures. For example, Hamiltonian paths in claw-free graphs and its sub classes have been explored [2]. Variants of Hamiltonian problems such as Hamiltonian path starting from a specific vertex, Hamiltonian path between a fixed pair of vertices, Hamiltonian connectedness, pancyclicity, etc., have also been explored in the literature.

On algorithmic perspective, the problem is NP-complete in general graphs, and in particular, special graph classes such as chordal [3], bipartite, chordal bipartite [4], planar [5], grid graphs [6], etc. On the other hand, polynomial-time results for the problem have been obtained for interval graphs [7], circular arc graphs [8], etc. It is important to note that although polynomial-time results are known for special graph classes, we still have a "thick complexity line" separating NP-complete instances and polynomial-time solvable instances. In this paper we revisit the Hamiltonian path problem in chordal graphs and present a tight hardness result. We attempt a micro level structural study for Hamiltonian path problem in split graphs and establish that Hamiltonian path problem in $K_{1,5}$-free split graph is NP-complete, which is a popular sub class of chordal graphs. Further, to make the borderline thin between NP-complete instances

© Springer International Publishing AG 2018
B. S. Panda and P. P. Goswami (Eds.): CALDAM 2018, LNCS 10743, pp. 30–44, 2018.
https://doi.org/10.1007/978-3-319-74180-2_3

and polynomial-time instances, we do a deeper investigation of the structure of $K_{1,4}$-free split graphs, which is a major contribution of this paper.

A split graph G is a $C_4, C_5, 2K_2$-free graph and the vertex set of G can be partitioned into a clique K and an independent set I. Such a split graph is denoted as $G(K \cup I, E)$. For a split graph $G(K \cup I, E)$, we assume K to be a maximum clique. For $S \subset V(G)$, $N(S) = \{u : u \notin S, v \in S, uv \in E(G)\}$. If $S = \{v\}$, $N(S)$ is also denoted as $N(v)$. For a split graph $G(K \cup I, E)$ and $S \subset K$ we define $N^I(S) = N(S) \cap I$. Accordingly, if $S = \{v\}$, $N^I(v) = N^I(S)$. $d^I(v) = |N^I(v)|$ and $\Delta^I = \max\{d^I(v) : v \in K\}$. For $S \subset V(G)$, $G - S$ represents the subgraph of G induced on the vertex set $V(G) \setminus S$. $c(G)$ represents the number of components in graph G. For a cycle or a path $C = (v_1, \ldots, v_n)$, by $\overrightarrow{C}$, we mean the visit of vertices in order $(v_1, \ldots, v_n)$. Similarly, by $\overleftarrow{C}$, we mean the visit of vertices in order $(v_n, \ldots, v_1)$. $u\overrightarrow{C}u$ represents the ordered vertices from u to v in C. For a path $P = (v_1, \ldots, v_{n \geq 1})$ of length n, for simplicity, we use P to denote the underlying set $V(P)$ and v_1, v_n are end vertices of P.

2 Polynomial-Time Results

We organize our results on Hamiltonian path as Hamiltonian path in $K_{1,3}$-free split graphs and Hamiltonian path in $K_{1,4}$-free split graphs. We present our results on $K_{1,4}$-free split graphs in a systematic way. That is, we shall present Hamiltonian path in $K_{1,4}$-free split graph with $\Delta^I = 1$, $\Delta^I = 2$ followed by $\Delta^I = 3$. We observe the following corollary to present our results.

Corollary 1 (of Claim A in [9]). *Let G be a connected $K_{1,4}$-free split graph with $v \in K, d^I(v) = 3$. For every vertex $w \in K \setminus \{v\}$, $N^I(v) \cap N^I(w) \neq \emptyset$.*

2.1 Results on $K_{1,3}$-free Split Graphs

Theorem 1. *Let G be a connected $K_{1,3}$-free split graph. G contains a Hamiltonian path if and only if G has at most 2 vertices $u, v \in I$ such that $d(u) = 1$, and $d(v) = 1$.*

Proof. If there exists at least three vertices $\{u, v, w\} \subseteq I$ such that $d(u) = d(v) = d(w) = 1$, then clearly G has no Hamiltonian path. For the sufficiency, we see the following cases.

Case 1: For every $u \in I$, if $d(u) \geq 2$, then G is 2-connected, and by [9], G has a Hamiltonian cycle. Thus G has a Hamiltonian path.

Case 2: If there exists only one vertex $u \in I$ with $d(u) = 1$, then observe that $G - u$ is 2-connected. By [9], there is a Hamiltonian cycle in $G - u$, which can be easily extended to a Hamiltonian path in G.

Case 3: There exists two vertices $u, v \in I$ with $d(u) = d(v) = 1$. If $I = \{u, v\}$, it is easy to see that there is a (u, v)-Hamiltonian path in G. If $I = \{u, v, w\}$, then we claim that $N(w) \cap N(u) = \emptyset$ and $N(w) \cap N(v) = \emptyset$. Suppose $N(w) \cap N(u) \neq \emptyset$, then let $N^I(u') = \{u, w\}$, $u' \in K$. Clearly, all the vertices

$x \in K \setminus \{u'\}$ are adjacent to w, otherwise $\{u', u, w, x\}$ induces a $K_{1,3}$. It follows that $K \cup \{w\}$ is a clique of larger size, contradicting the maximality of K. Similar arguments hold with respect to the vertex v, and hence $N(w) \cap N(v) = \emptyset$. Thus we conclude that $\Delta^I = 1$. From [9], if $|I| > 3$, since G is connected, $\Delta^I = 1$. Now we produce a Hamiltonian path in G with $|I| \geq 3$ as follows. Let $I = \{u, v, w_1, \ldots, w_k\}$, $k \geq 1$ such that for all w_i, $1 \leq i \leq k$, $d(w_i) > 1$. Let $x_i, y_i, 1 \leq i \leq k$ be any two elements in $N(w_i)$. Since $\Delta^I = 1$, note that for all $s, t \in I$, $N(s) \cap N(t) = \emptyset$. Let $P_i = (x_i, w_i, y_i)$, $1 \leq i \leq k$, $v' = N(v)$, $u' = N(u)$ and $\{z_1, \ldots, z_l\} = K \setminus \{x_1, \ldots, x_k, y_1, \ldots, y_k, u', v'\}$, then $P = (u, u', x_1, w_1, y_1, \ldots, x_i, w_i, y_i, \ldots, x_k, w_k, y_k, z_1, z_2, \ldots, z_l, v', v)$ is a Hamiltonian path in G. P can also be written as $(u, u', \overrightarrow{P_1}, \ldots, \overrightarrow{P_k}, z_1, z_2, \ldots, z_l, v', v)$. This completes a proof of Theorem 1. $\qquad\square$

2.2 Results on $K_{1,4}$-free Split Graphs

Theorem 2. *Let $G(K \cup I, E)$ be a connected $K_{1,4}$-free split graph with $\Delta^I = 1$. G contains a Hamiltonian path if and only if there exists at most 2 vertices $u, v \in I$ such that $d(u) = 1$, and $d(v) = 1$.*

Proof. The proof is similar to the proof of Case 3 in Theorem 1.

We shall define some special paths and cycles in a $K_{1,4}$-free split graph $G(K \cup I, E)$. We define the *restricted bipartite subgraph* H of G as follows. $V_a = \{u \in I : d(u) \leq 2\}$, $V_b = N(V_a)$, $V(H) = V_a \cup V_b$ and $E(H) = \{uv : u \in V_a, v \in V_b\}$. An induced cycle C in H is referred to as *short cycle* in H (as well as G) if $V(C) \subset V(G)$. An *I-I path* is a maximal path in H that starts and ends in I. Similarly *K-K path* and *I-K path* are maximal paths in H with end vertices in K and end vertices in I, K, respectively. A maximal *I-I path* P in H is referred to as *Short I-I path* if $V(P) \subset V(G)$.

Theorem 3. *Let $G(K \cup I, E)$ be a connected $K_{1,4}$-free split graph with $\Delta^I = 2$ and H be the restricted bipartite subgraph of G. G contains a Hamiltonian path if and only if the following holds true.*

1. *H has no short I-I path.*
2. *The number of I-K paths in H is at most 2.*

Proof. If there exists a short *I-I* path P in H, then note that $c(G - S) > |S| + 1$ where $S = P \cap K$, and there is no Hamiltonian path in G as per Chvatal's necessary condition [10]. It is easy to see that if the number of *I-K* paths in H is more than 2, then there is no spanning path in G that includes all the vertices in all such *I-K* paths. For sufficiency, we see the following. Since H is the restricted bipartite subgraph of G, H is a collection of maximal paths and short cycles. Moreover, any short cycle in H is also a maximal *I-K* path in H. We initialize a set $\mathbb{S}$ with the set of maximal paths in H. It follows that $\mathbb{S}$ has at most two *I-K* paths. Let $I' = I \setminus \bigcup_{\forall P \in \mathbb{S}} V(P)$ and $K' = K \setminus \bigcup_{\forall P \in \mathbb{S}} V(P)$. We now outline a procedure to update $\mathbb{S}$ in two stages, using which we construct a

Hamiltonian path in G. In the first stage, for every vertex $u \in K'$, which is by definition P_1, include P_1 in $\mathbb{S}$. Since $\Delta^I = 2$, observe that any vertex $v \in I'$ is not adjacent to any internal vertex of paths in $\mathbb{S}$. Thus such a vertex v is adjacent to the end vertices of paths in $\mathbb{S}$. In particular, v may be adjacent to some of the newly added P_1 in $\mathbb{S}$ during the first stage. Further, $d(v) \geq 3$ implies that v is adjacent to the end vertices of at least two different paths $Q_i, Q_j \in \mathbb{S}$. As a part of the second stage, we update $\mathbb{S}$ as follows. For every vertex $v \in I'$, we find paths Q_i, Q_j such that one of Q_i, Q_j is either a K-K path or P_1. The paths Q_i, Q_j are replaced with the path $(\overrightarrow{Q_i}, v, \overrightarrow{Q_j})$ in $\mathbb{S}$. Let $\mathbb{S}_f = \{Q_1, \ldots, Q_k\}$ be the resultant set of paths after completing the second stage. If there exists two I-K paths, then let it be Q_i, Q_j, $i < j$ and if there exists only one I-K path, then let it be Q_i. Then $(\overrightarrow{Q_i}, \overrightarrow{Q_1}, \ldots, \overrightarrow{Q}_{i-1}, \overrightarrow{Q}_{i+1}, \ldots, \overrightarrow{Q}_{j-1}, \overrightarrow{Q}_{j+1}, \ldots, \overrightarrow{Q_k}, \overrightarrow{Q_j})$ is a Hamiltonian path in G. This completes the sufficiency part and a proof of the theorem. $\square$

Definition: A connected $K_{1,4}$-free split graph G satisfies Property A if $|K| \geq |I| - 1 \geq 8$, G has no short I-I path, and the sum of the number of I-K paths and the number of short cycles is at most 2. In a $K_{1,4}$-free split graph G with $\Delta_G^I = 3$, we define $V_3 = \{v : v \in K, d^I(v) = 3\}$.

Consider a $K_{1,4}$-free split graph G with $\Delta_G^I = 3$. We shall now show that the number of short cycles in G is at most 1 and the length of short cycle is at most 8. Subsequently, if G satisfies Property A, then we produce a Hamiltonian path in G. Towards this attempt, we bring in a transformation which will transform an instance of $\Delta_G^I = 3$ into $\Delta_{G'}^I = 2$ instance. Our results are deep and investigates the structure of the restricted bipartite subgraph H' of G' to obtain a Hamiltonian path in G.

Lemma 1. *Let G be a connected $K_{1,4}$-free split graph with $\Delta^I = 3$. Then, the number of short cycles in G is at most one. Further, if G has a short cycle C_n, then $n \leq 8$.*

Proof. For a contradiction assume that there are at least two short cycles in G. Let C, D be any two short cycles in G. Since $\Delta_G^I = 3$, there exists $v \in V_3$. Clearly, there exist a vertex $v_1 \in N^I(v)$ such that v_1 is adjacent to all the vertices in $C \cap K$ and $D \cap K$. It follows that all the vertices in $K \setminus (C \cup D)$ are adjacent to v_1. Note that $K \cup \{v_1\}$ is a clique of larger size, which contradicts the maximality of clique K. For the second part, assume for a contradiction that there exists a short cycle $C_{n \geq 10}$. Consider the cycle C such that $V(C) = \{w_1, \ldots, w_j, x_1, \ldots, x_j\}$,

$$j \geq 5, \{w_1, \ldots, w_j\} \subset K, \{x_1, \ldots, x_j\} \subset I, E(C) = \bigcup_{i=1}^{j} \{w_i x_i, x_i w_{(i+1) \bmod j}\}.$$

Since $\Delta^I(G) = 3$, there exists $v \in V_3$. To complete our proof, we identify a vertex $v_1 \in N^I(v)$ as follows. If $v \notin C$, then from Corollary 1, there exists a vertex $v_1 \in N^I(v)$ such that $v_1 w_1 \in E(G)$. If $v \in C$, then without loss of generality, we assume $w_1 = v$. There exists $v_1 \in N^I(v)$ such that $v_1 \notin C$. We claim that the vertices $\{w_3, \ldots, w_{j-1}\}$ are adjacent to v_1, otherwise $N^I(w_1) \cup \{w_1, w_i\}$, $3 \leq i \leq j-1$ induces a $K_{1,4}$. Further $w_2 v_1 \in E(G)$, otherwise $N^I(w_4) \cup \{w_4, w_2\}$, induces a $K_{1,4}$. Also $w_j v_1 \in E(G)$, otherwise $N^I(w_3) \cup \{w_3, w_j\}$, induces a $K_{1,4}$. From Corollary 1, it follows that all the vertices in $K \setminus C$ are adjacent to v_1. Suppose there exists $w \in K \setminus C$ such that $wv_1 \notin E(G)$, then for any $u \in C \cap K$, $N^I(u) \cup \{u, w\}$ induces a $K_{1,4}$, a contradiction. Finally, $K \cup \{v_1\}$ is a larger clique, contradicting to the maximality of K. Therefore, no such $C_{n \geq 10}$ exists. Note that this is tight. This completes a proof of the lemma. $\square$

Definition: Let G be a connected $K_{1,4}$-free graph with $\Delta^I_G = 2$ satisfying property A and H be the restricted bipartite subgraph of G. By the constructive proof of Theorem 3, there exists a collection $\mathbb{S}_f$ of vertex disjoint paths containing all the vertices of G. Such a collection is termed as a *path collection* of H.

Let G be a $K_{1,4}$-free split graph with $\Delta^I = 3$, satisfying Property A. For a vertex $v \in V_3$ let $G' = G - N^I(v)$. Let H, H' be the restricted bipartite subgraphs of G, G', respectively and $\mathbb{S}_f$ be a path collection of H'. Clearly, H' is a subgraph of H. Let $\mathbb{P}_k$, $k \geq 1$ be the set containing all the maximal paths of length k in $\mathbb{S}_f$. Thus, $\mathbb{S}_f = \mathbb{P}_1 \cup \mathbb{P}_2, \ldots, \cup \mathbb{P}_k$, where $\mathbb{P}_j$ is the set of maximal paths of size j where for every $Q \in \mathbb{P}_j$, there does not exist $Q' \in \mathbb{S}_f$ such that $E(Q) \subset E(Q')$. Note that $\mathbb{S}_f$ has I-K paths (even length paths) and K-K paths (odd length paths). A K-K path $P_a \in \mathbb{S}_f$ is defined on the vertex set $V(P_a) = \{w_1, \ldots, w_j, x_1, \ldots, x_{j-1}\}$, $E(P_a) = \{w_i x_i : 1 \leq i \leq j-1\} \cup \{x_{k-1} w_k : 2 \leq k \leq j\}$ such that $\{w_1, \ldots, w_j\} \subseteq K$, $\{x_1, \ldots, x_{j-1}\} \subseteq I$. We denote such a path as $P_a = P(w_1, \ldots, w_j; x_1, \ldots, x_{j-1})$. Similarly, $P_b = P(w_1, \ldots, w_j; x_1, \ldots, x_j)$ represents an I-K path with $V(P_b) = \{w_1, \ldots, w_j, x_1, \ldots, x_j\}$, $\{w_1, \ldots, w_j\} \subseteq K$, $\{x_1, \ldots, x_j\} \subseteq I$ and $E(P_b) = \{w_i x_i : 1 \leq i \leq j\} \cup \{x_{k-1} w_k : 2 \leq k \leq j\}$.

Lemma 2. *Let G be a connected $K_{1,4}$-free split graph with $\Delta^I = 3$, satisfying Property A. If G has a short cycle C, then there exists a Hamiltonian path in G.*

Proof. Let $v \in V_3$, $N^I(v) = \{v_1, v_2, v_3\}$. Recall that H' is the restricted bipartite subgraph of $G - N^I(v)$ and $\mathbb{S}_f$ is a path collection of H'. From Lemma 1, there exists exactly one short cycle C in G. Let length of C be k and $P \in \mathbb{P}_k$ is such that $V(P) = V(C)$. We see the following cases depending on the presence of v in C.

Case 1: $v \notin C$. Consider a vertex $w \in C \cap K$. From Corollary 1, $N^I(v) \cap N^I(w) \neq \emptyset$. Thus there exists $v_1 \in N^I(v)$ such that $v_1 w \in E(G)$. Now we claim that there exists $x \in C \cap K$ such that $v_1 x \notin E(G)$. Suppose for a contradiction assume for every $x \in C \cap K$, $v_1 x \in E(G)$. It follows from Corollary 1 that all the vertices in $K \setminus C$ are adjacent to v_1 and $K \cup \{v_1\}$ is a larger clique, contradicting the maximality of K. Thus $v_1 x \notin E(G)$. Further, from Corollary 1, either $v_2 x \in E(G)$ or $v_3 x \in E(G)$. Without loss of generality, let $v_2 x \in E(G)$. Using the above vertices w, x, we claim that in the collection $\mathbb{S}_f$ of H', for $n \neq k$, $\mathbb{P}_{n \geq 4} = \emptyset$. Note that $|\mathbb{P}_k| = 1$. Suppose there exists a path $P_a \neq P$, $P_a \in \mathbb{P}_{n \geq 4}$, then there exists a vertex $u \in P_a \cap K$ such that in P_a, $d^I(u) = 2$. From Corollary 1, $N^I(u) \cap N^I(v) \neq \emptyset$. If $uv_1 \in E(G)$, then $N^I(x) \cup \{x, u\}$ induces a $K_{1,4}$, otherwise $N^I(w)\{w, u\}$ induces a $K_{1,4}$. If $d(v_3) = 1$, then since G satisfies Property A, there does not exist $z \in I \setminus (C \cup N^I(v))$ such that $d_G(z) = 1$. It follows that $\mathbb{P}_2 = \emptyset$.

This is true because G has already one short cycle and apart from that it can have at most one I-K path as per Property A. Since $d(v_3) = 1$, no such z exists. Let $\mathbb{P}_1 = \{\{v\}, \{w_1\}, \dots, \{w_k\}\}$ and $\mathbb{P}_3 = \{(w_{k+1}, x_1, w_{k+2}), (w_{k+3}, x_2, w_{k+4}), \dots\}$. We construct a path $Q = (w_1, w_2, \dots, w_k, w_{k+1}, x_1, w_{k+2}, w_{k+3}, x_2, w_{k+4}, \dots)$. As per the premise, $|K| \geq |I| - 1 \geq 8$. Thus $|I| \geq 9$, and $\mathbb{P}_3 \neq \emptyset$. It follows that Q is non-empty, further $|Q| \geq 5$. From Corollary 1, all the vertices in $Q \cap K$ are adjacent to both v_1 and v_2. Suppose there exists $s \in Q$ such that $v_1 s \notin E(G)$ or $v_2 s \notin E(G)$, then either $N^I(w) \cup \{w, s\}$ or $N^I(x) \cup \{x, s\}$ induces a $K_{1,4}$. Thus for every $s \in Q \cap K$, $v_1 s \in E(G)$ and $v_2 s \in E(G)$. Observe that $(v_3, v, v_2, \overrightarrow{Q}, v_1, w\overrightarrow{C})$ is a Hamiltonian path in G. If $d(v_3) > 1$, then we see the following. If $\mathbb{P}_2 \neq \emptyset$, since G satisfies Property A, $|\mathbb{P}_2| = 1$ i.e., $\mathbb{P}_2 = \{P_b\}$, $P_b = (y, z)$, $y \in K$. Further, there exists $w' \in K$ such that $v_3 w' \in E(G)$. Now we claim that $w' \notin C$. Suppose not, then there exists $w'' \in C \cap K$ such that $N^I(w'') \cup \{w'', w'\}$ induces a $K_{1,4}$. Similar to the argument with respect to the vertex s, for the vertex w', we argue that $v_1 w', v_2 w' \in E(G)$, and $\{w'\} \in \mathbb{P}_1$. Let $w' = w_i, i \leq k$. Then we construct a path $Q' = (w_1, w_2, \dots, w_{i-1}, w_{i+1}, \dots, w_k, w_{k+1}, x_1, w_{k+2}, w_{k+3}, x_2, w_{k+4}, \dots)$. Now we obtain $P_c = (z, y, w', v_3, v, v_2, \overrightarrow{Q'}, v_1, w\overrightarrow{C})$ as a Hamiltonian path in G. If $\mathbb{P}_2 = \emptyset$, then a $(w'\overrightarrow{P_c})$ is a Hamiltonian path in G.

Case 2: $v \in C$. Let $v_2, v_3 \in N^I(v) \cap C$. Then note that there exists $v_1 \in N^I(v)$ such that $v_1 \notin C$. Clearly, for all $w \in K \setminus C$, $wv_1 \in E(G)$. Since K is a maximal clique, it follows that there exists $x \in C \cap K$ such that $v_1 x \notin E(G)$. We now claim that $\mathbb{P}_{n \geq 4} = \emptyset$, $n \neq k$. Suppose there exists a path $P_a \neq P$, $P_a \in \mathbb{P}_{n \geq 4}$, then there exists a vertex $u \in P_a \cap K$ such that in P_a, $d^I(u) = 2$. We already observed that $v_1 u \in E(G)$. Further, $N^I(u) \cup \{u, x\}$ induces a $K_{1,4}$, a contradiction. Thus such a path P_a does not exist. If $\mathbb{P}_2 \neq \emptyset$, then let $(y, z) \in \mathbb{P}_2$, $y \in K$. Then $(\overrightarrow{C}v, v_1, \overrightarrow{Q}, y, z)$ is a Hamiltonian path in G. If $\mathbb{P}_2 = \emptyset$, then $(\overrightarrow{C}v, v_1, \overrightarrow{Q})$ is a Hamiltonian path in G. This completes the case analysis and a proof of the lemma. $\square$

We work on a $K_{1,4}$-free split graph G with $\Delta^I = 3$, satisfying Property A, with G', H' and $\mathbb{S}_f$ as defined previously. If G has a short cycle, then by Lemma 2, G has a Hamiltonian path. If G has no short cycles, then note that there exists at most 2 I-K paths in $\mathbb{S}_f$. For the following claims, we shall consider such a G with no short cycle. The structural study of paths in $\mathbb{S}_f$ is deep, which is the highlight of this paper. Now we shall present some claims to show the structural observations of paths in $\mathbb{S}_f$.

Claim 1. *If there exists a path $P_a \in \mathbb{P}_k, k \geq 10$ such that $P_a = P(w_1, \dots, w_{j'};$ $x_1, \dots, x_j), j + 1 \geq j' \geq j \geq 5$, then there exists $v_1 \in N^I(v)$ such that $v_1 w_i \in E(G), 2 \leq i \leq j$.*

Proof. First we show that for any two vertices $w_i, w_l \in K, 2 \leq i, l \leq j, |i - l| > 1$; $v_1 w_i, v_1 w_l \in E(G)$. By Corollary 1, clearly there exists $v_1 \in N(v)$ such that $v_1 w_i \in E(G)$. If $v_1 w_l \notin E(G)$, then by Corollary 1, $v_2 w_l \in E(G)$ or $v_3 w_l \in E(G)$ is true. It follows that $N^I(w_i) \cup \{w_i, w_l\}$ induces a $K_{1,4}$. Since $|\{w_2, \dots, w_j\}| \geq 4$, for every $2 \leq i \leq j$, $v_1 w_i \in E(G)$. $\square$

Claim 2. $\mathbb{P}_{i \geq 12} = \emptyset$.

Proof. Assume for a contradiction there exists a path $P_a \in \mathbb{P}_i, i \geq 12$. Let $P_a = P(w_1, \ldots, w_{j'}; x_1, \ldots, x_j)$, $j + 1 \geq j' \geq j \geq 6$. From Claim 1 there exists $v_1 \in N_G^I(v)$ such that $v_1 w_k \in E(G), 2 \leq k \leq j$. We now claim that $v_1 w_1 \in E(G)$. Suppose not, then from Corollary 1, $v_2 w_1 \in E(G)$ or $v_3 w_1 \in E(G)$. Further, $w_1 x_3 \in E(G)$, otherwise $N^I(w_3) \cup \{w_3, w_1\}$ or $N^I(w_4) \cup \{w_4, w_1\}$ has an induced $K_{1,4}$. Similarly, $w_1 x_5 \in E(G)$. It follows that $N^I(w_1) \cup \{w_1\}$ has an induced $K_{1,4}$, a contradiction. Thus $v_1 w_1 \in E(G)$. If P_a is an odd path, then similar arguments with respect to w_1 holds good for the vertex $w_{j'}$, and hence $v_1 w_{j'} \in E(G)$. Since the clique is maximum in G, there exists $s \in K$ such that $v_1 s \notin E(G)$. Further, there exists at least three vertices in $x_1, \ldots, x_j$ adjacent to s, otherwise, for some $2 \leq r \leq j$, $N_G^I(w_r) \cup \{w_r, s\}$ induces a $K_{1,4}$. Finally, from Corollary 1, either $v_2 s \in E(G)$ or $v_3 s \in E(G)$. It follows that $N_G^I(s) \cup \{s\}$ induces a $K_{1,4}$, a contradiction. Thus such a path P_a does not exist. This completes a proof of the claim. $\qquad\square$

Claim 3. *Let* $P_a = P(w_1, \ldots, w_{i'}; x_1, \ldots, x_i), i + 1 \geq i' \geq i \geq 2$, *and* $P_b = P(s_1, \ldots, s_{j'}; t_1, \ldots, t_j)$, $j + 1 \geq j' \geq j \geq 2$ *be arbitrary paths in* $\mathbb{S}_f$. *Then there exists* $v_1 \in N^I(v)$ *such that* $\forall\, 2 \leq l \leq i$, $v_1 w_l \in E(G)$, *and* $\forall\, 2 \leq m \leq j$, $v_1 s_m \in E(G)$.

Proof. From Corollary 1, there exists $v_1 \in N^I(v)$ such that $v_1 w_2 \in E(G)$. If $v_1 s_m \notin E(G)$, $2 \leq m \leq j$ then by Corollary 1, $v_2 s_m \in E(G)$ or $v_3 s_m \in E(G)$. It follows that $N^I(w_2) \cup \{w_2, s_m\}$ induces a $K_{1,4}$. Thus $v_1 s_m \in E(G)$. If path P_a has size more than 5, then for every $3 \leq l \leq i$, $v_1 w_l \in E(G)$. Suppose not, then by Corollary 1, $v_2 w_l \in E(G)$ or $v_3 w_l \in E(G)$. It follows that $N^I(s_2) \cup \{s_2, w_l\}$ induces a $K_{1,4}$. Therefore, we conclude that for all possible l, m; $v_1 w_l, v_1 s_m \in E(G)$, and the claim follows. $\qquad\square$

Corollary 2 *(of Claim 3).* *If* $P_a = P(w_1, \ldots, w_{l'}; x_1, \ldots, x_l), l + 1 \geq l' \geq l \geq 2$, $P_b = P(s_1, \ldots, s_{m'}; t_1, \ldots, t_m)$, $m + 1 \geq m' \geq m \geq 2$ *and* $P_c = P(q_1, \ldots, q_{n'}; r_1, \ldots, r_n)$, $n + 1 \geq n' \geq n \geq 2$ *are arbitrary paths in* $\mathbb{S}_f$, *then there exists* $v_1 \in N^I(v)$ *such that* $\forall\, 2 \leq i \leq l$, $v_1 w_i \in E(G)$, $\forall\, 2 \leq j \leq m$, $v_1 s_j \in E(G)$ *and* $\forall\, 2 \leq k \leq n$, $v_1 q_k \in E(G)$.

Claim 4. *If there exists* $P_a \in \mathbb{P}_{k \geq 8}$, *then there does not exist* P_b *such that* $|P_b| \geq 4$.

Proof. Assume for a contradiction that there exists such a path $P_b \in \mathbb{P}_j, j \geq 4$. Let $P_a = (w_1, \ldots, w_{l'}; x_1, \ldots, x_l)$, $l + 1 \geq l' \geq l \geq 4$ and $P_b = (s_1, \ldots, s_{r'}; t_1, \ldots, t_r)$, $r + 1 \geq r' \geq r \geq 2$. From Claim 3, $v_1 w_i \in E(G), 2 \leq i \leq l$ and $v_1 s_i \in E(G), 2 \leq i \leq r$. Now we claim $v_1 w_1 \in E(G)$. Otherwise, by Corollary 1, $v_2 w_1$ or $v_3 w_1$ is in $E(G)$. Observe that $w_1 x_3 \in E(G)$, otherwise $N^I(w_3) \cup \{w_3, w_1\}$ or $N^I(w_4) \cup \{w_4, w_1\}$ induces a $K_{1,4}$. Further, either $w_1 t_1 \in E(G)$ or $w_1 t_2 \in E(G)$, otherwise $N^I(s_2) \cup \{s_2, w_1\}$ induces a $K_{1,4}$. Now $N^I(w_1) \cup \{w_1\}$ induces a $K_{1,4}$, a contradiction and thus $v_1 w_1 \in E(G)$. If P_a is an odd path, then using similar argument, we establish $v_1 w_{l'} \in E(G)$. Now we claim that $v_1s_1 \in E(G)$. Otherwise, by Corollary 1, $s_1v_2 \in E(G)$ or $s_1v_3 \in E(G)$. Further, $s_1x_1 \in E(G)$ or $s_1x_2 \in E(G)$, otherwise $N^I(w_2) \cup \{w_2, s_1\}$ induces a $K_{1,4}$. Similarly, $s_1x_3 \in E(G)$ or $s_1x_4 \in E(G)$. Now $N^I(s_1) \cup \{s_1\}$ induces a $K_{1,4}$, a contradiction. Therefore, $v_1s_1 \in E(G)$. If P_b is an odd path, then using similar argument, we establish $v_1s_{r'} \in E(G)$. Since the clique is maximal, there exists a vertex $w' \in K$ such that $v_1w' \notin E(G)$. By Corollary 1, $w'v_2 \in E(G)$ or $w'v_3 \in E(G)$; without loss of generality, let $w'v_2 \in E(G)$. Also due to the similar reasoning for s_1, w' is adjacent to one among x_1, x_2, and w' is adjacent to one among x_3, x_4. Further, either $t_1w' \in E(G)$ or $t_2w' \in E(G)$, otherwise $N^I(s_2) \cup \{s_2, w'\}$ induces a $K_{1,4}$. Finally, $N^I(w') \cup \{w'\}$ induces a $K_{1,4}$, a contradiction. Therefore, P_b does not exist. This completes a proof of Claim 4.   □

Claim 5. *If there exists $P_a \in \mathbb{P}_{11}$, then G has a Hamiltonian path.*

Proof. Let $P_a = (w_1, \ldots, w_6; x_1, \ldots, x_5)$. From Claim 1, there exists a vertex say $v_1 \in N_G^I(v)$, such that $v_1w_i \in E(G)$, $2 \leq i \leq 5$. From the proof of the previous claim, $v_1w_1, v_1w_6 \in E(G)$. Since the clique is maximal, there exists $w' \in K$, such that $w'v_1 \notin E(G)$. By Corollary 1, $w'v_2 \in E(G)$ or $w'v_3 \in E(G)$. Without loss of generality, let $w'v_2 \in E(G)$. We claim $w'x_2 \in E(G)$ and $w'x_4 \in E(G)$, otherwise for some $2 \leq i \leq 5$, $N^I(w_i) \cup \{w_i, w'\}$ induces a $K_{1,4}$. One among v_2, x_2, x_4 is adjacent to w_1, otherwise $N^I(w') \cup \{w_1, w'\}$ induces a $K_{1,4}$. Similar argument holds good with respect to the vertex w_6. From Claim 4, $\mathbb{P}_j = \emptyset$, $j \geq 4$. If $\mathbb{P}_2 \neq \emptyset$, since G satisfies Property A, note that at most two vertices of I have degree 1. If $d(v_3) = 1$, then $|\mathbb{P}_2| \leq 1$. Let $P_b \in \mathbb{P}_2$. Let $\overrightarrow{Q}$ represents an ordering of paths in $\mathbb{P}_1 \cup \mathbb{P}_3$, excluding the paths $\{v\}$, $\{w'\}$. $P = (v_3, v, v_2, w', x_2\overrightarrow{P_a}w_6, v_1, w_1\overrightarrow{P_a}w_2, \overrightarrow{Q}, \overrightarrow{P_b})$ or $(\overrightarrow{P}w_2, \overrightarrow{Q})$ is a Hamiltonian path in G. If $d(v_3) > 1$, then $|\mathbb{P}_2| \leq 2$ and let $P_c \in \mathbb{P}_2$, $P_c \neq P_b$. Note that there exists $w'' \in K$ such that $v_3w'' \in E(G)$ and w'' is adjacent to at least two vertices in $\{x_1, \ldots, x_5\}$, otherwise for some $2 \leq i \leq 5$, $N^I(w_i) \cup \{w_i, w''\}$ induces a $K_{1,4}$. Thus $\{w''\} \in \mathbb{P}_1$. Let $\overrightarrow{Q'}$ represents an ordering of paths in $\mathbb{P}_1 \cup \mathbb{P}_3$, excluding the paths $\{v\}$, $\{w'\}$, $\{w''\}$. $P' = (\overleftarrow{P_b}, w'', v_3, v, v_2, w', x_2\overrightarrow{P_a}w_6, v_1, w_1\overrightarrow{P_a}w_2, \overrightarrow{Q'}, \overrightarrow{P_c})$ is a Hamiltonian path in G. If $|\mathbb{P}_2| < 2$, then a observe that $(\overrightarrow{P'}w_2, \overrightarrow{Q'})$ or $(w''\overrightarrow{P'})$ or $(w''\overrightarrow{P'}w_2, \overrightarrow{Q'})$ is a spanning path of G. This completes a proof of the claim. □

Claim 6. *If there exists $P_a \in \mathbb{P}_{10}$, then G has a Hamiltonian path.*

Proof. Due to page constraints, proof of this claim is included in [11].   □

From here onwards, for producing Hamiltonian paths in the proof of claims, we obtain a special path, termed as *desired path*, which is a path containing the vertices $\{v_1, v_2, v_3\}$ and all (at most two) the I-K paths in $\mathbb{S}_f$ along with some K-K paths. Let $\overrightarrow{Q}$ be an ordering of the paths in $\mathbb{S}_f$ which are not included in the desired path. Depending on the adjacency of the first vertex of $\overrightarrow{Q}$ to $N^I(v)$, $(\overrightarrow{P_a}, v_i, \overrightarrow{Q}, \overrightarrow{P_b})$ is a Hamiltonian path in G, where $(\overrightarrow{P_a}, v_i, \overrightarrow{P_b})$, $i \in \{1, 2, 3\}$ is the desired path, with $|\overrightarrow{P_a}| \geq 0$, $|\overrightarrow{P_b}| \geq 0$.

Claim 7. *If there exists $P_a \in \mathbb{P}_9$, then G has a Hamiltonian path.*

Proof. Let $P_a = (w_1, \ldots, w_5; x_1, \ldots, x_4)$. There exists a vertex in $v_1 \in N^I(v)$ such that $v_1 w_2 \in E(G)$. Note that $v_1 w_4 \in E(G)$, otherwise $N^I(w_2) \cup \{w_2, w_4\}$ induces a $K_{1,4}$. Recall from Claim 4, $\mathbb{P}_{j \geq 4} = \emptyset$. Now we claim that for an arbitrary path $P \in \mathbb{P}_2 \cup \mathbb{P}_3$ and $s \in K$ be an end vertex of P, then $v_1 s \in E(G)$. Suppose not, then $v_2 s \in E(G)$ or $v_3 s \in E(G)$. Further, s is adjacent to either x_1 or x_2, otherwise $N^I(w_2) \cup \{w_2, s\}$ induces a $K_{1,4}$. Similarly, s is adjacent to one of x_3, x_4, otherwise $N^I(w_4) \cup \{w_4, s\}$ induces a $K_{1,4}$. Therefore, $N^I(s) \cup \{s\}$ induces a $K_{1,4}$, a contradiction. Thus $v_1 s \in E(G)$. Note that the above argument is true for any end vertex $s \in K$ of every such paths in $\mathbb{P}_2 \cup \mathbb{P}_3$. Since the clique is maximal, there exists a non-adjacency for v_1 in K, and based on the non-adjacency, we see the following cases as shown in Table 1.

Due to page constraints, for the following 3 claims, proofs are included in [11].

Claim 8. *If there exists $P_a \in \mathbb{P}_8$, then G has a Hamiltonian path.*

Table 1. Case analysis for the proof of Claim 7

Case	Arguments
Case 1: $v_1 w_1 \notin E(G)$ or $v_1 w_5 \notin E(G)$	Without loss of generality, we shall assume $v_1 w_1 \notin E(G)$. Note that one of v_2, v_3 is adjacent to w_1, without loss of generality, let $v_2 w_1 \in E(G)$. Further, note that w_1 is adjacent to one of x_3, x_4, otherwise, $N^I(w_4) \cup \{w_4, w_1\}$ induces a $K_{1,4}$
Case 1.1: $d(v_3) = 1$	From Property A, at most 2 vertices of I have degree 1. Thus, $|\mathbb{P}_2| \leq 1$. Let $P_b \in \mathbb{P}_2$. Further, from Property A, $|I| \geq 9$. Therefore, there exists $P_c \in \mathbb{P}_3$. Let $\overrightarrow{Q}$ be an ordering of the paths in $\mathbb{P}_1 \cup \mathbb{P}_3$ excluding paths $\{v\}$ and P_c. Then $(v_3, v, v_2, w_1 \overrightarrow{P_a} w_5, \overrightarrow{Q}, \overrightarrow{P_c}, v_1, \overrightarrow{P_b})$ is a Hamiltonian path in G
Case 1.2: $d(v_3) > 1$	Recall $d^I(w_1) = 3$. We claim v_3 is not adjacent to $s \in K$ such that s is an end vertex of any path in $\mathbb{P}_2 \cup \mathbb{P}_3$. Recall that $v_1 s \in E(G)$. Suppose $v_3 s \in E(G)$, then $N^I(w_1) \cup \{w_1, s\}$ induces a $K_{1,4}$. Thus $v_3 s \notin E(G)$. It follows that v_3 is adjacent to w_3 or w_5 or a vertex in $\mathbb{P}_1$. Observe that $|\mathbb{P}_2| \leq 2$. To complete the proof, we shall assume that $|\mathbb{P}_2| = 2$. Proof of other two cases are similar. Let $P_b, P_d \in \mathbb{P}_2$, $y \in P_b \cap K$
Case 1.2.1	$v_3 w_5 \in E(G)$ or $v_3 w_3 \in E(G)$ or $v_3 z \in E(G)$, $z \notin P_a$. Proof is included in [11]
Case 2:	$v_1 w_3 \notin E(G)$ or $v_1 t \notin E(G)$, $t \notin P_a$. Proof is included in [11]

Claim 9. *If there exists $P_a, P_b \in \mathbb{P}_7 \cup \mathbb{P}_6$, then there does not exist $P_c \in \mathbb{S}_f$ such that $P_c \neq P_a$, $P_c \neq P_b$ and $|P_c| \geq 4$. Further, $|\mathbb{P}_7| + |\mathbb{P}_6| \leq 2$.*

Claim 10. *If there exists $P_a, P_b \in \mathbb{P}_7 \cup \mathbb{P}_6$, then G has a Hamiltonian path.*

Claim 11. *If there $|\mathbb{P}_7| + |\mathbb{P}_6| = 1$, then $|\mathbb{P}_5| + |\mathbb{P}_4| \leq 1$.*

Proof. Let $P_a = P(w_1, \ldots, w_l; x_1, \ldots, x_3)$, $l \in \{3, 4\}$, $P_b = P(s_1, \ldots, s_m; t_1, \ldots, t_2)$, $m \in \{2, 3\}$. Assume for a contradiction that there exists a path $P_c = P(q_1, \ldots, q_n; r_1, \ldots, r_2)$, $n \in \{2, 3\}$. From Corollary 2, there exists $v_1 \in N^I(v)$ such that $v_1 w_2, v_1 w_3, v_1 s_2, v_1 q_2 \in E(G)$. We claim that $v_1 w_1, v_1 s_1, v_1 q_1 \in E(G)$. Suppose $v_1 w_1 \notin E(G)$, then w_1 is adjacent to either v_2 or v_3, and one each from $P_b \cap I$ and $P_c \cap I$. Therefore, $N^I(w_1) \cup \{w_1\}$ induces a $K_{1,4}$, a contradiction. Similar argument holds good for the other edges. If P_a, P_b, P_c are odd paths, then similar to the previous argument, the end vertices w_l, s_m, q_n are adjacent to v_1. Since the clique is maximal, there exists $w' \in K$ such that $v_1 w' \notin E(G)$. From the previous argument, $w' \notin \{w_1, \ldots, w_l, s_1, \ldots, s_m, q_1, \ldots, q_n\}$. By Corollary 1, either $v_2 w' \in E(G)$ or $v_3 w' \in E(G)$. Further, we argue that $w' x_2 \in E(G)$, otherwise $N^I(w_2) \cup \{w_2, w'\}$ or $N^I(w_3) \cup \{w_3, w'\}$ induces a $K_{1,4}$. Observe that w' is adjacent to one of $\{t_1, t_2\}$, otherwise $N^I(s_2) \cup \{s_2, w'\}$ induces a $K_{1,4}$. Similarly, w' is adjacent to one of $\{r_1, r_2\}$. Now, $N^I(w') \cup \{w'\}$ induces a $K_{1,4}$, which is a final contradiction to the existence of such a path P_c. This completes a proof of the claim. $\square$

Claim 12. *If there exists $P_a \in \mathbb{P}_7 \cup \mathbb{P}_6$, $P_b \in \mathbb{P}_5 \cup \mathbb{P}_4$, then G has a Hamiltonian path.*

Proof. Let $P_a = P(w_1, \ldots, w_l; x_1, \ldots, x_3)$, $l \in \{3, 4\}$, $P_b = P(s_1, \ldots, s_m; t_1, \ldots, t_2)$, $m \in \{2, 3\}$. From Corollary 2, there exists $v_1 \in N^I(v)$ such that $v_1 w_2, v_1 w_3, v_1 s_2 \in E(G)$. We argue that $v_1 w_1 \in E(G)$. Suppose not, then from Corollary 1, w_1 is adjacent to either v_2 or v_3. Further, w_1 is adjacent to one of $\{x_2, x_3\}$, otherwise $N^I(w_3) \cup \{w_3, w_1\}$ induces a $K_{1,4}$. w_1 is also adjacent to one of $\{t_1, t_2\}$, otherwise $N^I(s_2) \cup \{s_2, w_1\}$ induces a $K_{1,4}$. It follows that $N^I(w_1) \cup \{w_1\}$ induces a $K_{1,4}$, a contradiction. Thus $v_1 w_1 \in E(G)$. If P_a is an odd path, then similar arguments holds good with respect to the other end vertex of P_a and $v_1 w_l \in E(G)$. From Claim 11, there are no paths of size 4 or more other than P_a, P_b in $\mathbb{S}_f$. Now we claim that for an arbitrary path $P \in \mathbb{P}_2 \cup \mathbb{P}_3$ and $s \in K$ be an end vertex of P, then $v_1 s \in E(G)$. Suppose not, then $v_2 s \in E(G)$ or $v_3 s \in E(G)$. Further, s is adjacent to x_2, otherwise $N^I(w_2) \cup \{w_2, s\}$ or $N^I(w_3) \cup \{w_3, s\}$ induces a $K_{1,4}$. Similarly, s is adjacent to one of t_1, t_2, otherwise $N^I(s_2) \cup \{s_2, s\}$ induces a $K_{1,4}$. From the above arguments it follow that $N^I(s) \cup \{s\}$ induces a $K_{1,4}$, a contradiction. Therefore, $v_1 s \in E(G)$. Note that the above argument is true for any end vertex $s \in K$ of every such paths in $\mathbb{P}_2 \cup \mathbb{P}_3$. Since the clique is maximal, there exists $y \in K$ such that $v_1 y \notin E(G)$. Now we see the following cases depending on the length of paths P_a, P_b, and in each case, we also see the possibility of adjacency of y.

Case 1: $|P_a| = 7, |P_b| = 5$ and $v_1 y \notin E(G)$.

Case 1.1: $y \in \{s_1, s_3\}$, without loss of generality, let $s_1 v_1 \notin E(G)$. From Corollary 1, $s_1 v_2 \in E(G)$ or $s_1 v_3 \in E(G)$. Without loss of generality, let $s_1 v_2 \in E(G)$. Note that $s_1 x_2 \in E(G)$, otherwise $N^I(w_2) \cup \{w_2, s_1\}$ or $N^I(w_3) \cup \{w_3, s_1\}$ induces a $K_{1,4}$. Consider $d(v_3) = 1$. Since there are at most 2 vertices in I of degree 1, $|\mathbb{P}_2| \leq 1$, let $P_c \in \mathbb{P}_2$. We obtain $(v_3, v, v_2, s_1 \overrightarrow{P_b}, \overrightarrow{P_a}, v_1, \overrightarrow{P_c})$ as a desired path in G. If $d(v_3) > 1$, then we observe the following. Since there are at most 2 vertices in I of degree 1, $|\mathbb{P}_2| \leq 2$, let $P_c, P_d \in \mathbb{P}_2$. Observe that w_1, w_4 are adjacent to a vertex in $N^I(s_1)$. Thus $d^I(w_1) = d^I(w_4) = 3$. We now claim that there does not exist a vertex $s \in K$ such that s is an end vertex of a path in $\mathbb{P}_2 \cup \mathbb{P}_3$ and $v_3 s \in E(G)$. Suppose, for such a vertex s, let $v_3 s \in E(G)$, then $N^I(s_1) \cup \{s_1, s\}$ induces a $K_{1,4}$. From the above observations we conclude that v_3 is adjacent to either s_3 or a vertex in $\mathbb{P}_1$. If $v_3 s_3 \in E(G)$, then note that $s_3 x_2 \in E(G)$, otherwise for some $w \in \{w_2, w_3, s_1\}$, $N^I(w) \cup \{w, s_3\}$ induces a $K_{1,4}$. Further, $w_1 x_2, w_4 x_2 \in E(G)$. Note that $d(v_2) = d(v_3) = d(t_1) = d(t_2) = 2$. Since G has no short cycles, there exists $w' \in \mathbb{P}_1$ such that w' is adjacent to one among $\{v_2, v_3, t_1, t_2\}$. In each cases, we obtain a desired path P as follows.

If $w' v_2 \in E(G)$, then $P = (\overrightarrow{P_c}, w', v_2, v, v_3, s_3 \overleftarrow{P_b} s_1, \overrightarrow{P_a}, v_1, \overrightarrow{P_d})$.
If $w' v_3 \in E(G)$, then $P = (\overrightarrow{P_c}, w', v_3, s_3 \overleftarrow{P_b} s_1, v_2, v, v_1, \overrightarrow{P_a}, \overrightarrow{P_d})$.
If $w' t_1 \in E(G)$, then $P = (\overrightarrow{P_c}, w', t_1, s_1, v_2, v, v_3, s_3, t_2, s_2, v_1, \overrightarrow{P_a}, \overrightarrow{P_d})$.
If $w' t_2 \in E(G)$, then $P = (\overrightarrow{P_c}, w', t_2, s_3, v_3, v, v_2, s_1, t_1, s_2, v_1, \overrightarrow{P_a}, \overrightarrow{P_d})$.
For a vertex $z \in \mathbb{P}_1$, if $v_3 z \in E(G)$, then $P = (\overrightarrow{P_c}, z, v_3, v, v_2, s_1 \overrightarrow{P_b} s_3, \overrightarrow{P_a}, v_1, \overrightarrow{P_d})$.

Case 1.2: $y \in \mathbb{P}_1$. In this case we shall assume that $v_1 s_1, v_1 s_3 \in E(G)$. From Corollary 1, $y v_2 \in E(G)$ or $y v_3 \in E(G)$. Without loss of generality, let $y v_2 \in E(G)$. Consider $d(v_3) = 1$. Since there are at most 2 vertices in I of degree 1, $|\mathbb{P}_2| \leq 1$, let $P_c \in \mathbb{P}_2$. We obtain $(v_3, v, v_2, y, \overrightarrow{P_b}, \overrightarrow{P_a}, v_1, \overrightarrow{P_c})$ as a desired path in G. On the other hand, if $d(v_3) > 1$, then we observe the following. Since there are at most 2 vertices in I of degree 1, $|\mathbb{P}_2| \leq 2$, let $P_c, P_d \in \mathbb{P}_2$. If $v_3 s_1 \in E(G)$, then $(\overrightarrow{P_c}, s_3 \overrightarrow{P_b} s_1, v_3, v, v_2, y, \overrightarrow{P_a}, v_1, \overrightarrow{P_d})$ is a desired path in G. If $v_3 s_3 \in E(G)$, then $(\overrightarrow{P_c}, s_1 \overrightarrow{P_b} s_3, v_3, v, v_2, y, \overrightarrow{P_a}, v_1, \overrightarrow{P_d})$ is a desired path in G. If $v_3 z \in E(G)$ for some $z \in \mathbb{P}_1$, then we see the following. Observe that $y x_2 \in E(G)$ and there exists a vertex in $P_b \cap I$ adjacent to y. Therefore, $d^I(y) = 3$ and $z \neq y$. We obtain $(\overrightarrow{P_c}, z, v_3, v, v_2, y, \overrightarrow{P_b}, \overrightarrow{P_a}, v_1, \overrightarrow{P_d})$ as a desired path in G.

One could produce desired paths in all the other possibilities, and a detailed analysis is shown in [11].

Claim 13. *If there exists $P_a \in \mathbb{P}_7$ and there does not exist $P \in \mathbb{S}_f$ such that $P \neq P_a$ and $|P| \geq 4$, then G has a Hamiltonian path.*

Proof. Let $P_a = (w_1, \ldots, w_4; , x_1, \ldots, x_3)$. From Corollary 1, the vertices w_2, w_3 are adjacent to at least one of the vertices in $N^I(v)$. Depending on this adjacency, we see the following two cases.

Case 1: There exists $v_1 \in N^I(v)$ such that $v_1 w_2, v_1 w_3 \in E(G)$. Consider a path $Q \in \mathbb{P}_2 \cup \mathbb{P}_3$, and $s \in Q \cap K$. We first claim that s is adjacent to either v_1 or x_2. Suppose that $v_1 s, x_2 s \notin E(G)$. Note that $v_2 s \in E(G)$ or $v_3 s \in E(G)$. If $sx_1 \notin E(G)$, then $N^I(w_2) \cup \{w_2, s\}$ induces a $K_{1,4}$. Therefore, $sx_1 \in E(G)$ and similarly, $sx_3 \in E(G)$, otherwise $N^I(w_3) \cup \{w_3, s\}$ induces a $K_{1,4}$. It follows that $N^I(s) \cup \{s\}$ induces a $K_{1,4}$. Therefore all the end vertices of all such paths are adjacent to either v_1 or x_2. Since the clique is maximal, there exists $v' \in K$ such that $v_1 v' \notin E(G)$. We further classify based on the possibilities of v' as follows.

Case 1.1: $v' \in P_a$; i.e., w_1 or w_4 is non-adjacent to v_1. Without loss of generality, let $v_1 w_4 \notin E(G)$. Note that w_4 is adjacent to either v_2 or v_3. Without loss of generality, let $v_2 w_4 \in E(G)$.

Case 1.1.1: $d(v_3) = 1$. Since $|I| \geq 9$, there exists $P_d, P_e \in \mathbb{P}_3$; $P_d = P(s_1, s_2; t_1)$, and $P_e = P(q_1, q_2; r_1)$. Further, note that $|\mathbb{P}_2| \leq 1$. If $|\mathbb{P}_2| = 1$, then let $P_b \in \mathbb{P}_2$. Recall that the vertices s_1, s_2, q_1, q_2 are adjacent to at least one of v_1, x_2. We obtain a desired path P as follows.

If $s_1 v_1, q_1 v_1 \in E(G)$, then $P = (v_3, v, v_2, \overleftarrow{P_a}, \overleftarrow{P_d}, v_1, \overrightarrow{P_e}, \overrightarrow{P_b})$.
If $s_1 x_2, q_1 x_2 \in E(G)$, then $P = (v_3, v, v_2, w_4, x_3, w_3, v_1, w_2, x_1, w_1, \overrightarrow{P_d}, x_2, \overrightarrow{P_e}, \overrightarrow{P_b})$.
If $s_1 v_1, q_1 x_2 \in E(G)$, then $P = (v_3, v, v_2, w_4, x_3, w_3, v_1, \overrightarrow{P_d}, \overleftarrow{P_e}, x_2 \overrightarrow{P_a}, \overrightarrow{P_b})$.
If $s_1 x_2, q_1 v_1 \in E(G)$, then $P = (v_3, v, v_2, w_4, x_3, w_3, v_1, \overrightarrow{P_e}, \overleftarrow{P_d}, x_2 \overrightarrow{P_a}, \overrightarrow{P_b})$.

Case 1.1.2: $d(v_3) > 1$. Note that w_4 is adjacent to a vertex in $N^I(w_2)$. In particular, either $w_4 x_1 \in E(G)$ or $w_4 x_2 \in E(G)$. Thus the only vertex to which v_3 is adjacent in P_a is w_1. Consider $v_3 w_1 \in E(G)$. Since $|I| \geq 9$, there exists $P_d \in \mathbb{P}_3$; $P_d = P(s_1, s_2; t_1)$. Note that $|\mathbb{P}_2| \leq 2$. If $|\mathbb{P}_2| = 2$, then let $P_b, P_c \in \mathbb{P}_2$. Let $y_1 = P_b \cap K$, $y_2 = P_c \cap K$. Recall that y_1, y_2, s_1, s_2 are adjacent to either v_1 or x_2. Let $C = (w_2, x_1, w_1, v_3, v, v_2, w_4, x_3, w_3, x_2, w_2)$ and $C' = (w_2, x_1, w_1, v_3, v, v_2, w_4, x_3, w_3, v_1, w_2)$. We obtain desired path P as follows. If $s_1 v_1 \in E(G)$ then we observe the following. If $y_1 x_2 \in E(G)$, then $P = (\overrightarrow{P_b}, x_2, w_2 \overrightarrow{C'} v_1, \overrightarrow{P_d}, \overrightarrow{P_c})$. Note $y_2 x_2 \in E(G)$ is a symmetric case. If $y_1 v_1, y_2 v_1 \in E(G)$, then note that y_1 is adjacent to a vertex s in $N^I(w_4)$. Note $N^I(w_4) \subset C$. We obtain $P = (\overrightarrow{P_b}, s \overrightarrow{C}, \overrightarrow{P_d}, v_1, \overrightarrow{P_c})$.

If $s_1 x_2 \in E(G)$ and $y_1 v_1 \in E(G)$, then $P = (\overrightarrow{P_b}, v_1, w_2 \overrightarrow{C} x_2, \overrightarrow{P_d}, \overrightarrow{P_c})$. Note that $s_1 x_2, y_2 v_1 \in E(G)$ is a symmetric case. If $s_1 x_2, y_1 x_2, y_2 x_2 \in E(G)$, then note that y_1 is adjacent to a vertex s in $N^I(v)$. Note $N^I(v) \subset C'$. We obtain $P = (\overrightarrow{P_b}, s \overrightarrow{C'}, \overrightarrow{P_d}, x_2, \overrightarrow{P_c})$. Now we consider the case in which v_3 is adjacent to a vertex in $\mathbb{P}_2 \cup \mathbb{P}_3$. If $v_3 y_1 \in E(G)$, then we observe the following. Observe that either $y_1 v_1 \in E(G)$ or $y_1 x_2 \in E(G)$. If $w_4 x_1 \in E(G)$, then $N^I(y_1) \cup \{y_1, w_4\}$ induces a $K_{1,4}$. Therefore, $w_4 x_2 \in E(G)$. Further, all the end vertices of paths in $\mathbb{P}_2 \cup \mathbb{P}_3$ which are in K are adjacent to x_2. We obtain $(\overrightarrow{P_b}, v_3, v, v_2, w_4, x_3, w_3, v_1, w_2, x_1, w_1, \overrightarrow{P_d}, x_2, \overrightarrow{P_c})$ as a desired path.

If $v_3 s_1 \in E(G)$, then similar to the arguments for with respect to the vertex y_1, all the end vertices of paths in $\mathbb{P}_2 \cup \mathbb{P}_3$ which are in K are adjacent to x_2. We obtain $(\overrightarrow{P_b}, x_2, s_2, t_1, s_1, v_3, v, v_2, w_4, x_3, w_3, v_1, w_2, x_1, w_1, \overrightarrow{P_c})$ as a desired path. Now we shall consider the case in which v_3 is adjacent to a vertex in $\mathbb{P}_1$; for $w' \in \mathbb{P}_1$, let $v_3 w' \in E(G)$. We obtain desired path P as follows.

If $y_1 x_2, s_1 x_2 \in E(G)$, then $P = (\overrightarrow{P_b}, x_2, \overrightarrow{P_d}, w', v_3, v, v_2, w_4, x_3, w_3, v_1, w_2, x_1, w_1, \overrightarrow{P_c})$.

If $y_1 v_1, s_1 v_1 \in E(G)$, then $P = (\overrightarrow{P_b}, v_1, \overrightarrow{P_d}, w', v_3, v, v_2, w_4 \overleftarrow{P_a} w_1, \overrightarrow{P_c})$.

If $y_1 x_2, s_1 v_1 \in E(G)$, then $P = (\overrightarrow{P_b}, x_2 \overleftarrow{P_a} w_1, \overrightarrow{P_d}, v_1, w_3, x_3, w_4, v_2, v, v_3, w', \overrightarrow{P_c})$.
Note $y_1 v_1, s_1 x_2 \in E(G)$ is a symmetric case.

Proof. when $v' \notin P_a$ and Case 2, which is if there exists $v_1, v_2 \in N^I(v)$ such that $v_1 w_2, v_2 w_3 \in E(G)$, are included in [11].   □

Claim 14. *If there exists $P_a \in \mathbb{P}_6$ and there does not exists $P \in \mathbb{S}_f$ such that $P \neq P_a$ and $|P| \geq 4$, then G has a Hamiltonian path.*

Proof. The proof is similar to the proof of Claim 13.   □

Claim 15. *If $\mathbb{P}_j = \emptyset, j \geq 6$, and $\mathbb{P}_5 \cup \mathbb{P}_4 \neq \emptyset$, then $|\mathbb{P}_5| + |\mathbb{P}_4| \leq 2$.*

Proof. Let $P_a = P(w_1, \ldots, w_l; x_1, x_2)$, $l \in \{2, 3\}$, $P_b = P(s_1, \ldots, s_m; t_1, t_2)$, $m \in \{2, 3\}$. Assume for a contradiction that there exists a path $P_c = P(q_1, \ldots, q_n; r_1, r_2)$, $n \in \{2, 3\}$. From Corollary 2, there exists $v_1 \in N^I(v)$ such that $v_1 w_2, v_1 s_2, v_1 q_2 \in E(G)$. We claim that $v_1 w_1, v_1 s_1, v_1 q_1 \in E(G)$. Suppose $v_1 w_1 \notin E(G)$, then w_1 is adjacent to one among $\{v_2, v_3\}$, and one each from $P_b \cap I$ and $P_c \cap I$. Therefore, $N^I(w_1) \cup \{w_1\}$ induces a $K_{1,4}$, a contradiction. Similar argument holds good for the other edges and thus $v_1 s_1, v_1 q_1 \in E(G)$. If P_a, P_b, P_c are odd paths, then similar to the previous argument, the end vertices w_3, s_3, q_3 are adjacent to v_1. Since the clique is maximal, there exists $w' \in K$ such that $v_1 w' \notin E(G)$. From the previous arguments, $w' \notin \{w_1, \ldots, w_l, s_1, \ldots, s_m, q_1, \ldots, q_n\}$. By Corollary 1, either $v_2 w' \in E(G)$ or $v_3 w' \in E(G)$. Further, we argue that w' is adjacent to one of $\{x_1, x_2\}$, one of $\{t_1, t_2\}$ and one of $\{r_1, r_2\}$. Now, $N^I(w') \cup \{w'\}$ induces a $K_{1,4}$, which is a final contradiction to the existence of such a path P_c. This completes a proof of the claim.   □

Claim 16. *If there exists $P_a, P_b \in \mathbb{P}_5 \cup \mathbb{P}_4$ and there does not exists $P \in \mathbb{S}_f$ such that $P \neq P_a$, $P \neq P_b$ and $|P| \geq 4$, then G has a Hamiltonian path.*

Proof. Case analysis is similar to Claim 10.   □

Claim 17. *If there exists $P_a \in \mathbb{P}_5 \cup \mathbb{P}_4$, and there does not exists $P \in \mathbb{S}_f$ such that $P \neq P_a$, $P \neq P_b$ and $|P| \geq 4$, then G has a Hamiltonian path.*

Claim 18. *If $\mathbb{P}_{j \geq 4} = \emptyset$, then G has a Hamiltonian path.*

Proof. For the above two claims, case analysis are similar to Claim 13.   □

Theorem 4. *Let G be a $K_{1,4}$-free split graph with $|K| \geq |I| - 1 \geq 8$. Then G has a Hamiltonian path if and only if G has no short I-I path, and the sum of the number of I-K paths and the number of short cycles is at most 2. Further, finding such a cycle is polynomial-time solvable.*

Proof. Necessity is trivial. Sufficiency follows from all the results mentioned in this section.   □

3   Hardness Result

Akiyama et al. [12] proved the NP-completeness of Hamiltonian cycle in planar bipartite graphs with maximum degree 3. In [9], Hamiltonian cycle problem in planar bipartite graphs with maximum degree 3 is reduced to Hamiltonian cycle problem in $K_{1,5}$-free split graph. An in depth analysis of the reduction reveals that the reduced instances are split graphs with $\Delta^I \leq 3$. We show a polynomial-time reduction from Hamiltonian cycle problem in split graphs with $\Delta^I \leq 3$ to Hamiltonian path problem in split graphs with $\Delta^I \leq 3$ as follows. Further note that such graphs are sub class of $K_{1,5}$-free split graphs. For a given instance of split graph G with $\Delta^I \leq 3$, having partitions K and I, we create m instances of $K_{1,5}$-free split graphs $G_j, 1 \leq j \leq m$ with partitions K_j and I_j where $m = |E = \{uv : u \in K, v \in I\}|$. That is, corresponding to each edge uv in G such that $u \in K, v \in I$, G_j is constructed as follows:

$$K_j = K \cup \{z\}, I_j = I \cup \{s, t\}, E' = \{zw : w \in K\},$$
$$E(G_j) = E(G) \cup E' \cup \{zt, zv, us\} \setminus \{uv\}.$$

In the next theorem, we prove that Unless $P = NP$, Hamiltonian path problem in $K_{1,5}$-free split graph has no polynomial-time algorithm using the above reduction.

Theorem 5. *Unless $P = NP$, there is no polynomial-time algorithm for Hamiltonian path problem in $K_{1,5}$-free split graphs.*

Proof. Due to page constraints, proof is included in [11].   □

Conclusions: We produced a dichotomy on the Hamiltonian path problem in split graphs. A natural extension is to study longest path problem and minimum-leaf spanning tree problem which are generalizations of Hamiltonian path problem.

References

1. Broersma, H.J.: On some intriguing problems in hamiltonian graph theory - a survey. Discrete Math. **251**(1), 47–69 (2002)
2. Gould, R.J.: Advances on the Hamiltonian problem - a survey. Graphs Comb. **19**(1), 7–52 (2003)
3. Bertossi, A.A., Bonuccelli, M.A.: Hamiltonian circuits in interval graph generalizations. Inf. Process. Lett. **23**(4), 195–200 (1986)
4. Muller, H.: Hamiltonian circuits in chordal bipartite graphs. Discrete Math. **156**(1), 291–298 (1996)
5. Garey, M.R., Johnson, D.S., Tarjan, R.E.: Planar hamiltonian circuit problem is NP-complete. SIAM J. Comput. **5**(4), 704–714 (1976)
6. Gordon, V.S., Orlovich, Y.L., Werner, F.: Hamiltonian properties of triangular grid graphs. Discrete Math. **308**(24), 6166–6188 (2008)
7. Keil, J.M.: Finding Hamiltonian circuits in interval graphs. Inf. Process. Lett. **20**(4), 201–206 (1985)
8. Shih, W.K., Chern, T.C., Hsu, W.L.: An $O(n^2 \log n)$ algorithm for the Hamiltonian cycle problem on circular-arc graphs. SIAM J. Comput. **21**(6), 1026–1046 (1992)
9. Renjith, P., Sadagopan, N.: Hamiltonicity in split graphs - a dichotomy. In: Gaur, D., Narayanaswamy, N.S. (eds.) CALDAM 2017. LNCS, vol. 10156, pp. 320–331. Springer, Cham (2017). https://doi.org/10.1007/978-3-319-53007-9_28
10. West, D.B.: Introduction to Graph Theory. Prentice Hall of India, New Delhi (2003)
11. Renjith, P., Sadagopan, N.: Hamiltonian path in split graphs - a dichotomy. In: arXiv (2017)
12. Akiyama, T., Nishizeki, T., Saito, N.: NP-completeness of the Hamiltonian cycle problem for bipartite graphs. J. Inf. Process. **3**(2), 73–76 (1980)

A Fully Polynomial Time Approximation Scheme for Refutations in Weighted Difference Constraint Systems

Bugra Caskurlu[1], Matthew Williamson[2(✉)], K. Subramani[3],
Vahan Mkrtchyan[4], and Piotr Wojciechowski[3]

[1] TOBB University of Economics and Technology, Ankara, Turkey
caskurlu@gmail.com
[2] Marietta College, Marietta, OH, USA
williamm@marietta.edu
[3] West Virginia University, Morgantown, WV, USA
k.subramani@mail.wvu.edu, pwojciec@mix.wvu.edu
[4] University of Verona, Verona, Italy
vahanmkrtchyan2002@ysu.am

Abstract. This paper is concerned with the design and analysis of approximation algorithms for the problem of finding the least weight refutation in a weighted difference constraint system (DCS). In a weighted DCS (WDCS), a positive weight is associated with each constraint. Every infeasible DCS has a refutation, which attests to its infeasibility. The length of a refutation is the number of constraints used in the derivation of a contradiction. Associated with a DCS **D** is its constraint network **G**. **D** is infeasible if and only if **G** has a simple, negative cost cycle. It follows that the shortest refutation of **D** corresponds to the length of the shortest negative cost cycle in **G**. The constraint network of a WDCS is represented by a constraint network, where each edge contains both a cost and a positive, integral length. In the case of a WDCS, the weight of a refutation is defined as the sum of the lengths of the edges corresponding to the refutation. The problem of finding the minimum weight refutation in a WDCS is called the weighted optimal length resolution refutation (WOLRR) problem and is known to be **NP-hard**. In this paper, we describe a pseudo-polynomial time algorithm for the WOLRR problem and convert it into a fully polynomial time approximation scheme (FPTAS). We also generalize our FPTAS to determine the optimal length refutation of a class of constraints called Unit Two Variable per Inequality (UTVPI) constraints.

Keywords: Difference constraint systems
"No"-certificate · Approximation algorithms · Graph theory
Negative cost cycle · Certification

© Springer International Publishing AG 2018
B. S. Panda and P. P. Goswami (Eds.): CALDAM 2018, LNCS 10743, pp. 45–58, 2018.
https://doi.org/10.1007/978-3-319-74180-2_4

1 Introduction

This paper is concerned with the design and analysis of approximation algorithms for determining the least weight refutation in a weighted difference constraint system (DCS). Every infeasible DCS has a refutation that verifies its infeasibility. For a DCS, the refutation is a subset of the difference constraints such that its conjunction results in a contradiction of the form $0 \leq -b$, $b > 0$. The length of a refutation is the number of difference constraints in the subset that proves the infeasibility of the DCS.

For each DCS $\mathbf{D}$, there exists a corresponding difference constraint network $\mathbf{G}$. $\mathbf{D}$ is infeasible if and only if $\mathbf{G}$ contains a simple, negative cost cycle. The length of a negative cost cycle is the number of edges in the negative cost cycle. The shortest negative cost cycle is defined as the negative cost cycle having the fewest number of edges. It follows that the refutation in $\mathbf{D}$ with the fewest number of constraints corresponds to the length of the shortest negative cost cycle in $\mathbf{G}$.

If each difference constraint in a DCS has unit length, then the problem of determining the length of the refutation with the fewest number of constraints is called the optimal length resolution refutation (OLRR) problem. The OLRR problem is motivated by a number of applications, as discussed in [6], including program verification [5], real-time scheduling [3], and incremental shortest paths in weighted networks [1]. The first polynomial time algorithm for this problem was proposed in [6] and runs in $O(n^3 \cdot \log K)$ time, where n is the number of vertices in $\mathbf{G}$, and K is the OLRR. The current fastest algorithm runs in $O(m \cdot n \cdot K)$ time [7], where m is the number of edges in $\mathbf{G}$.

In this paper, we are interested in a weighted DCS (WDCS), where a positive weight is associated with each constraint. We represent the constraint network of a WDCS $\mathbf{D}$ as a constraint network $\mathbf{G}$, where each edge has both a cost and a positive, integral length. Note that the term "weight" is used for a WDCS, while the term "length" is used for the difference constraint network. In the case of a WDCS, the weight of a refutation is defined as the sum of the lengths of the edges in the corresponding negative cost cycle in $\mathbf{G}$. The problem of finding the minimum weight refutation in a WDCS is called the weighted optimal length resolution refutation (WOLRR) problem. This problem is known to be **NP-hard** [6].

In this paper, we present a pseudo-polynomial time algorithm for computing the WOLRR for a WDCS. Our algorithm applies a dynamic programming approach for computing the WOLRR. If L is the largest edge length in $\mathbf{G}$, our algorithm runs in $O(n^4 \cdot L)$ time. In addition, we present a fully polynomial-time approximation scheme (FPTAS) for the WOLRR problem. This algorithm transforms $\mathbf{G}$ into a simpler network by removing specific edges and then applying the pseudo-polynomial time WOLRR algorithm. The FPTAS runs in $O(n^6 \cdot \frac{1}{\epsilon})$ time, and the approximation ratio is bounded by $(1 + \epsilon)$, where $\epsilon > 0$.

We extend our FPTAS to finding optimal length refutations in Unit Two Variable Per Inequality (UTVPI) constraint systems. These constraints generalize difference constraint systems. We provide both a pseudo-polynomial time algorithm and an FPTAS for finding the weighted optimal length resolution refutation (WOLRR) of a weighted UTVPI constraint system (WUCS).

2   Formal Problem Statement

A constraint of the form $\Sigma_{i=1}^{n} a_i \cdot x_i \leq b$ is called a *linear inequality* or *linear constraint*, where $a_i, b \in \mathbb{R}$. A conjunction of linear inequalities is called a *linear inequality system* or a *linear constraint system*. This system is represented in matrix form as: $\mathbf{A} \cdot \mathbf{x} \leq \mathbf{b}$.

A constraint of the form $x_i - x_j \leq b_{ij}$ is called a *difference constraint*. A conjunction of difference constraints is called a *difference constraint system* (DCS). If $\mathbf{A} \cdot \mathbf{x} \leq \mathbf{b}$ is a DCS, then there exists a vector $\mathbf{x} \in \mathbb{R}^n$ such that $\mathbf{A} \cdot \mathbf{x} \leq \mathbf{b}$ if and only if there exists a vector $\mathbf{x} \in \mathbb{Z}^n$ such that $\mathbf{A} \cdot \mathbf{x} \leq \mathbf{b}$, provided that $\mathbf{b}$ is integral. This is because the constraint matrix of a DCS is totally unimodular [4]. Therefore, the linear and integer programs are indistinguishable from the perspective of non-emptiness of the constraint polyhedron when $\mathbf{b}$ is integral. We should note this is not always the case since linear programming is polynomial time solvable, while integer programming is **NP-hard**.

Given a DCS $\mathbf{A} \cdot \mathbf{x} \leq \mathbf{b}$, we can construct the corresponding *difference constraint network* $\mathbf{G} = \langle \mathbf{V}, \mathbf{E}, \mathbf{b} \rangle$ as follows:

1. For each variable x_i, we create vertex v_i.
2. For each difference constraint $x_i - x_j \leq b_{ij}$, we add the directed edge $e_{ji} = (v_j, v_i)$ with cost b_{ij}.

In the difference constraint network $\mathbf{G}$, $\mathbf{V}$ is the vertex set with n vertices, $\mathbf{E}$ is the edge set with m edges, and $\mathbf{b} : \mathbf{E} \to \mathbb{R}$ is the cost function that assigns a real number to each edge in $\mathbf{E}$.

Every refutation in a DCS corresponds to a simple negative cost cycle in the corresponding constraint network [6]. Thus, any negative cost cycle is a "no"-certificate for the satisfiability of the DCS. It therefore follows that the refutation with the fewest number of constraints corresponds to the simple negative cost cycle with the fewest number of edges. The problem of finding the size of such a negative cost cycle is known as the optimal length resolution refutation (OLRR) problem [6].

In this paper, we are concerned with weighted difference constraints. A constraint $\mathcal{C}$ of the form $x_i - x_j \leq b_{ij}$ is called a *weighted difference constraint* if $\mathcal{C}$ is associated with a weight $l_{ij} > 0$, where $\mathbf{l} : \mathbf{E} \to \mathbb{Z}^+$ is the weight function. Observe that if $\mathbf{E}' \subseteq \mathbf{E}$ is a set of edges in $\mathbf{G}$, then $l(\mathbf{E}') = \sum_{e_{ij} \in \mathbf{E}'} l_{ij}$ is defined as the sum of the lengths of all edges in $\mathbf{E}'$. A conjunction of weighted difference constraints is called a *weighted difference constraint system* (WDCS). Constructing the corresponding constraint network of a WDCS $\mathbf{G} = \langle \mathbf{V}, \mathbf{E}, \mathbf{b}, \mathbf{l} \rangle$ is similar to constructing the constraint network of a DCS. The key difference is that for each constraint $\mathcal{C} : x_i - x_j \leq b_{ij}$ with weight l_{ij}, we add the edge $e_{ji} = (v_j, v_i)$ with cost b_{ij} and *length* $l(\mathcal{C}) = l_{ij}$. We denote the length of a path P_{ij} from vertex v_i to vertex v_j as $l(P_{ij})$. Note that for constraint networks, we use the term "length" rather than "weight."

We already know that if a DCS is unsatisfiable, then there must exist a simple negative cost cycle in the corresponding constraint network [6]. The same applies for a WDCS. Therefore, the refutation with the smallest total weight corresponds to the negative cost cycle ("no"-certificate) with the smallest total length. We call the length of such a negative cost cycle the *weighted optimal length resolution refutation* (WOLRR).

Using the terminology above, we define the WOLRR problem as follows: *Given a WDCS* $D : A \cdot x \leq b$, *where the weight of each constraint is a positive integer, find the weight of a refutation having the smallest total weight.*

Alternatively, based on the equivalence between difference constraints and constraint networks, we define the WOLRR problem as: *Given a network* $G = \langle V, E, b, l \rangle$, *where* b *is the set of real edge costs and* l *is the set of positive integral edge lengths, find the length of a negative cost cycle having the smallest total length.*

3 A Pseudo-Polynomial Time Algorithm

In this section, we present a pseudo-polynomial time algorithm for computing the WOLRR in a WDCS. Recall that an algorithm runs in pseudo-polynomial time if the running time is bounded by both the *size* (i.e., number of bits) and *magnitude* (i.e., value) of the input. Note that pseudo-polynomial time algorithms may run in exponential time in the worst case scenario. However, they may run in polynomial time if the input is bounded by a polynomial function.

Consider the difference constraint network $G = \langle V, E, b, l \rangle$. Our approach applies the pseudo-polynomial time algorithm described in [2]. This algorithm computes the shortest path from source vertex v_1 to all other vertices $v_j \in V$ for networks with positive integral edge costs and having a transition time at most T. Note that [2] denotes the cost of an edge with l_{ij}, while we use b_{ij}. Furthermore, [2] uses the term "delay" (denoted as t_{ij}), while we use the term "length" (denoted as l_{ij}) to define the same property.

Assume the vertices are enumerated from 1 to n, where v_1 denotes the source vertex. Let $L(j, t)$ denote the cost of the shortest path from vertex v_1 to vertex v_j with length at most t. We compute $L(j, t)$ using the following dynamic program [2]:

$$L(j,t) = \begin{cases} 0 & j = 1 \quad\quad t = 0, \ldots, T \\ \infty & j = 2, \ldots, n \; t = 0 \\ \min\left\{ L(j, t-1), \min_{k | t_{kj} \leq t, e_{kj} \in E} \{L(k, t - t_{kj}) + b_{kj}\} \right\} & j = 2, \ldots, n \; t = 1, \ldots, T. \end{cases}$$

Our algorithm modifies the dynamic program for networks with real edge costs. We use the notation $D(j, t)$ rather than $L(j, t)$. This is to differentiate our modified dynamic program from the dynamic program in [2]. We initialize $D(j, t)$ to 0 when $j = 1$. However, we compute $D(j, t) = \min\{D(j, t-1), \min_{k | t_{kj} \leq t, e_{kj} \in E}\{D(k, t - t_{kj}) + b_{kj}\}\}$ when $j = 1, \ldots n$. Note that our algorithm does not apply to networks where the edge lengths may be zero. Otherwise, $D(j, t)$ could be defined in terms of itself if $t_{kj} = 0$.

After computing $D(j, t)$ for all $v_j \in V$ and a single value of t, we check if $D(1, t) < 0$. If this is true, then there exists a negative cost cycle from vertex

v_1 to itself with length t. Otherwise, we repeat the computation for $t+1$, where $t+1 \leq T$.

To compute the WOLRR, we apply the above dynamic program for all vertices. For each source vertex v_s, let $D_s(j,t)$ be the shortest path from v_s to v_j with length t. We compute $D_s(j,t)$ for all values of s, j, and a single value of t. We then check if $D_s(s,t) < 0$ for any $v_s \in \mathbf{V}$. If this is true, we immediately halt the algorithm and return t as the WOLRR. Otherwise, we repeat the calculations for $t+1$, where $t+1 \leq T$. If L denotes the largest length of any edge in $\mathbf{G}$, then we set $T = n \cdot L$, which is the largest possible length for any negative cost cycle.

The above observations are summarized in Algorithms 3.1 and 3.2. Observe that Algorithm 3.2 gives us only the weight of the shortest refutation. The actual negative cost cycle can be obtained by using a predecessor subgraph.

Function SINGLE-VERTEX-WOLRR($\mathbf{G}, v_s, t$)
1: Enumerate the vertices such that v_s is v_1.
2: **for** $(j = 1$ to $n)$ **do**
3: $\quad D_1(j,t) = \min \left\{ D_1(j,t-1), \min_{k|l_{kj} \leq t, e_{kj} \in \mathbf{E}} \{ D_l(k, t - l_{kj}) + b_{kj} \} \right\}$.
4: **if** $(D_1(1,t) < 0)$ **then**
5: $\quad$ **return** (**True**).
6: **return** (**False**).

Algorithm 3.1: Single Vertex WOLRR Algorithm

Function PSEUDO-WOLRR($\mathbf{G}$)
1: $n = |\mathbf{V}|$.
2: $L = \max_{v_i, v_j \in \mathbf{V}} \{ l_{ij} \}$.
3: $T = n \cdot L$.
4: **for** (each vertex $s \in \mathbf{V}$) **do**
5: $\quad$ **for** $(t = 0$ to $T)$ **do**
6: $\quad\quad D_s(s,t) = 0$.
7: $\quad\quad$ **for** $(v_j \in \mathbf{V} - \{s\})$ **do**
8: $\quad\quad\quad D_s(j,t) = \infty$.
9: **for** $(t = 1$ to $T)$ **do**
10: $\quad$ **for** (each vertex $v_s \in \mathbf{V}$) **do**
11: $\quad\quad$ SINGLE-VERTEX-WOLRR($\mathbf{G}, v_s, t$).
12: $\quad\quad$ **if** (SINGLE-VERTEX-WOLRR returned **True**) **then**
13: $\quad\quad\quad$ **return** ("The WOLRR is t").

Algorithm 3.2: Pseudo-Polynomial Time Algorithm for WOLRR

3.1   Analysis

From [2], we know that Algorithm 3.1 takes $O(m)$ time because we scan each edge exactly once.

We now analyze the running time of Algorithm 3.2. The **for** loop at line 4 clearly has $O(n)$ iterations. Observe that the **for** loop at line 5 takes $O(T) = O(n \cdot L)$ time, where L is the largest length among all edges. Furthermore, the **for** loop at line 7 takes $O(n)$ time. Therefore, the **for** loop at line 4 runs in $O(n^2 \cdot L)$ time.

To analyze the **for** loop at line 9, observe that line 11 takes $O(m)$ time since this is Algorithm 3.1. The **for** loop at line 10 has $O(n)$ iterations, and the **for** loop at line 9 has $O(T) = O(n \cdot L)$ iterations. Therefore, the **for** loop at line 9 takes $O(n \cdot L \cdot n \cdot m) = O(m \cdot n^2 \cdot L) = O(n^4 \cdot L)$ time.

3.2   Correctness

We now prove the correctness of our pseudo-polynomial time algorithm. We first address the correctness of the dynamic program. Observe that [2] proves that the dynamic program correctly computes the shortest paths from vertex v_j to all other vertices with length t for $j = 2, \ldots, n$ and $t = 1, \ldots, T$, where $T = n \cdot L$. The key difference with our algorithm is that we include $j = 1$ in the computation. Thus, it must be the case that the dynamic program correctly computes the shortest paths from the source vertex $v_j = v_1$ to all other vertices.

This implies that when we update the value of $D(j, t)$ for $j = 1$, $D(j, t)$ represents the cost of the shortest cycle containing vertex v_1 whose length is at most t. The smallest t, for which $D_s(s, t) < 0$, represents the length of the shortest negative cost cycle $\mathbf{C}$ containing v_s. We need to show that $\mathbf{C}$ is a simple negative cost cycle.

Suppose $\mathbf{C}$ is not a simple negative cost cycle, and there exists a vertex $v \in \mathbf{V}$ that appears in $\mathbf{C}$ more than once. Consider the path along $\mathbf{C}$ from v to itself. This path forms a cycle, which we denote as $\mathbf{C}_1$. Furthermore, $\mathbf{C} \backslash \mathbf{C}_1$ forms a second cycle, denoted as $\mathbf{C}_2$. Observe that the total cost of $\mathbf{C}$ is the sum of the costs of $\mathbf{C}_1$ and $\mathbf{C}_2$. Likewise, the total length of $\mathbf{C}$ is the sum of the lengths of $\mathbf{C}_1$ and $\mathbf{C}_2$.

Since $\mathbf{C}$ has a negative cost, at least $\mathbf{C}_1$ or $\mathbf{C}_2$, or both, must also have a negative cost. Without loss of generality, assume that $\mathbf{C}_1$ is the negative cost cycle. Since all edge lengths are strictly positive, the total length of $\mathbf{C}_1$ is less than the total length of $\mathbf{C}$. This contradicts the fact that $\mathbf{C}$ is the negative cost cycle with the smallest length. Therefore, $\mathbf{C}$ must be a simple negative cost cycle.

4   A Fully Polynomial-Time Approximation Scheme

In this section, we present a fully polynomial time approximation scheme (FPTAS) for computing the WOLRR of a DCS.

4.1   Preprocessing Phase

The first phase of our algorithm converts $\mathbf{G}$ into a simpler network by erasing a carefully selected subset of edges. This phase preserves the WOLRR of $\mathbf{G}$.

Function PRE-PROCESS(**G**)

1: Let A be a vector of edges initialized as $\mathcal{A} = \emptyset$.
2: **while** (**G** has a negative cost cycle) **do**
3:     Let e_{ij} denote the edge of **G** with the largest length.
4:     Remove e_{ij} from **G**.
5:     Add e_{ij} to $\mathcal{A}$.
6:     Let e_{uv} be the last edge added to $\mathcal{A}$.
7:     **for** (each edge e_{st} in $\mathcal{A}$ such that $l_{st} \leq n \cdot l_{uv}$) **do**
8:         Add e_{st} back to **G**.

Algorithm 4.1: Preprocessing Step

Algorithm 4.1 removes the edges of **G** one-by-one in descending order with respect to the lengths until **G** does not have a negative cost cycle. Let e_{uv} be the last edge removed in this manner. Observe that the length of any negative cost cycle in **G** is at least l_{uv}. This is because any negative cost cycle has to contain at least one edge whose length is at least l_{uv}. Therefore, l_{uv} is a lower bound for the WOLRR of **G**.

Consider the moment immediately before the algorithm removes e_{uv} from **G**. l_{uv} is an upper bound for the lengths of the remaining edges in **G** at that time moment since the algorithm removes the edges in descending order with respect to their lengths. Since **G** has a negative cost cycle at that moment, and a simple cycle can have at most n edges, $(n \cdot l_{uv})$ is an upper bound for the WOLRR of **G**. In other words, if $|OPT|$ is the length of the negative cost cycle with the smallest length in **G**, then $|OPT| \leq n \cdot l_{uv}$.

Algorithm 4.1 then inserts the edges with length at most $(n \cdot l_{uv})$ back into **G**. This means that when the algorithm terminates, the only edges that are pruned are the ones whose lengths are more than $(n \cdot l_{uv})$. Note that the transformation made by Algorithm 4.1 on **G** preserves the WOLRR. Since one can check the existence of a negative cost cycle on a constraint network in $O(m \cdot n)$ time, the running time of Algorithm 4.1 is $O(m^2 \cdot n) = O(n^5)$.

4.2   An FPTAS for WOLRR

We next present the main part of our algorithm. Let $\mathbf{G} = \langle \mathbf{V}, \mathbf{E}, \mathbf{b}, \mathbf{l} \rangle$ be the constraint network after the preprocessing step, and let $\epsilon > 0$. We let $P = \frac{\epsilon \cdot l_{uv}}{n}$, where $\epsilon > 0$ is arbitrarily chosen. For each edge e_{ij} remaining in **G**, we set l'_{ij} to be $\left\lceil \frac{l_{ij}}{P} \right\rceil$. We then apply Algorithm 3.2 on $\mathbf{G}' = \langle \mathbf{V}, \mathbf{E}, \mathbf{b}, \mathbf{l}' \rangle$, and the resulting WOLRR is our approximation. The above observations are summarized in Algorithm 4.2.

Let OPT denote the negative cost cycle with the smallest length in **G**. Let OPT' denote the negative cost cycle with the smallest length after running Algorithm 4.2. Let $|OPT|$ and $|OPT'|$ denote the lengths of their respective negative cost cycles. Our algorithm returns $|OPT'|$ at termination. In order to prove that our algorithm is an FPTAS, we will show that $|OPT'| \leq (1 + 2 \cdot \epsilon) \cdot |OPT|$.

Function WOLRR-FPTAS(**G**)

1: PRE-PROCESS().
2: Let **G** be the resulting constraint network.
3: Let $P = \frac{\epsilon \cdot l_{uv}}{n}$.
4: **for** (each edge $e_{ij} \in$ **E**) **do**
5: $\quad l'_{ij} = \left\lceil \frac{l_{ij}}{P} \right\rceil$.
6: Define: $\mathbf{G}' = \langle \mathbf{V}, \mathbf{E}, \mathbf{b}, \mathbf{l}' \rangle$
7: Let OPT' denote the resulting negative cost cycle with the smallest length from running PSEUDO-WOLRR(**G**').
8: **return** ($|OPT'|$).

Algorithm 4.2: FPTAS for WOLRR

Clearly, this will prove our claim since $\epsilon > 0$ is chosen arbitrarily and 2 is a constant.

Recall that for each edge $e_{ij} \in$ **E**, we have $l'_{ij} = \left\lceil \frac{l_{ij}}{P} \right\rceil < \frac{l_{ij}}{P} + 1$. We claim that $l_{ij} < P \cdot l'_{ij} + P$. If $l_{ij} \geq P \cdot l'_{ij} + P$, then $\frac{l_{ij}}{P} \geq l'_{ij} + 1$, and therefore, $l'_{ij} = \left\lceil \frac{l_{ij}}{P} \right\rceil \geq \frac{l_{ij}}{P} \geq l'_{ij} + 1$, which is a contradiction.

Let $l'(\mathbf{C}) = \sum_{e_{ij} \in \mathbf{C}} l'_{ij}$ be defined as the sum of the scaled and rounded lengths of the edges in **C**. If we add the above inequalities for all edges e_{ij} that lie in OPT', we have $l(OPT') < P \cdot l'(OPT') + P \cdot n$. Here we used the fact that OPT' contains at most n edges. Now, observe that OPT' is a negative cost cycle with the smallest length in **G**'. Hence, it must be the case that $l'(OPT') \leq l'(OPT)$. Thus, we will have $l(OPT') < P \cdot l'(OPT) + P \cdot n$. Taking into account that $l'_{ij} < \frac{l_{ij}}{P} + 1$ and $\epsilon \cdot l_{uv} = n \cdot P$, we get

$$|OPT'| = l(OPT') < P \cdot l'(OPT) + P \cdot n < P \cdot (\frac{l(OPT)}{P} + n) + P \cdot n$$
$$= l(OPT) + 2 \cdot P \cdot n = |OPT| + 2 \cdot \epsilon \cdot l_{uv}.$$

Recall that e_{uv} is the last edge added to **G** such that the absence of e_{uv} would result in **G** having no negative cost cycles. This means that any negative cost cycle in **G** must include an edge of length at least l_{uv}. Hence, the length of any negative cost cycle in **G** must be at least l_{uv}. Therefore, $l_{uv} \leq |OPT|$. Thus, we have $|OPT| + 2 \cdot \epsilon \cdot l_{uv} \leq |OPT| + 2 \cdot \epsilon \cdot |OPT| = (1 + 2 \cdot \epsilon) \cdot |OPT|$. Therefore, we can conclude that $|OPT'| \leq (1 + 2 \cdot \epsilon) \cdot |OPT|$.

We now analyze the running time of Algorithm 4.2. As previously stated, Line 1 takes $O(n^5)$ time. The **for** loop at line 4 takes $O(m) = O(n^2)$ time. For line 7, recall that Algorithm 3.2 takes $O(n^4 \cdot L)$ time, where L is the length of the largest edge length. However, in this case, the pseudo-polynomial time algorithms takes $O(n^4 \cdot L')$ time, where $L' = \left\lceil \frac{L}{P} \right\rceil$. Hence, the total running time is $O(n^4 \cdot L')$.

We distinguish two cases. In the first case, $\frac{L}{P} < 1$. This means that $O\left(n^4 \cdot \left\lceil \frac{L}{P} \right\rceil\right) = O\left(n^4\right)$. In the second case, $\frac{L}{P} \geq 1$. This implies that $\left\lceil \frac{L}{P} \right\rceil \leq \left\lfloor \frac{L}{P} \right\rfloor + 1 \leq \frac{L}{P} + 1 \leq 2 \cdot \frac{L}{P}$. Therefore,

$$O\left(n^4 \cdot \left\lceil \tfrac{L}{P} \right\rceil\right) \leq O\left(n^4 \cdot \tfrac{L}{P}\right) = O\left(n^4 \cdot \tfrac{n \cdot L}{\epsilon \cdot l_{uv}}\right) = O\left(n^4 \cdot n \tfrac{L}{\epsilon \cdot l_{uv}}\right)$$
$$\leq O\left(n^5 \cdot \tfrac{n \cdot l_{uv}}{\epsilon \cdot l_{uv}}\right) = O\left(n^6 \cdot \tfrac{1}{\epsilon}\right)$$

Observe that if $\epsilon < 1$, then $O(n^6 \cdot \tfrac{1}{\epsilon})$ dominates the running time of our algorithm. Since the running time is polynomial in both $(1/\epsilon)$ and the size of the input instance, the above algorithm is an FPTAS.

5   The WOLRR Problem in UTVPI Constraint Systems

In this section, we generalize the algorithms discussed in previous sections to determine optimal length refutations in UTVPI constraint systems (UCS). We recall that a UTVPI constraint has the form: $a_i \cdot x_i + a_j \cdot x_j \leq b_{ij}$, where $a_i, a_j \in \{0, 1, -1\}$. Clearly, such constraints are more general than difference constraints, since in the latter we must have $a_i = -a_j$.

5.1   Constructing the Constraint Network

Our algorithms utilize the constraint network introduced in [8]. Let $\mathbf{U} : \mathbf{A} \cdot \mathbf{x} \leq \mathbf{b}$ be the UTVPI constraint system. We can construct the corresponding UTVPI constraint network $\mathbf{G} = \langle \mathbf{V}, \mathbf{E}, \mathbf{b} \rangle$ as follows: For each variable x_i, we create a vertex $v_i \in \mathbf{V}$. For each constraint, we create an edge in $\mathbf{E}$ using the following rules:

1. A constraint of the form $x_i - x_j \leq b_{ij}$ is represented as an undirected "gray" edge, $(v_j \overset{b_{ij}}{\blacksquare} v_i)$, or $(v_i \overset{b_{ij}}{\blacksquare} v_j)$, with cost b_{ij}.
2. A constraint of the form $-x_i - x_j \leq b_{ij}$ is represented by an undirected "black" edge, $(v_i \overset{b_{ij}}{\blacksquare} v_j)$, with cost b_{ij}.
3. A constraint of the form $x_i + x_j \leq b_{ij}$ is represented by an undirected "white" edge, $(v_i \overset{b_{ij}}{\square} v_j)$, with cost b_{ij}.

We then add vertex v_0 to $\mathbf{G}$. This vertex allows us to include absolute constraints which are of the form $a_i \cdot x_i \leq b_i$, where $a_i \in \{0, 1, -1\}$. Each absolute constraint $x_i \leq b_i$ is replaced by a pair of constraints $x_i + x_0 \leq b_i$ and $x_i - x_0 \leq b_i$. The corresponding edges $(v_0 \overset{b_i}{\square} v_i)$ and $(v_0 \overset{b_i}{\blacksquare} v_i)$ are added to $\mathbf{G}$.

Unlike difference constraint networks, a cycle in a UTVPI constraint network may consist of edges and vertices that occur more than once. An edge reduction represents the addition of the two corresponding UTVPI constraints such that the resulting constraint is also a UTVPI constraint. An edge reduction can be thought of as determining the edge equivalent to a two-edge path. If a path can be reduced to an edge of type t, where $t \in \{\square, \blacksquare, \blacksquare, \blacksquare\}$, then we say that the path has type t. The shortest path of type t between v_i and v_j is a path of type t from v_i and v_j with minimum cost. Furthermore, a *negative cost gray cycle* is a path that can be reduced to an edge $(v_i \overset{b_i}{\blacksquare} v_i)$ (or $(v_i \overset{b_i}{\blacksquare} v_i)$) such that $b_i < 0$.

UTVPI constraint systems may require refutations that use constraints more than once. In fact, every infeasible UTVPI constraint system has a refutation where each constraint is used at most twice [8]. [8] also shows that a UTVPI constraint system has a refutation of length l if and only if the corresponding constraint network has a negative cost gray cycle of length l. Thus, the optimal length refutation can be obtained by finding the negative cost gray cycle with the smallest length.

This paper is concerned with the Weighted OTLR (WOLRR) problem. In this problem, each UVTPI constraint $a_i \cdot x_i + a_j \cdot x_j \leq b_{ij}$ also has a positive weight l_{ij}, where $\mathbf{l} : \mathbf{E} \to \mathbb{Z}^+$ is the weight function. Constructing the constraint network is similar to constructing the constraint network of a UCS except we assign to each edge e_{ij} a length l_{ij}. Note that the term "length" is used for the constraint network rather than "weight." The refutation with the smallest total weight corresponds to the negative cost gray cycle with the smallest total length. We call the length of such a negative cost gray cycle the WOLRR.

5.2 Modifying the Pseudo-Polynomial Time Algorithm to Handle UTVPI Constraints

In this subsection, we describe how we modify Algorithm 3.2 to find the WOLRR of a weighted UTVPI constraint system. Recall that in Sect. 3, we modified the dynamic program described by Goel [2] for networks with real edge costs. To utilize the dynamic program for UTVPI constraints, we make the following additional modifications:

1. We apply the dynamic program for each edge type in the weighted UCS. Let $D^{(\square)}(j,t)$, $D^{(\blacksquare)}(j,t)$, $D^{(\square)}(j,t)$, and $D^{(\square)}(j,t)$ denote the shortest path from vertex v_1 to v_j with length at most t for the respective edge types.
2. We initialize $D^{(\square)}(j,t)$, $D^{(\blacksquare)}(j,t)$, $D^{(\square)}(j,t)$, and $D^{(\square)}(j,t)$ to 0 when $j = 1$.
3. We let v_r, instead of v_k, denote the neighboring vertex of v_j.
4. We compute the minimum value for $j = 1, \ldots n$ instead of $j = 2, \ldots, n$. This allows us to compute the shortest path from vertex v_1 to itself.
5. Instead of computing $\min\{D(j, t-1), \min_{k|t_{kj} \leq t, e_{kj} \in \mathbf{E}}\{D(k, t - t_{kj}) + b_{kj}\}\}$, we apply the dynamic programs from [8]. We formally define our dynamic programs below.

$$D^{(\square)}(j,t) = \begin{cases} 0 & j=1 \quad t=0 \\ \infty & j=2,\ldots,n \quad t=0 \\ \min \begin{cases} D^{(\square)}(j, t-1) \\ \min_{r|l_{rj}\leq t} \begin{cases} D^{(\square)}(r, t - l_{rj}) + b(v_r \blacksquare v_j) \\ D^{(\square)}(r, t - l_{rj}) + b(v_r \square v_j) \end{cases} \end{cases} & \begin{matrix} j=1,\ldots,n \quad t=1,\ldots,T. \end{matrix} \end{cases}$$

$$D^{(\,\blacksquare\,)}(j,t) = \begin{cases} 0 & j = 1 \quad\quad t = 0 \\ \infty & j = 2,\dots,n \;\; t = 0 \\ \min\begin{cases} D^{(\,\blacksquare\,)}(j,t-1) \\ \min_{r\,|\,l_{rj}\le t}\begin{cases} D^{(\,\blacksquare\,)}(r,t-l_{rj}) + b(v_r \,\blacksquare\, v_j) \\ D^{(\,\blacksquare\,)}(r,t-l_{rj}) + b(v_r \,\square\, v_j) \end{cases} & j = 1,\dots,n \;\; t = 1,\dots,T. \end{cases} \end{cases}$$

$$D^{(\,\blacksquare\,)}(j,t) = \begin{cases} 0 & j = 1 \quad\quad t = 0 \\ \infty & j = 2,\dots,n \;\; t = 0 \\ \min\begin{cases} D^{(\,\blacksquare\,)}(j,t-1) \\ \min_{r\,|\,l_{rj}\le t}\begin{cases} D^{(\,\blacksquare\,)}(r,t-l_{rj}) + b(v_r \,\blacksquare\, v_j) \\ D^{(\,\blacksquare\,)}(r,t-l_{rj}) + b(v_r \,\blacksquare\, v_j) \end{cases} & j = 1,\dots,n \;\; t = 1,\dots,T. \end{cases} \end{cases}$$

$$D^{(\,\blacksquare\,)}(j,t) = \begin{cases} 0 & j = 1 \quad\quad t = 0 \\ \infty & j = 2,\dots,n \;\; t = 0 \\ \min\begin{cases} D^{(\,\blacksquare\,)}(j,t-1) \\ \min_{r\,|\,l_{rj}\le t}\begin{cases} D^{(\,\blacksquare\,)}(r,t-l_{rj}) + b(v_r \,\blacksquare\, v_j) \\ D^{(\,\square\,)}(r,t-l_{rj}) + b(v_r \,\blacksquare\, v_j) \end{cases} & j = 1,\dots,n \;\; t = 1,\dots,T. \end{cases} \end{cases}$$

After computing $D^{(type)}(j,t)$ for all $j \in V$, for all $type \in \{\, \square\,, \,\blacksquare\,, \,\blacksquare\,, \,\blacksquare\,\}$, and a single value of t, we check if $D^{(\,\blacksquare\,)}(1,t) < 0$ (or $D^{(\,\blacksquare\,)}(1,t) < 0$). If this is true, then there exists a negative cost gray cycle from vertex v_1 to itself with length t. Otherwise, we repeat the computation for $(t+1) \le T$.

As with Algorithm 3.2, we apply the dynamic programs for all vertices. For each source vertex v_s, $D_s^{(type)}(j,t)$ is the shortest path from v_s to v_j with length t, where $type \in \{\, \square\,, \,\blacksquare\,, \,\blacksquare\,, \,\blacksquare\,\}$. We compute $D_s^{(type)}(j,t)$ for all values of v_s, v_j, $type$, and a single value of t. We then check if $D_s^{(\,\blacksquare\,)}(s,t) < 0$ (or $D_s^{(\,\blacksquare\,)}(s,t) < 0$) for any $v_s \in V$. If this is true, then t is the WOLRR. Otherwise, we repeat the calculations for $(t+1) \le T$. If L denotes the largest edge length of any edge in $\mathbf{G}$, then we set $T = (2 \cdot n + 2) \cdot L$, which is the largest possible length for any negative cost gray cycle.

The above observations are summarized in Algorithms 5.1 and 5.2.

Function SINGLE-VERTEX-WOLRR-UTVPI$(\mathbf{G}, v_s, t)$

1: Enumerate the vertices such that v_s is v_1.
2: **for** $j = 1$ to n **do**
3:     **for** $(type \in \{\, \square\,, \,\blacksquare\,, \,\blacksquare\,, \,\blacksquare\,\})$ **do**
4:         Compute $D_1^{(type)}(j,t)$ as described above.
5: **if** $(D_1^{(\,\blacksquare\,)}(j,t) < 0)$ **then**
6:     **return** (**True**).
7: **return** (**False**).

Algorithm 5.1: Single Vertex WOLRR Algorithm for UTVPI Constraints

Function PSEUDO-WOLRR-UTVPI(**G**)
1: $n = |\mathbf{V}|$.
2: $L = \max_{v_i, v_j \in \mathbf{V}} \{l_{ij}\}$.
3: $T = (2 \cdot n + 2) \cdot L$.
4: **for** (each vertex $v_s \in \mathbf{V}$) **do**
5:    **for** ($t = 0$ to T) **do**
6:       **for** ($type \in \{\square, \blacksquare, \blacksquare, \blacksquare\}$) **do**
7:          $D_s^{(type)}(s,t) = 0$.
8:          **for** ($v_j \in \mathbf{V} - \{v_s\}$) **do**
9:             $D_s^{(type)}(j,t) = \infty$.
10: **for** ($t = 1$ to T) **do**
11:    **for** (each vertex $v_s \in \mathbf{V}$) **do**
12:       SINGLE-VERTEX-WOLRR-UTVPI(**G**, v_s, t).
13:       **if** (SINGLE-VERTEX-WOLRR-UTVPI returned **True**) **then**
14:          **return** ("The WOLRR is t").

Algorithm 5.2: Pseudo-Polynomial Time Algorithm for WOLRR in UTVPI Constraints

Analysis. From [2], we know that Algorithm 5.1 takes $O(m)$ time because we scan each edge exactly once. In Algorithm 5.2, observe that lines 10 to 14 dominate the running time. Since these steps are identical to lines 9 to 13 in Algorithm 3.2, and those steps take $O(m \cdot n^2 \cdot L) = O(n^4 \cdot L)$ time, Algorithm 5.1 takes $O(n^4 \cdot L)$ time.

Correctness. We now prove the correctness of our pseudo-polynomial time algorithm. We first address the correctness of the dynamic program in Algorithm 5.1. Consider the dynamic program for computing $D_1^{(\square)}(j,t)$. When computing the shortest white path from v_j to all other vertices (i.e., $t = 1, \ldots, T$), we compute the following dynamic program:

$$D^{(\square)}(j,t) = \min \begin{cases} D^{(\square)}(j,t-1) \\ \min_{r|l_{rj} \le t} \begin{cases} D^{(\square)}(r, t - l_{rj}) + b(v_r \,\blacksquare\, v_j) \\ D^{(\blacksquare)}(r, t - l_{rj}) + b(v_r \,\square\, v_j) \end{cases} \end{cases}$$

From [8], we know that the above dynamic program correctly computes the shortest white paths in a UCS. When we include the case where $t = 0$, we have the same dynamic program from Sect. 3 except it applies to a WUCS instead of a WDCS. Since Sect. 3.2 proves the correctness of the dynamic program in Sect. 3, it must be the case that our dynamic program correctly computes the shortest white paths in a WUCS. Similar arguments can be used to prove that $D_1^{(\blacksquare)}(j,t)$, $D_1^{(\blacksquare)}(j,t)$, and $D_1^{(\blacksquare)}(j,t)$ correctly compute their respective shortest paths in a WUCS.

This implies that when we update the values of $D_1^{(type)}(j,t)$ for $j = 1$ and $type \in \{\square, \blacksquare, \blacksquare, \blacksquare\}$, $D_1^{(\blacksquare)}(j,t)$ represents the cost of the shortest gray cycle containing vertex v_1 whose length is at most t. The smallest t, for which $D_s^{(\blacksquare)}(s,t) < 0$, represents the length of the shortest negative cost gray cycle $\mathbf{C}_s$ containing v_s. The smallest length among all $\mathbf{C}_i$ for all $v_i \in \mathbf{V}$ must give the WOLRR.

5.3   Modifying the FPTAS for the WOLRR Problem

We now describe how we modify our FPTAS in Sect. 4.2 to detect the WOLRR. We find that our modified FPTAS is nearly identical to Algorithm 4.2 with a few changes:

1. In Line 7 of Algorithm 4.1, we change $l_{ij} \le n \cdot l_{uv}$ to $l_{ij} \le (2 \cdot n + 2) \cdot l_{uv}$. This is because a negative cost gray cycle has at most $(2 \cdot n + 2)$ edges. Hence, $((2 \cdot n + 2) \cdot l_{uv})$ is an upper bound for the WOLRR.
2. In Line 3 of Algorithm 4.2, we change $\frac{\epsilon \cdot l_{uv}}{n}$ to $\frac{\epsilon \cdot l_{uv}}{2 \cdot n + 2}$. This is because the largest negative cost gray cycle has $(2 \cdot n + 2)$ edges.
3. In Line 7 of Algorithm 4.2, we execute PSEUDO-WOLRR-UTVPI($\mathbf{G}'$) instead of PSEUDO-WOLRR($\mathbf{G}'$).

Observe that the running time of Algorithm 5.2 is the same as the running time of Algorithm 3.2. Thus, using the same analysis for the FPTAS in Sect. 4.2, our FPTAS for the WOLRR runs in $O(n^6 \cdot \frac{1}{\epsilon})$ time.

Acknowledgments. This work was done while the first author was at West Virginia University. The first author was supported in part by the National Science Foundation through Award CNS-0849735 and the Air Force Office of Scientific Research through Award FA9550-12-1-0199. The third author was supported in part by the National Science Foundation through Awards CCF-1305054 and CNS-0849735, and the Air Force Office of Scientific Research through Award FA9550-12-1-0199. The fourth author was supported in part by the Air Force Office of Scientific Research through Award FA9550-12-1-0199. The fifth author was supported in part by the National Science Foundation through Award CCF-1305054, and by NASA through the West Virginia Space Grant. We thank Ashish Goel for useful conversations.

References

1. Demtrescu, C., Italiano, G.F.: A new approach to dynamic all pairs shortest paths. J. ACM **51**(6), 968–992 (2004)
2. Goel, A., Ramakrishnan, K.G., Kataria, D., Logothetis, D.: Efficient computation of delay-sensitive routes from one source to all destinations. In: Proceedings of the IEEE Conference on Computer Communications. Twentieth Annual Joint Conference of the IEEE Computer and Communications Society, INFOCOM 2001 (Cat. No. 01CH37213), vol. 2, pp. 854–858 (2001)
3. Han, C.C., Lin, K.J.: Job scheduling with temporal distance constraints. Technical report, UIUCDCS-R-89-1560, University of Illinois at Urbana-Champaign, Department of Computer Science (1989)

4. Nemhauser, G.L., Wolsey, L.A.: Integer and Combinatorial Optimization. Wiley, New York (1999)
5. Seshia, S.A., Lahiri, S.K., Bryant, R.E.: A hybrid SAT-based decision procedure for separation logic with uninterpreted functions. In: DAC, pp. 425–430 (2003)
6. Subramani, K.: Optimal length resolution refutations of difference constraint systems. J. Autom. Reason. (JAR) **43**(2), 121–137 (2009)
7. Subramani, K., Williamson, M., Gu, X.: Improved algorithms for optimal length resolution refutation in difference constraint systems. Form. Asp. Comput. **25**(2), 319–341 (2013)
8. Subramani, K., Wojciechowski, P.J.: A combinatorial certifying algorithm for linear feasibility in UTVPI constraints. Algorithmica **78**(1), 166–208 (2017)

Probabilistic Properties of Highly Connected Random Geometric Graphs

Bodo Manthey and Victor M. J. J. Reijnders[✉]

University of Twente, Enschede, The Netherlands
{b.manthey,v.m.j.j.reijnders}@utwente.nl

Abstract. In this paper, we study the probabilistic properties of reliable networks of minimum costs in d-dimensional Euclidean space. We study reliability in terms of k-edge-connectivity in graphs. We show that this problem fits into Yukich's framework for Euclidean functionals for arbitrary k, dimension d and distant-power gradient p with $p < d$. With this framework results on convergence and concentration of the value of optimal solutions of random inputs follow. These results are then extended to optimal k-edge-connected power assignment graphs, where we assign transmit power to vertices, and two vertices are connected if they both have sufficient transmit power. This variant models wireless networks. Finally, we devise a partitioning heuristic to find approximate solutions quickly, and we analyze its performance in the framework of smoothed analysis.

Keywords: Random geometric graphs · Average-case analysis
Connectivity of graphs

1 Introduction

The design of fault tolerant networks is an important issue in today's research, due to their numerous applications. The goal is to find cheap and reliable networks with some specific characteristics. Reliability is generally expressed in terms of the connectivity of the network. Applications for these type of problems can be found in the design of reliable communication and transportation networks [2,10].

Wireless ad hoc networks have also received significant attention in recent studies [6,9]. Instead of direct connections between nodes, communication can also take place by relaying through intermediate nodes. Here we assign a transmission power to each node. As the transmission range is directly related to the power used by a node, the goal is to find a fault tolerant network with minimum total power usage. Possible applications are environmental monitoring, emergency disaster relief where wiring is difficult, and networks of vehicles [3,9].

In this paper, we study the probabilistic properties of the value of the optimal solution within Yukich's framework for Euclidean optimization problems. This yields results both about convergence and concentration of the value of optimal

© Springer International Publishing AG 2018
B. S. Panda and P. P. Goswami (Eds.): CALDAM 2018, LNCS 10743, pp. 59–72, 2018.
https://doi.org/10.1007/978-3-319-74180-2_5

solutions as a function of the number n of vertices, dimension d of the underlying Euclidean space, distance-power gradient p, and connectivity requirement k.

Finding a cheapest k-edge-connected network is NP-hard [8], and so is finding a minimal power wireless network [5]. As we still want to have reasonably good solutions in acceptable computation time, we need to find a good approximation algorithm. Partitioning algorithms are a simple, easy-to-implement type of heuristics that show good performance on Euclidean optimization problems [4]. We devise a partitioning heuristic for the design of optimal k-edge-connected networks. Furthermore, we analyze it in the framework of smoothed analysis [14, 19] in order to explain its performance.

The rest of this paper is organized as follows. In Sect. 2 we give the relevant definitions. We summarize related work in Sect. 3. The properties of k-edge-connected graphs are in Sect. 4. The partitioning heuristic and its smoothed analysis are presented in Sect. 5. We extend these results to k-edge-connected power graphs in Sect. 6. We conclude with some open problems.

2 Definitions

Let $V \subseteq \mathbb{R}^d$ be a finite set of vertices, where $d \in \mathbb{N}$ is an arbitrary constant. In the rest of the paper, $n = |V|$ is the number of vertices. For two nodes $u, v \in V$, let (u, v) denote the edge connecting u and v, and let $|(u, v)|$ denote the Euclidean distance between u and v.

A graph $G = (V, E)$ is called k-edge-connected if G is connected after removal of any set of at most $k - 1$ of its edges. An alternative characterization of k-edge-connectedness is that there exist k edge-disjoint paths between every pair of vertices. In this paper, we also call any complete graph k-edge-connected, even if it contains fewer than k vertices. (Otherwise, no k-edge-connected graphs below a certain size exist, which would cause technical issues.)

In this paper, we study k-edge-connected graphs of minimal costs, where costs are defined as the sum of all edge lengths. The cost of an edge (u, v) is $|(u, v)|^p$, where $p > 0$ denotes the distant-power gradient. For a given graph $G = (V, E)$, the costs of this graph is the sum of its edge costs, i.e., $\sum_{(u,v) \in E} |(u, v)|^p$. Then $k\text{-EC}^p(V)$ is the minimum cost of any k-edge-connected graph on V with costs computed with distant-power gradient p. In the remainder of this paper, $k \in \mathbb{N}$, $d \in \mathbb{N}$, and $p > 0$ are assumed to be fixed constants.

Besides the model above, where we pay per edge, and which could be viewed as modeling wired networks, we also consider a model where we assign transmit power $\text{PA}(v)$ to the vertices v. Two vertices u and v are connected by an edge if both have sufficient transmit power, i.e., if $\text{PA}(u), \text{PA}(v) \geq |(u, v)|^p$. The costs of such a power assignment is the sum of all transmit powers, i.e., $\sum_{v \in V} \text{PA}(v)$. The graph resulting of a power assignment is called the corresponding power assignment graph.

$k\text{-ECPA}^p(V)$ denotes the minimum costs of any power assignment whose corresponding power assignment graph is k-edge-connected. Both $k\text{-ECPA}^p$ and $k\text{-EC}^p$ are Euclidean functionals. This means that they map a finite point set to a non-negative real number, are translation invariant, and scaling all points by a factor of $\alpha > 0$ changes the costs by a factor of α^p [20].

Following Yukich [20], for a Euclidean functional L^p, we write $L^p(V, R)$ to denote the functional on $V \cap R$, where R is some hyperrectangle. Usually, $R = [0,1]^d$, and we omit R if it is clear from the context.

In order to fit $k\text{-EC}^p$ and $k\text{-ECPA}^p$ into Yukich's framework for Euclidean optimization problems [20], we have to define corresponding canonical boundary functionals, an idea first articulated in Redmond's thesis [16]. Roughly speaking, in these functionals, the entire boundary of the rectangle is considered as one additional vertex that can be used. To distinguish between a functional and its boundary functional, we refer to the former as the original functional.

Given a hyperrectangle R and a point set $V \subseteq R$, a boundary graph is a graph with nodes V plus the boundary ∂R of R as additional node. A vertex v is connected to the boundary ∂R by adding edge (v, v_∂) where $v_\partial = \arg\min_{w \in \partial R} |(v, w)|$. A boundary graph is called k-edge-connected if the graph restricted to V is k-edge-connected, or if the graph on $V \cup \{\partial R\}$ is k-edge-connected. Here, any edge connecting $v \in V$ to ∂R counts as up to k independent edges. We denote by $k\text{-EC}_B^p$ the boundary functional corresponding to $k\text{-EC}^p$. This means that $k\text{-EC}_B^p(V, R)$ is the minimum total length of a k-edge-connected graph in terms of summed edge lengths on $V \cup \partial R$ in d-dimensional rectangle R with pth power-weighted edges. Similarly, $k\text{-ECPA}_B^p$ is the boundary functional of $k\text{-ECPA}^p$. Here v is connected to the boundary if $\text{PA}(v) \geq |(v, v_\partial)|^p$.

3   Related Work

For a survey about properties of k-edge-connected graphs, we refer to Kammer and Täubig [12]. Finding k-edge-connected networks of minimum costs is NP-hard for $k \geq 2$, both in the classical and the power assignment model [5,8]. Therefore, a considerable amount of research has been focused on approximation algorithms.

Khuller and Vishkin [13] proved that the problem of finding a minimum-cost k-edge connected graph can be approximated within a factor of 2 approximation algorithm. Czumaj and Lingas [7] gave a polynomial-time approximation scheme for this problem for the Euclidean case, where points are contained in $\mathbb{R}^d$ for some fixed d.

Althaus et al. [1] devised several approximation algorithms for the problem of finding a connected power assignment graph. Santi et al. [18] studied the connectivity of power assignment graphs in the Euclidean case for $d \in \{1, 2, 3\}$ under the restriction that every vertex is assigned the same power r. They derived bounds for r to achieve connectivity with high probability. De Graaf and Manthey [9] analyzed connectivity in power assignment graphs. They proved properties similar to our results of Sect. 4 for simple connectivity.

Călinescu and Wan [6] analyzed several approximation algorithms for finding cheap k-edge-connected power assignment graphs and obtained an approximation ratio of $2k$ for this problem.

4   Properties of k-EC^p

In this section we show that k-EC^p fits into Yukich's framework for Euclidean functionals [20]. We make heavy use of the following lemma, which states that the union of two k-edge-connected graphs with non-empty intersection is also k-edge-connected.

Lemma 4.1 (Matula [15]). *Let $G_1 = (V_1, E_1)$ and $G_2 = (V_2, E_2)$ be k-edge-connected graphs with $V_1 \cap V_2 \neq \emptyset$. Then the graph $H = (V_1 \cup V_2, E_1 \cup E_2)$ is k-edge-connected.*

4.1   Yukich's Framework

First, we prove that k-EC^p is a geometrically subadditive functional. This shows that the function value of a whole set is not larger – up to an additive error term – than the sum of the function values of the sets in a partition.

Lemma 4.2. *For $p \geq 1$, k-EC^p is geometrically subadditive, i.e. for all finite sets V, all rectangles R and partitions of R into rectangles R_1 and R_2 we have*

$$k\text{-EC}^p(V, R) \leq k\text{-EC}^p(V, R_1) + k\text{-EC}^p(V, R_2) + C_1(\text{diam } R)^p,$$

where $C_1 = C_1(d, p)$ is a constant.

Proof. Consider the graphs (V_1, E_1) and (V_2, E_2) with $V_i = V \cap R_i$ that realize the optimal solutions of k-EC$^p(V, R_1)$ and k-EC$^p(V, R_2)$, respectively. Without loss of generality, we assume that $|V_1| \geq |V_2|$. We distinguish three cases.

1. $|V_1|, |V_2| \geq k + 1$. We join (V_1, E_1) and (V_2, E_2) by k vertex-disjoint edges $e_1, \ldots, e_k$. This results in a k-edge-connected graph on V. We have $|e_i| \leq$ diam R. Thus, the costs of this k-edge-connected graph is bounded by

$$k\text{-EC}^p(V, R) \leq k\text{-EC}^p(V, R_1) + k\text{-EC}^p(V, R_2) + \sum_{i=1}^{k} |e_i|^p$$

$$\leq k\text{-EC}^p(V, R_1) + k\text{-EC}^p(V, R_2) + kd^{p/2}(\text{diam } R)^p.$$

2. $|V_1| \geq k + 1$ and $|V_2| \leq k$. We know that (V_2, E_2) is complete. For each vertex $v_i \in V_2$ we add k edges $e_{i_1}, \ldots e_{i_k}$ with endpoints in V_1. Each pair of vertices from V_1 has at least k edge-disjoint paths in E_1, and by adding e_{i_j} each pair in V has k edge-disjoint paths as well. At most k^2 edges have been added this way, and we can bound the costs of the k-edge-connected graph thus obtained in a similar way as in the first case.

3. $|V_1|, |V_2| \leq k$. Both (V_1, E_1) and (V_2, E_2) are complete, and we know that a complete graph is always k-edge-connected. So if we make the combined graph complete as well, it is k-edge-connected. We need to add at most k^2 edges this way, and k-$\text{EC}^p(V, R)$ can be bounded as in the previous cases. $\quad\square$

Subadditivity gives an upper bound on the growth of k-EC^p as a function of $|V|$.

Lemma 4.3 (growth bound). *Let $0 < p < d$. Then there exists a constant $C = C(d, p)$ such that for all cubes $R \subset \mathbb{R}^d$ and all $V \subset R$, we have*

$$k\text{-}\text{EC}^p(V, R) \leq C(\operatorname{diam} R)^p |V|^{(d-p)/d}.$$

Proof. This lemma follows directly by combining Lemma 4.2 and a result by Yukich [20, Lemma 3.3]. $\quad\square$

Unfortunately, as the Euclidean functionals considered by Yukich [20], k-EC^p is not superadditive. If it were superadditive, then together with subadditivity this makes the functional nearly additive in the sense that

$$k\text{-}\text{EC}^p(V, R) \approx k\text{-}\text{EC}^p(V, R_1) + k\text{-}\text{EC}^p(V, R_2).$$

We could then approximate the optimal solution value of the whole set by the sum of optimal solutions on its partitions. The way to superadditivity is via the canonical boundary functional of k-EC_B^p, which is superadditive according to the following lemma.

Lemma 4.4. *For $p \geq 1$, k-EC_B^p is superadditive, i.e. for all finite sets V, all rectangles R and all partitions of R into rectangles R_1 and R_2 we have*

$$k\text{-}\text{EC}_B^p(V, R) \geq k\text{-}\text{EC}_B^p(V, R_1) + k\text{-}\text{EC}_B^p(V, R_2).$$

Proof. Let V and R together with a partition of R into rectangles R_1 and R_2 be given. Let $V_i = V \cap R_i$ for $i \in \{1, 2\}$. Let E be the edge set of a boundary k-edge-connected graph on V or $V \cup \{\partial R\}$ of minimum costs. For each edge $(u_1, u_2) \in E$ with $u_i \in V_i$, we add two edges: One edge connecting u_1 to the closest point ∂u_1 on the boundary of R_1, and one edge connecting u_2 to the closest point ∂u_2 on the boundary of R_2. Now, he two induced subgraphs are k-edge-connected. The sum of the costs of these two graphs does not exceed the costs of k-$\text{EC}^p(V, R)$ by the triangle inequality and because $p \geq 1$. $\quad\square$

Next, we show that k-EC^p and k-EC_B^p are pointwise close. This yields that both are approximately subadditive and superadditive.

Lemma 4.5. *For $1 \leq p < d$, k-EC^p is pointwise close to k-EC_B^p, i.e. for all finite sets $V \subset [0, 1]^d$ we have*

$$\left| k\text{-}\text{EC}^p(V) - k\text{-}\text{EC}_B^p(V) \right| = o(|V|^{(d-p)/d}).$$

Proof. Let $V \subseteq [0,1]^d$ be a finite set of points. Clearly, $k\text{-EC}^p(V) \geq k\text{-EC}^p_B(V)$. Thus, we only have to prove $k\text{-EC}^p(V) \leq k\text{-EC}^p_B(V) + o(|V|^{(d-p)/p})$. If $|V| \leq k$, then this holds as $|V|$ is constant. Thus, we assume that $|V| \geq k+1$ from now on. We first need the following claim.

Claim 4.6. *Let $V \subseteq [0,1]^d$, $|V| = n$, and $1 \leq p < d$. Consider a graph $G = (V, E)$ that realizes the optimal solution of $k\text{-EC}^p_B(V)$. Then the sum of the p-th powers of the lengths of the edges connecting vertices in V with the boundary of $[0,1]^d$ is bounded by $O(n^{(d-p-1)/(d-1)})$.*

Proof. The proof follows ideas from Yukich [20] and depends on a dyadic subdivision of $[0,1]^d$. Let Q_0 by the cube of edge length $1/3$ and centered within $[0,1]^d$. Let Q_1 be the cube of edge length $2/3$, also centered within $[0,1]^d$. We partition $Q_1 - Q_0$ into subcubes of edge length $1/6$. The number of such subcubes is bounded by $C6^{d-1}$ for some constant $C = C(d)$.

We continue with this subdivision recursively. This means that at the j-th stage we define cube Q_j of edge length $1 - 2(3 \cdot 2^j)^{-1}$ and partition $Q_j - Q_{j-1}$ into subcubes of edge length $(3 \cdot 2^j)^{-1}$. The number of such subcubes is bounded from above by $C3^{d-1}2^{j(d-1)}$. We carry out this recursion until the ℓ-th stage, where ℓ is the unique integer satisfying $2^{(\ell-1)(d-1)} \leq n \leq 2^{\ell(d-1)}$.

This procedure produces nested cubes $Q_1 \subseteq Q_2 \subseteq \cdots \subseteq Q_\ell$. It produces a dyadic covering of the cube until the moat $[0,1]^d - Q_\ell$ has a width of $O(n^{-1/(d-1)})$. We use these properties to prove Claim 4.6 as follows.

This dyadic subdivision partitions the largest cube Q_ℓ into at most $\sum_{j=0}^{k} C3^{d-1}2^{j(d-1)} \leq Cn$ subcubes, each with an edge length equal to the distance between the subcube and the boundary of $[0,1]^d$. Furthermore, by partitioning each subcube into $(k2^y)^d$ congruent subcubes, where y is the least integer satisfying $2^y \geq d^{1/2}$, we obtain a partition $\mathcal{P}$ of Q_ℓ consisting of at most Cn subcubes with the property that k times the diameter of each subcube is less than the distance to the boundary.

Suppose $V \subseteq [0,1]^d$ is a finite point set and $G = (V, E)$ is the graph realizing the optimal solution of $k\text{-EC}^p_B(V)$. We observe that in G, each subcube Q in $\mathcal{P}$ contains at most k points in V which are rooted to the boundary. If there were more than k points in $V \cap Q$ rooted to the boundary, we can do the following. We know that these points will not have edges directly between them as they already have k edge-disjoint paths between them, and edges between them would then not all be in G. So we can take one of them and connect it to all the other points rooted to the boundary (which are at least k), while removing the connection to the boundary. As the diameter of each subcube is less than $1/k$ times the distance to the boundary, this relinking gives us a cheaper solution, contradicting the optimality of G.

The sum of the p-th powers of the lengths of the edges connecting vertices in $V \cap (Q_j - Q_{j-1})$ with the boundary is thus bounded by the product of the number of subcubes in $Q_j - Q_{j-1}$ and the p-th power of the common diameter of the subcubes, namely

$$C3^{d-1}2^{j(d-1)} \cdot (3 \cdot 2^j)^{-p} := A(d, p, j).$$

Summing over all $1 \leq j \leq \ell$ gives a bound for the sum of the p-th power of the lengths of the edges connecting points to the boundary in $V \cap Q_\ell$:

$$\sum_{j=1}^{\ell} A(d,p,j) \leq \begin{cases} C \max\{n^{(d-p-1)/(d-1)}, \log n\} & \text{if } 1 \leq p \leq d-1 \\ C & \text{if } d-1 < p < d. \end{cases} \tag{1}$$

The $\log n$ term is needed to cover the case $p = d-1$. The sum of the p-th powers of the lengths of the edges connecting vertices in $V \cap ([0,1]^d - Q_\ell)$ with the boundary is at most the product of $n = |V|$ and the p-th power of the width of the moat $[0,1]^d - Q_\ell$, i.e. at most

$$n \cdot Cn^{-p/(d-1)} = Cn^{(d-p-1)/(d-1)}. \tag{2}$$

Combining (1) and (2) proves the claim.   □

We can now continue with the proof of Lemma 4.5. Consider $U \subset V$ the set of all vertices connected to the boundary. Let $\mathcal{B} \subseteq \partial[0,1]^d$ be the set of points on the boundary to which vertices are connected. Then $|U| \geq |\mathcal{B}|$. As we want to remove all edges to the boundary to get to a solution for our original functional, we have to add new edges to maintain k-edge-connectivity. Recall that edges to the boundary can count as up to k edges. As in the proof of Claim 4.6, we know there are no edges between the points of U. To get a good bound on the increase of costs that incurs by adding these edges, we use the following lemma.

Lemma 4.7. *Fix $1 \leq p < d$ and let $G = (V, E)$ be a k-edge-connected graph realizing the optimal solution for k-$EC^p(V, [0,1]^d)$. Then there exists a constant $c = c(k, d)$ such that the degree of every vertex $v \in V$ is bounded by c.*

Proof. Let us assume to the contrary that there exists a vertex $v \in V$ for which the degree is not bounded by c. We then divide $[0,1]^d$ into cones originating from v and with the property that for every two points x and y in a cone we have $\angle(x, y, v) < \pi/3$. In this way, the number of cones we create is finite since d is constant, and every point is covered by a cone. We consider a cone C with an unbounded number of points connected to v and look at the point y that is furthest from v and connected to v. Such a cone has to exist as the number of cones is bounded. Let us consider two cases.

In the first case, the degree of y is greater than c as well, and y is connected to all vertices in C that are also connected to v. This means both are connected to more than k vertices in C. In that case we can remove the edge between v and y as we would still have more than k edge-disjoint paths between v and y (and other points will not be affected). Removing an edge would only lower the cost for the graph, so this would contradict the optimality of the solution.

In the other case, we can find a vertex to which y is not connected, but v is. Let us call this vertex z and consider the triangle $\Delta(v, z, y)$. As we know that $\angle(z, v, y) < \pi/3$, either $\angle(z, y, v) > \pi/3$ or $\angle(v, z,)y > \pi/3$ (or both). Using that $|(v, y)| \geq |(v, z)|$, we can see that $|(y, z)| < |(v, y)|$. If we then replace the edge from v to y by the edge from y to z, the number of edge-disjoint paths from y to

z or v cannot decrease. But as $|(y, z)| < |(v, y)|$, we also have $|(y, z)|^p < |(v, y)|^p$. Thus, we have lowered the cost and still have a k-edge-connected graph. This again would contradict the optimality of G.   □

With this lemma we can continue the proof of Lemma 4.5. We can also get a bound on the length of each edge we need to add. By using the triangle inequality for $p = 1$ we get $|(u, v)| \le |(u, u_\partial)| + |(u_\partial, v_\partial)| + |(v_\partial, v)|$. Using the triangle inequality for powers of metrics, we obtain

$$\sum_{(u,v) \in M} |(u, v)|^p \le 2^{p-1} \left(\sum_{(u,\cdot) \in M} |(u, u_B)|^p + \sum_{(u,v) \in M} |(u_B, v)|^p \right) \tag{3}$$

$$\le 4^{p-1} \left(\sum_{(u,\cdot) \in M} |(u, u_B)|^p + \sum_{(u,v) \in M} |(u_B, v_B)|^p + \sum_{(\cdot,v) \in M} |(v_B, v)|^p \right)$$

$$\le 4^p \sum_{(u,\cdot) \in M} |(u, u_B)|^p + 4^{p-1} k\text{-EC}^p(\mathcal{B}, [0, 1]^d)$$

$$\le 4^p C_1 \sum_{u \in U} |(u, u_B)|^p + 4^{p-1} C_2 |V|^{(d-p-1)/(d-1)} \qquad \text{(by Lemma 4.3)}$$

$$\le 4^p C_1 C_3 |V|^{(d-p-1)/(d-1)} + o(|V|^{(d-p)/d}) \qquad \text{(by Claim 4.6)}$$

$$\le o(|V|^{(d-p)/d}).$$

Now we have changed a graph that achieves the optimal solution for $k\text{-EC}_B^p(V, [0, 1]^p)$ into a k-edge-connected graph. The cost of this new graph provides an upper bound for $k\text{-EC}^p(V)$. Observing that we have increased the costs by at most $o(|V|^{(d-p)/d})$ concludes the proof.   □

We have shown geometric subadditivity, superadditivity, and pointwise closeness. The last property that we need is smoothness. Roughly speaking, smoothness means that adding or removing a few vertices does not change the function value by much.

Theorem 4.8. *For $1 \le p < d$, $k\text{-EC}^p$ is smooth, i.e. for all finite sets $U, V \subseteq [0, 1]^d$ we have*

$$\left| k\text{-EC}^p(U \cup V) - k\text{-EC}^p(U) \right| = O(|V|^{(d-p)/d}).$$

Proof. Subadditivity (Lemma 4.2) and Lemma 4.3 together yield

$$k\text{-EC}^p(U \cup V) \le k\text{-EC}^p(U) + k\text{-EC}^p(V) + O(1) \le k\text{-EC}^p(U) + O(|V|^{(d-p)/d}).$$

It remains to be shown that

$$k\text{-EC}^p(U, [0, 1]^d) - k\text{-EC}^p(U \cup V, [0, 1]^d) \le O(|V|^{(d-p)/d}).$$

We start with a graph $G = (U \cup V, E)$ that realizes $k\text{-EC}^p(U \cup V)$. After removal of V, we can modify the remaining graph to obtain a k-edge-connected graph on U without increasing the cost by more than $O(|V|^{(d-p)/d})$. Let $N_V \subseteq U$ be the set of direct neighbors of vertices in V. By Lemma 4.7 we know that $|N_V| \leq c|V|$ for some constant c, that only depends on k and d. Let $m = |N_V|$.

We now compute a k-edge-connected graph T of minimum cost on N_V. By Lemma 4.3, the cost of this graph is bounded by $O(m^{(d-p)/d})$. Let F be the set of edges of the subgraph of G induced by U. Clearly, the graph with edge set $T \cup F$ has weight at most $k\text{-EC}^p(U \cup V, [0,1]^d) + O(|V|^{(d-p)/d})$. It remains to be proved that it is k-edge-connected. Fix any two vertices $u, v \in V$. We distinguish three cases:

1. Both u and v are in N_V. As T is k-edge-connected graph, there are k edge-disjoint paths connecting u to v.
2. Only one of u and v is in N_V. Without loss of generality, let $v \in N_V$ and $u \in U \backslash N_V$. Consider k edge-disjoint paths $P_1, \ldots, P_k$ from u to v in $(U \cup V, E)$. Let q_i be the first vertex of N_V that P_i reaches. The nodes $q_1, \ldots, q_k$ are not necessarily distinct, and we can have $q_i = v$. However, since T is k-edge-connected, there exist k edge-disjoint paths within T connecting q_i to v for each $i \in \{1, \ldots, k\}$.
3. Both u and v are not in N_V. Take any $x \in N_V$, then Item 2 yields that there are k edge-disjoint paths from u to x and from x to v. By Lemma 4.1, we know there are also k edge-disjoint paths from u to v. $\square$

Smoothness for the boundary functional follows with an almost identical proof.

Theorem 4.9. *For $1 \leq p < d$, $k\text{-EC}_B^p$ is smooth.*

4.2   Limit Theorems

For the theorems on the convergence of optimal solutions, we need the notion of complete convergence. Let X_n, for $n \in \mathbb{N}$, be a sequence of random variables. Then X_n converges completely (c.c.) to a constant C if and only if for all $\epsilon > 0$ we have

$$\sum_{n=1}^{\infty} P(|X_n - C| > \epsilon) < \infty.$$

This notion of convergence was first introduced by Hsu and Robbins [11].

As we have geometric subadditivity, superadditivity, pointwise closeness, and smoothness, several limit theorems directly follow. These are given in this section. The results show that the functional on random points is highly concentrated around its expected value.

Theorem 4.10. *Let V be a set of n points drawn independently and uniformly from $[0,1]^d$. Fix $1 \leq p < d$ and $k \in \mathbb{N}$. Then there exists a positive constant $\alpha = \alpha(k\text{-EC}^p, d, k)$ such that*

$$\lim_{n \to \infty} \frac{k\text{-EC}^p(V)}{n^{(d-p)/p}} = \alpha \quad c.c., \; and$$

$$\lim_{n \to \infty} \frac{k\text{-EC}_B^p(V)}{n^{(d-p)/p}} = \alpha \quad c.c.$$

The following famous limit theorem is due to Rhee [17]. This theorem states that the solution value is not far from its expected value when we look at randomly places vertices.

Theorem 4.11. *Fix $d \geq 2$, $1 \leq p < d$ and $k \in \mathbb{N}$. Let V be a set of n points drawn independently and uniformly from $[0,1]^d$. Then there exists constants $c_1 = c_1(k\text{-EC}^p, d)$ and $c_2 = c_2(k\text{-EC}^p, d)$ such that for all $t > 0$ we have*

$$\mathbb{P}\left(\left|k\text{-EC}^p(V) - \mathbb{E}[k\text{-EC}^p(V)]\right| > t\right) \leq c_1 \exp\left(\frac{-c_2 t^{2d/(d-p)}}{n}\right).$$

5   Partitioning Heuristics

Partitioning heuristics are a generic approach to design heuristics for Euclidean optimization problems: The d-dimensional Euclidean space is divided into a number of cells such that each cell contains only a small number of points. This allows es to compute quickly optimal solutions for the set of points in each cell. Finally, the solutions of the individual cells are combined to obtain a solution of the whole set of points. We describe this more formally in the following Algorithm 5.1.

Algorithm 5.1. (Partitioning Scheme)
Input: *set $V \subseteq [0,1]^d$ of n points and number $s > k$.*

1. *Partition $[0,1]^d$ into $\ell = \sqrt[d]{n/s}$ stripes of dimension $d-1$ such that each stripe contains exactly $n/\ell = (n^{d-1}s)^{1/d}$ points.*
2. *Keep partitioning each $i+1$-dimensional stripe into ℓ stripes of dimension i such that each stripe contains exactly $n/\ell^i = (n^{d-i}s^i)^{1/d}$ points. Stop at $i = 1$ so that each 2-dimensional stripe is partitioned into ℓ cells with $n/\ell^d = s$ points. In this way we end up with $\ell^d = n/s$ cells.*
3. *Compute a graph achieving the optimal solution of k-EC^p for each cell.*
4. *Join the graphs to obtain a k-edge-connected graph on V as follows: Choose k points of each cell. Connect these k points to the k points of an adjacent cell such that the graph becomes k-edge-connected.*

Overall, we obtain the following upper bound on the approximation performance.

Theorem 5.2. *Let $s > k$, and let $1 \leq p < d$. The partitioning heuristic (Algorithm 5.1) for k-EC^p can be implemented to run in time $2^{O(s^2)} + O(n)$. Furthermore, let $\text{PSE}^p(V)$ denote the cost of the k-edge-connected graph computed by Algorithm 5.1. Then $\text{PSE}^p(V) \leq k\text{-EC}^p(V) + O\left((n/s)^{\frac{d-p}{d}}\right)$.*

Proof. First, the graph that we get as an output from Algorithm 5.1 is k-edge-connected because of Lemma 4.1. Thus, a feasible solution is computed. Second, a simple brute-force search shows that an optimal k-edge-connected graph of s points can be computed in time $2^{O(s^2)}$. The joining can easily be done in linear time. Third, let $\text{PSE}^p(V)$ be the cost of the k-edge-connected graph on V computed by Algorithm 5.1. By using subadditivity, $\text{PSE}^p(V)$ is bounded from above by $k\text{-EC}^p(V) + C_1$ plus the costs of joining the optimal solutions in the cells (Step 4). Here, $C_1 = C_1(d, p) > 0$ is the constant of Lemma 4.2.

The joining (Step 4) can be implemented to yield additional costs of at most $O((n/s)^{(d-p)/d})$ using Theorem 4.3). We do this by lacing cells together in a snakelike succession. We create a minimal length matching of k vertices between two succeeding cells. It is easy to check that this yields a k-edge-connected graph on the cells we lace together. This yields the last claim of the theorem. $\square$

The running-time of the partitioning algorithm is polynomial for $s = \Omega(n/\sqrt{\log n})$. The above theorem does not yield an approximation ratio in the worst case. If the value of an optimum solution is small compared to the additive error term, then the approximation guarantee is poor. However, typically, this is not the case, and partitioning heuristics work quite well on typical instances.

In order to explain this, partitioning heuristics for Euclidean optimization problems have been analyzed in the framework of smoothed analysis for Euclidean optimization problems introduced by Bläser et al. [4]. They use the so-called one-step model to construct semi-random instances: Let $\phi \geq 1$ be a perturbation parameter. For each of the n points, an adversary specifies a probability density function $[0, 1]^d \to [0, \phi]$. Then the points are drawn independently according to their respective probability density function. The smoothed performance is then the maximum expected performance that the adversary can achieve by choosing the density functions. The parameter ϕ limits the power of the adversary: If $\phi = 1$, then the adversary can only choose the uniform distribution on the unit hypercube $[0, 1]^d$. For larger ϕ, the adversary can concentrate more and more probability mass and, thus, is able to specify (worst-case) instances more accurately.

For ease of presentations, we restrict ourselves to the case $d = 2$ and $p = 1$ in the remainder of this section, because Bläser et al. [4] also stated their results only for $d = 2$. It is quite straightforward to generalize the results to larger values of d, but it seems to be non-trivial to generalize them to $p > 1$. We obtain the following smoothed analysis result.

Theorem 5.3. *For $p = 1$, $d = 2$, and $s = \Theta(\sqrt{\log n})$, Algorithm 5.1 has polynomial running-time and achieve a smoothed approximation ratio of $1 + O\left(\sqrt{\frac{\phi}{\sqrt{\log n}}}\right)$.*

Proof. The polynomial running-time follows from the discussion above. The approximation ratio follows from a result by Bläser et al. [4, Theorem 3.8] together with Theorem 5.2. $\square$

6   Extension to k-ECPA

The results for $k\text{-EC}^p$ are easily copied to $k\text{-ECPA}^p$ by making some small adjustments to the proofs. We start off with subadditivity.

Lemma 6.1. *For $p \geq 1$, $k\text{-ECPA}^p$ is a geometric subadditive functional.*

Proof. We follow the proof of Lemma 4.2. We need to increase $PA(v)$ accordingly for the edges that need to be added in the power assignment graph. We define $d_{uv} = \max\{0, |(u, v)| - PA(u)\}$. This denotes the increase in power needed for u to reach v. Then we can add an edge (u, v) by a change of power equal to $(PA(u) + d_{uv})^p + (PA(v) + d_{vu})^p - PA(u)^p - PA(v)^p \leq 2(\text{diam } R)^p$. The rest of the proof follows.   □

Proving superadditivity of k-ECPA$_B^p$ is even easier than proving it of k-EC$_B^p$ as the power of vertices that lose a connection is already large enough to reach to boundary.

Lemma 6.2. *For $p \geq 1$, k-ECPA$_B^p$ is superadditive.*

Obtaining pointwise closeness of k-ECPA^p and k-ECPA$_B^p$ is a bit more difficult as not all theorems we used in the proof of Lemma 4.5 hold for these functionals. We first state corresponding auxiliary results.

Claim 6.3. *Let $V \subset [0, 1]^d$, $|V| = n$, and $1 \leq p < d$. Consider a graph realising the optimal solution of k-ECPA$_B^p(V, [0, 1]^d)$. The sum of the p-th powers of the lengths of the edges connecting vertices in V with the boundary of $[0, 1]^d$ is bounded by $O(n^{(d-p-1)/(d-1)})$.*

Proof. The proof is similar to that of Claim 4.6, except that for power assignment, in case more than k points are rooted to the boundary, all of them are also connected to each other (as the diameter of the subcube is less than the distance to the boundary). Without changing the solution, we can remove one of the roots to the boundary. This also gives us that we only have to take into account at most k points rooted to the boundary in each subcube. The rest of the proof follows.   □

As we know the degree of vertices in an optimal power assignment graph can be unbounded [9], we cannot show Lemma 4.7 for k-ECPA^p as well. We do however have the following lemma:

Lemma 6.4. *Fix $1 \leq p < d$, let $V \subset [0, 1]^d$ be V is a finite subset and let R be a d-dimensional rectangle. Then we have that k-ECPA$^p(V, R) \leq 2\,k$-EC$^p(V, R)$.*

Proof. Consider a graph $G = (V, E)$ achieving an optimal solution for k-EC$^p(V, R)$. Then for each vertex $v \in V$ we take the longest edge from v and we set power $PA(v)$ to the length of this edge. Now all edges that were in E, are also in the graph resulting from power assignment PA. So the power assignment graph resulting from power assignment PA is also k-edge-connected, and has costs no more than twice k-EC$^p(V, R)$. An optimal solution for k-ECPA$^p(V, R)$ cannot have costs higher than PA, so k-ECPA$^p(V, R) \leq 2k$-EC$^p(V, R)$ follows. □

These now make sure we can obtain pointwise closeness.

Lemma 6.5. *For $1 \leq p < d$, k-ECPA^p is pointwise close to k-ECPA$_B^p$.*

Proof. The proof for power assignments is the same as the proof of Lemma 4.5, except that use Lemma 6.4 and Eq. (3) to get the following result for a k-edge-connected power assignment graph on U

$$k\text{-ECPA}^p(U, [0,1]^d) \le 2\,k\text{-EC}^p(U, [0,1]^d) = 2 \sum_{(u,v) \in M} \mathrm{d}(u,v)^p \le o(|V|^{(d-p)/d}),$$

where M is the set of edges used create a k-edge-connected graph on U. This gives us for k-ECPA^p that we have $k\text{-ECPA}^p(V, [0,1]^d) \le k\text{-ECPA}_B^p(V, [0,1]^d) + o(|V|^{(d-p)/p})$, and therefore k-ECPA$_B^p$ and k-ECPA^p are pointwise close. □

If we try to extend the proof of Theorem 4.8 to power assignments, we get into trouble with the possible unbounded degree of power assignment. So instead of trying to bound the number of vertices connected to one vertex, we bound the number of k-edge-connected components connected to one vertex. To do this, we use another lemma [9, Lemma 3.2]. This way, we get smoothness for k-ECPA^p.

Theorem 6.6. *For $1 \le p < d$, k-ECPA^p is smooth.*

Now the limit theorems also hold for k-ECPA^p (Theorems 4.10 and 4.11), as well as the results for the partitioning algorithm (Theorems 5.2 and 5.3).

7 Conclusions and Open Problems

In this paper, we have looked at fault tolerant networks in terms of k-edge-connected graphs. We studied both a standard (wired) model and a model for wireless networks. We analyzed the corresponding Euclidean functionals k-EC^p and k-ECPA^p on random inputs. We fitted k-EC^p into Yukich's framework for Euclidean functionals, and obtained probabilistic results for k-EC^p. With Yukich's framework we derived several concentration results. We derived a partitioning heuristic for k-EC^1, for which we proved an additive approximation guarantee. We analyzed its approxmiation ratio in the framework of smoothed analysis. Finally, we transferred results to k-ECPA^p.

We conclude this paper with a few open problems for future research.

In this paper we have looked only at k-edge-connected graphs. But we can also consider connectivity in terms of k-vertex-connected graphs. While subadditivity, superadditivity of the corresponding boundary functional, and pointwise closeness is relatively straightforward to prove, we feel that the main technical difficulty in proving smoothness is the lack of a counterpart of Lemma 4.1.

Another possible extension would be the case $p \ge d$, which would require closeness in mean and smoothness in mean, because pointwise closeness and smoothness do not hold in this setting.

References

1. Althaus, E., Călinescu, G., Mandoiu, I.I., Prasad, S., Tchervenski, N., Zelikovsky, A.: Power efficient range assignment for symmetric connectivity in static ad hoc wireless networks. Wireless Netw. **12**(3), 287–299 (2006)

2. Bendali, F., Diarrassouba, I., Mahjoub, A.R., Didi Biha, M., Mailfert, J.: A branch-and-cut algorithm for the k-edge connected subgraph problem. Networks **55**(1), 13–32 (2010)
3. Bettstetter, C.: On the minimum node degree and connectivity of a wireless multi-hop network. In: Proceedings of the 3rd ACM International Symposium on Mobile Ad Hoc Networking and Computing (MobiHoc), pp. 80–91. ACM (2002)
4. Bläser, M., Manthey, B., Rao, B.V.R.: Smoothed analysis of partitioning algorithms for euclidean functionals. Algorithmica **66**(2), 397–418 (2013)
5. Clementi, A.E.F., Penna, P., Silvestri, R.: On the power assignment problem in radio networks. Mob. Netw. Appl. **9**(2), 125–140 (2004)
6. Calinescu, G., Wan, P.-J.: Range assignment for high connectivity in wireless ad hoc networks. In: Pierre, S., Barbeau, M., Kranakis, E. (eds.) ADHOC-NOW 2003. LNCS, vol. 2865, pp. 235–246. Springer, Heidelberg (2003). https://doi.org/10.1007/978-3-540-39611-6_21
7. Czumaj, A., Lingas, A.: On approximability of the minimum-cost k-connected spanning subgraph problem. In: Tarjan, R.E., Warnow, T.J. (eds.) Proceedings of the 10th Annual ACM-SIAM Symposium on Discrete Algorithms (SODA), pp. 281–290. ACM/SIAM (1999)
8. Garey, M.R., Johnson, D.S.: Computers and Intractability: A Guide to the Theory of NP-Completeness (1979)
9. de Graaf, M., Manthey, B.: Probabilistic analysis of power assignments. Random Struct. Algorithms **51**(3), 483–505 (2017)
10. Grötschel, M., Monma, C.L., Stoer, M.: Polyhedral approaches to network surviv-ability. In: Roberts, F., Hwang, F., Monma, C.L. (eds.) Proceedings of the Work-shop on Reliability of Computer and Communication Networks. Series in Discrete Mathematics and Theoretical Computer Science, vol. 5, pp. 121–141. American Mathematical Society (1991)
11. Hsu, P.L., Robbins, H.: Complete convergence and the law of large numbers. Proc. Nat. Acad. Sci. **33**(2), 25–31 (1947)
12. Kammer, F., Täubig, H.: Connectivity. In: Brandes, U., Erlebach, T. (eds.) Net-work Analysis. LNCS, vol. 3418, pp. 143–177. Springer, Heidelberg (2005). https://doi.org/10.1007/978-3-540-31955-9_7
13. Khuller, S., Vishkin, U.: Biconnectivity approximations and graph carvings. J. ACM **41**(2), 214–235 (1994)
14. Manthey, B., Röglin, H.: Smoothed analysis: analysis of algorithms beyond worst case. IT – Inf. Technol. **53**(6), 280–286 (2011)
15. Matula, D.W.: The cohesive strength of graphs. In: Chartrand, G., Kapoor, S.F. (eds.) The Many Facets of Graph Theory. LNM, vol. 110, pp. 215–221. Springer, Heidelberg (1969). https://doi.org/10.1007/BFb0060120
16. Redmond, C.: Boundary rooted graphs and euclidean matching algorithms. Ph.D. thesis, Lehigh University, Bethlehem, PA, USA (1993)
17. Rhee, W.T.: A matching problem and subadditive euclidean functionals. Ann. Appl. Probab. **3**(3), 794–801 (1993)
18. Santi, P., Blough, D.M., Vainstein, F.: A probabilistic analysis for the range assign-ment problem in ad hoc networks. In: Proceedings of the 2nd ACM International Symposium on Mobile Ad Hoc Networking and Computing, pp. 212–220. ACM (2001)
19. Spielman, D.A., Teng, S.H.: Smoothed analysis: an attempt to explain the behavior of algorithms in practice. Commun. ACM **52**(10), 76–84 (2009)
20. Yukich, J.E.: Probability Theory of Classical Euclidean Optimization Problems. LNM, vol. 1675. Springer, Heidelberg (1998). https://doi.org/10.1007/BFb0093472

On Indicated Coloring of Some Classes of Graphs

P. Francis, S. Francis Raj$^{(\boxtimes)}$, and M. Gokulnath

Department of Mathematics, Pondicherry University,
Puducherry 605014, India
selvafrancis@gmail.com, francisraj_s@yahoo.com,
gokulnath.math@gmail.com

Abstract. Indicated coloring is a slight variant of the game coloring which was introduced by Grzesik [6]. In this paper, we obtain structural characterization of connected $\{P_5, K_4, Kite, Bull\}$-free graphs which contains an induced C_5 and connected $\{P_6, C_5, K_{1,3}\}$-free graphs that contains an induced C_6. Also, we prove that these graphs are k-indicated colorable for all $k \geq \chi(G)$. In addition, we show that complete expansion of C_5 is k-indicated colorable for all $k \geq \chi(G)$ and as a consequence, we exhibit that $\{P_2 \cup P_3, C_4\}$-free graphs, $\{P_5, C_4\}$-free graphs are k-indicated colorable for all $k \geq \chi(G)$. This partially answers one of the questions which was raised by Grzesik [6].

Keywords: Game chromatic number · Indicated chromatic number
P_5-free graphs

2000 AMS Subject Classification: 05C75

1 Introduction

All graphs considered in this paper are simple, finite and undirected. For $S, T \subseteq V(G)$, let $\langle S \rangle$ denote the subgraph induced by S in G and let $[S, T]$ denote the set of all edges with one end in S and the other end in T. Let $\mathcal{F}$ be a family of graphs. We say that G is $\mathcal{F}$-free if it contains no induced subgraph which is isomorphic to a graph in $\mathcal{F}$. For two vertex-disjoint graphs G_1 and G_2, the join of G_1 and G_2, denoted by $G_1 + G_2$, is the graph whose vertex set $V(G_1 + G_2) = V(G_1) \cup V(G_2)$ and the edge set $E(G_1 + G_2) = E(G_1) \cup E(G_2) \cup \{xy : x \in V(G_1), y \in V(G_2)\}$. We write $H \sqsubseteq G$, if H is an induced subgraph of G. Next, the coloring number of a graph G, denoted by $\mathrm{col}(G)$, is defined by $\mathrm{col}(G) = 1 + \max_{H \subseteq G} \delta(H)$.

A game coloring of a graph is a coloring of the vertices in which two players Ann and Ben are alternatively coloring the vertices of the graph G properly by using a fixed set of colors C. The first player Ann is aiming to get a proper coloring of the whole graph, where as the second player Ben is trying to prevent the realization of this project. If all the vertices are colored then Ann wins the

© Springer International Publishing AG 2018
B. S. Panda and P. P. Goswami (Eds.): CALDAM 2018, LNCS 10743, pp. 73–80, 2018.
https://doi.org/10.1007/978-3-319-74180-2_6

game, otherwise Ben wins (that is, at that stage of the game there appears a block vertex. A *block* vertex means an uncolored vertex which has all colors from C on its neighbors). The minimum number of colors required for Ann to win the game on a graph G irrespective of Ben's strategy is called the game chromatic number of the graph G and it is denoted by $\chi_g(G)$. There has been a lot of papers on game coloring. See for instance, [11–13]. In indicated coloring, the role of Ann and Ben is as follows: in each round the first player Ann selects a vertex and the second player Ben colors it properly, using a fixed set of colors. The aim of Ann as in game coloring is to achieve a proper coloring of the whole graph G, while Ben tries to "block" some vertex. The smallest number of colors required for Ann to win the game on a graph G is known as the indicated chromatic number of G and is denoted by $\chi_i(G)$. Clearly from the definition we see that $\omega(G) \leq \chi(G) \leq \chi_i(G) \leq \Delta(G)+1$. For a graph G, if Ann has a winning strategy while using k colors, then we say that G is k-indicated colorable.

In [13], Zhu has asked the following question for game coloring. If Ann has a winning strategy using k colors, will Ann have a winning strategy using $k + 1$ colors? The same question was asked by Grzesik for indicated coloring. The question can be equivalently stated as "Whether G is k-indicated colorable for every $k \geq \chi_i(G)$". There has been already some partial answers to this question. For instance see [5, 8, 10]. In this paper, we obtain structural characterization of connected $\{P_5, K_4, Kite, Bull\}$-free graphs which contains an induced C_5 and connected $\{P_6, C_5, K_{1,3}\}$-free graphs which contains an induced C_6. Also, we prove that these graphs are k-indicated colorable for all $k \geq \chi(G)$. In addition, we show that $\mathbb{K}[C_5]$, the complete expansion of C_5, is k-indicated colorable for all $k \geq \chi(G)$ and as a consequence, we exhibit that $\{P_2 \cup P_3, C_4\}$-free graphs, $\{P_5, C_4\}$-free graphs are k-indicated colorable for all $k \geq \chi(G)$.

2 Structural Characterization of Some Free Graphs and Their Indicated Coloring

For a bipartite graph G, the game chromatic number can be arbitrarily large when compared to the indicated chromatic number. In [6], Grzesik has shown that $\chi_i(G) = 2$.

Theorem 2.1 ([6]). *Every bipartite graph is k-indicated colorable for every* $k \geq 2$.

Next, let us recall the definition of complete expansion and independent expansion of a graph G. Let G be a graph on n vertices $v_1, v_2, \ldots, v_n$, and let $H_1, H_2, \ldots, H_n$ be n vertex-disjoint graphs. An expansion $G(H_1, H_2, \ldots, H_n)$ of G is the graph obtained from G by

(i) replacing each v_i of G by H_i, $i = 1, 2, \ldots, n$, and
(ii) by joining every vertex in H_i with every vertex in H_j whenever v_i and v_j are adjacent in G.

For $i \in \{1, 2, \ldots, n\}$, if $H_i = K_{m_i}$, then $G(H_1, H_2, \ldots, H_n)$ is said to be a complete expansion of G and is denoted by $\mathbb{K}[G](m_1, m_2, \ldots, m_n)$ or $\mathbb{K}[G]$. For $i \in \{1, 2, \ldots, n\}$, if $H_i = \overline{K_{m_i}}$, then $G(H_1, H_2, \ldots, H_n)$ is said to be an independent expansion of G and is denoted by $\mathbb{I}[G](m_1, m_2, \ldots, m_n)$ or $\mathbb{I}[G]$.

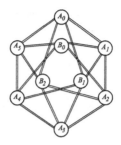

Fig. 1. $\{P_6, C_5, K_{1,3}\}$-free graph contains an induced C_6

The study of P_5-free graphs and P_6-free graphs has been of interest for a lot of coloring parameters. For instance see, $[2, 4, 7, 9]$. In this direction, we have obtained the structural characterization of a connected $\{P_6, C_5, K_{1,3}\}$-free graph that contains an induced C_6.

Theorem 2.2. *If G is a connected $\{P_6, C_5, K_{1,3}\}$-free graph which contains an induced C_6 then G is isomorphic to the graph given in Fig. 1. Here $V(G) = (\cup_{i=0}^{5} A_i) \cup (\cup_{j=0}^{2} B_j)$ and the circle denote the complete subgraph induced by the sets A_i and B_j and the double line between any two sets denote the join of the two sets.*

Proof. Let G be a connected $\{P_6, C_5, K_{1,3}\}$-free graph that contains an induced $C_6 \cong \langle \{v_0, v_1, v_2, v_3, v_4, v_5\} \rangle = \langle N_0 \rangle$, and let $N_i = \{x \in V(G) : \mathrm{dist}(x, N_0) = i\}$, $i \geq 1$.

Claim 1: If $x \in N_1$, then $\langle N(x) \cap N_0 \rangle \cong P_3$ or $2K_2$.

For $x \in N_1$, the possibilities for $\langle N(x) \cap N_0 \rangle$ are $K_1, K_2, P_3, P_4, P_5, 2K_1, 3K_1,$ $2K_2, K_1 \cup K_2, K_1 \cup P_3$ and C_6. Here (a) if $\langle N(x) \cap N_0 \rangle \cong K_1$ or K_2, then $P_6 \sqsubseteq G$, (b) if $\langle N(x) \cap N_0 \rangle \cong P_4$ or $K_1 \cup K_2$, then $C_5 \sqsubseteq G$, (c) if $\langle N(x) \cap N_0 \rangle \cong P_5$ or C_6 (or) $K_1 \cup P_3$ (or) $3K_1$ (or) $2K_1$, then $K_{1,3} \sqsubseteq G$, a contradiction. Finally, if $\langle N(x) \cap N_0 \rangle \cong P_3$ or $2K_2$, we see that neither P_6 nor C_5 (nor) $K_{1,3}$ is an induced subgraph of $\langle N_0 \cup N_1 \rangle$. Thus $\langle N(x) \cap N_0 \rangle \cong P_3$ or $2K_2$.

Throughout this proof, for any integer i, v_i means $v_{i \pmod 6}$ and A_i means $A_{i \pmod 6}$. For $0 \leq i \leq 5$, let $A_i = \{x \in N_1 : N(x) \cap N_0 = \{v_{i-1}, v_i, v_{i+1}\}\} \cup \{v_i\}$ and $B_i = \{x \in N_1 : N(x) \cap N_0 = \{v_{i-2}, v_{i-1}, v_{i+1}, v_{i+2}\}\}$.

Claim 2: $\langle \cup_{i=0}^{5} A_i \rangle \cong \mathbb{K}[C_6]$.

For every i, $0 \le i \le 5$, we have (a) $\langle A_i \rangle$ is complete (suppose if there exist vertices $x, y \in A_i$ such that $xy \notin E(G)$, then $\langle \{v_{i+1}, v_{i+2}, x, y\} \rangle \cong K_{1,3} \sqsubseteq G$), (b) $[A_i, A_{i+1}]$ is complete, (if not, there exist vertices $x \in A_i$ and $y \in A_{i+1}$ such that $xy \notin E(G)$, and hence $\langle \{x, v_i, y, v_{i+2}, v_{i+3}, v_{i+4}\} \rangle \cong P_6 \sqsubseteq G$), (c) $[A_i, A_{i+2}] = \emptyset$, (suppose if there exist vertices $x \in A_i$ and $y \in A_{i+2}$ such that $xy \in E(G)$, then $\langle \{x, y, v_{i+3}, v_{i+4}, v_{i+5}\} \rangle \cong C_5 \sqsubseteq G$), (d) $[A_i, A_{i+3}] = \emptyset$, (otherwise as shown previously, we can find $x \in A_i$ and $y \in A_{i+3}$ such that $xy \in E(G)$, and $\langle \{x, v_{i-1}, v_{i+1}, y\} \rangle \cong K_{1,3} \sqsubseteq G$). Thus from (a), (b), (c) and (d), it can be seen that $\langle \cup_{i=0}^{5} A_i \rangle \cong \mathbb{K}[C_6]$.

Claim 3: $\langle B_i \rangle$ is complete, for $i = 0, 1, 2, 3, 4, 5$.

Here, if there exist vertices $x, y \in B_i$ such that $xy \notin E(G)$, then $\langle \{v_{i-1}, v_i, x, y\} \rangle \cong K_{1,3} \sqsubseteq G$, a contradiction.

Claim 4: $[B_i, B_{i+1}] = \emptyset$, for $i = 0, 1, 2, 3, 4, 5$.

Suppose if there exist vertices $x \in B_i$ and $y \in B_{i+1}$ such that $xy \in E(G)$, then $\langle \{x, y, v_{i+1}, v_{i-2}\} \rangle \cong K_{1,3} \sqsubseteq G$, a contradiction.

Claim 5: $[A_i, B_i] = \emptyset$, $i = 0, 1, 2, 3, 4, 5$.

On the contrary, if there exist vertices $x \in A_i$ and $y \in B_i$ such that $xy \in E(G)$, then $\langle \{y, x, v_{i-2}, v_{i+2}\} \rangle \cong K_{1,3} \sqsubseteq G$, a contradiction.

Claim 6: $[A_i, B_{i+1}]$ is complete, for $i = 0, 1, 2, 3, 4, 5$.

If not, there exist vertices $x \in A_i$ and $y \in B_{i+1}$ such that $xy \notin E(G)$. Here $\langle \{x, v_{i-1}, y, v_{i+2}, v_{i+1}\} \rangle \cong C_5 \sqsubseteq G$, a contradiction.

Claim 7: $[A_i, B_{i+2}]$ is complete, for $i = 0, 1, 2, 3, 4, 5$.

It is easy to observe that if there exist vertices $x \in A_i$ and $y \in B_{i+2}$ such that $xy \notin E(G)$, then $\langle \{x, v_{i-1}, v_{i-2}, y, v_{i+1}\} \rangle \cong C_5 \sqsubseteq G$, a contradiction.

Claim 8: $N_i = \emptyset$, for all i, $i \ge 2$.

It is enough to show that $N_2 = \emptyset$. Suppose $N_2 \ne \emptyset$, then there exists a vertex $x \in N_2$. Since G is connected, there exists a vertex $y \in A_j$ or $y \in B_j$ for some $j \in \{0, 1, \ldots, 5\}$ such that $xy \in E(G)$. Then $\langle \{y, v_{j-1}, v_{j+1}, x\} \rangle \cong K_{1,3} \sqsubseteq G$, a contradiction. Thus $V(G) = N_0 \cup N_1$.

Note that $B_j = B_{j+3}$ for every $j \in \{0, 1, 2\}$. From all these Claims, we see that G will be isomorphic to the graph shown in Fig. 1. □

Corollary 2.3. *If G is a connected $\{P_6, C_5, \overline{P_5}, K_{1,3}\}$-free graph that contains an induced C_6 then $G \cong \mathbb{K}[C_6]$.*

$(\overline{P_2 \cup P_3})$ Dart   Kite   House   Bull

Fig. 2. Some special graphs

Next, we have obtain the structural characterization of a connected $\{P_5, K_4, Kite, Bull\}$-free graphs that contains an induced C_5. Here, the graphs Kite and Bull are shown in Fig. 2. The proof of Theorem 2.4 is similar to that given in Theorem 2.2 but with a little more involvement.

Theorem 2.4. *If G is a connected $\{P_5, K_4, Kite, Bull\}$-free graph that contains an induced C_5, then $V(G) = V_1 \cup V_2 \cup V_3$ such that (1) $\langle V_2 \rangle$ is a complete bipartite graph with bipartition B and S, (2) $\langle V_1 \cup V_3 \rangle$ is disjoint union of $\mathbb{I}[C_5]$'s and bipartite graphs, (3) $[V_1, B]$ is complete, $[V_1, S] = [V_1, V_3] = [V_3, B] = \emptyset$ and (4) there exists $x^* \in S$ such that $[x^*, V_3]$ is complete.*

Even though the graph G shown in Fig. 1 looks simple, it looks challenging to obtain the indicated chromatic number of G. So, we have only considered the indicated coloring of $\mathbb{K}[C_6]$.

Proposition 2.5. *For $1 \leq i \leq 6$, let m_i's be positive integers. Then the graph $G = \mathbb{K}[C_6](m_1, m_2, m_3, m_4, m_5, m_6)$ is k-indicated colorable for all $k \geq \chi(G)$.*

Without much difficulty, by using Theorem 2.9, Corollary 2.3 and Proposition 2.5 one can get Corollary 2.6.

Corollary 2.6. *If G is a $\{P_6, C_5, \overline{P_5}, K_{1,3}\}$-free graph that contains an induced C_6, then G is k-indicated colorable for all $k \geq \chi(G)$.*

By using the structural characterization given in Theorem 2.4, one can easily find the chromatic number of $\{P_5, K_4, Kite, Bull\}$-free graph that contain an induced C_5.

Corollary 2.7. *If G is a $\{P_5, K_4, Kite, Bull\}$-free graph that contain an induced C_5, then $\chi(G) = 3$ if and only if $G \cong \mathbb{I}[C_5]$, otherwise $\chi(G) = 4$.*

Now let us consider the indicated coloring of $G = \mathbb{I}[C_n](m_1, m_2, \ldots, m_n)$. By presenting the vertices of any induced C_n cyclically in G and then by presenting the remaining vertices in any order, Ann will get a winning strategy for G using k colors, for all $k \geq \chi(G)$.

Theorem 2.8. *For $1 \leq i \leq n$, let m_i's be positive integers. Then the graph $G = \mathbb{I}[C_n](m_1, m_2, \ldots, m_n)$ is k-indicated colorable for all $k \geq \chi(G)$.*

Let us recall the result on indicated coloring of union of two graphs.

Theorem 2.9 ([10]). *Let* $G = G_1 \cup G_2$. *If* G_1 *is* k_1-*indicated colorable for every* $k_1 \geq \chi_i(G_1)$ *and* G_2 *is* k_2-*indicated colorable for every* $k_2 \geq \chi_i(G_2)$, *then* $\chi_i(G) = \max\{\chi_i(G_1), \chi_i(G_2)\}$ *and* G *is* k-*indicated colorable for all* $k \geq \chi_i(G)$.

Now, let us consider the indicated coloring of $\{P_5, K_4, Kite, Bull\}$-free graphs which contains an induced C_5.

Theorem 2.10. *Let* G *be a* $\{P_5, K_4, Kite, Bull\}$-*free graph which contains an induced* C_5. *Then* G *is* k-*indicated colorable for all* $k \geq \chi(G)$.

Proof. By Theorem 2.9, it is enough to prove the result for a connected $\{P_5, K_4, Kite, Bull\}$-free graph that contains an induced C_5. Let G be such a graph. Suppose $\chi(G) = 3$, then $G \cong \mathbb{I}[C_5]$. Thus by Theorem 2.8, G is k-indicated colorable for all $k \geq \chi(G)$. Suppose not, $\chi(G) = 4$. Then G is isomorphic to the graph mentioned in Theorem 2.4. Let $\{1, 2, \ldots, k \geq 4\}$ be the set of colors. We shall show that G is k-indicated colorable. Let Ann start by presenting x^* and a vertex $b \in B$. Without loss of generality, let the color used by Ben for b and x^* be 1 and 2 respectively. Since $[b, V_1]$ is complete and $[x^*, V_3]$ is complete, the set of available colors for V_1 and V_2 are $\{2, 3, \ldots, k\}$ and $\{1, 3, 4, \ldots, k\}$ respectively. Since $[V_1, V_3] = \emptyset$, $\langle V_1 \cup V_3 \rangle$ is a disjoint union of $\mathbb{I}[C_5]$'s and bipartite graphs, by Theorems 2.1 and 2.8, $\langle V_1 \cup V_3 \rangle$ is l-indicated colorable for all $l \geq 3$. That is, Ann has a winning strategy for $\langle V_1 \rangle$ while using the colors $\{2, 3, \ldots, k\}$ and a winning strategy for $\langle V_3 \rangle$ while using the colors $\{1, 3, 4, \ldots, k\}$. After presenting the vertices of V_1 and V_3 by using these winning strategies, Ann will present the remaining vertices of B and S in any order. Clearly, the color of the vertices b and x^*, namely 1 and 2 are available for the uncolored vertices of B and S respectively. Thus Ann wins the game on G with k colors, $k \geq 4$.   □

3   Indicated Coloring of $\mathbb{K}[C_5]$ and Some of its Consequences

Let us start this Section by recalling two of the results which were proved in [10].

Theorem 3.1 ([10]). *Any graph* G *is* k-*indicated colorable for all* $k \geq col(G)$.

Theorem 3.2 ([10]). *Let* $G = G_1 + G_2$. *If* G_1 *is* k_1-*indicated colorable for every* $k_1 \geq \chi_i(G_1)$ *and* G_2 *is* k_2-*indicated colorable for every* $k_2 \geq \chi_i(G_2)$, *then* $\chi_i(G) = \chi_i(G_1) + \chi_i(G_2)$ *and* G *is* k-*indicated colorable for all* $k \geq \chi_i(G)$.

Next, let us recall the structural characterization of $\{P_2 \cup P_3, C_4\}$-free graphs, $\{P_5, C_4\}$-free graphs and $\{P_5, (\overline{P_2 \cup P_3}), \overline{P_5}, Dart\}$-free graphs which contains an induced C_5. The graphs $(\overline{P_2 \cup P_3})$ and Dart are shown in Fig. 2.

Theorem 3.3 ([3]). *If* G *is a connected* $\{P_2 \cup P_3, C_4\}$-*free graph, then* G *is chordal or there exists a partition* (V_1, V_2, V_3) *of* $V(G)$ *such that* (1) $\langle V_1 \rangle \cong \overline{K_m}$, *for some* $m \geq 0$, (2) $\langle V_2 \rangle \cong K_t$, *for some* $t \geq 0$, (3) $\langle V_3 \rangle$ *is isomorphic to a graph obtained from one of the basic graphs* G_t $(1 \leq t \leq 17)$ *shown in Fig. 3 by expanding each vertex indicated in circle by a complete graph (of order* ≥ 1), (4) $[V_1, V_3] = \emptyset$ *and* (5) $[V_2, V_3 \backslash S]$ *is complete.*

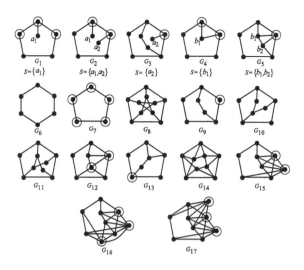

Fig. 3. Basic graphs used in Theorem 3.3 ($S = \emptyset$ for G_i, $6 \leq i \leq 17$)

Theorem 3.4 ([4]). *Let G be a connected $\{P_5, C_4\}$-free graph. Then $V(G) = V_1 \cup V_2$ such that*

(i) $\langle V_1 \rangle$ is a P_5-free graph which is also chordal.
(ii) If $V_2 \neq \emptyset$, then $\langle V_2 \rangle = A_1 \cup A_2 \cup \cdots \cup A_l$ where each A_i is a $\mathbb{K}[C_5]$, for every $i \in \{1, 2, \ldots, l\}$ for some $l \geq 1$. Also, $\langle N(A_i) \rangle$ is a complete subgraph of V_1 and $[A_i, N(A_i)]$ is complete.

Theorem 3.5 ([1]). *If G is a connected $\{P_5, (\overline{P_2 \cup P_3}), \overline{P_5}, Dart\}$-free graph that contains an induced C_5, then G is either isomorphic to $C_5(S_1, S_2, S_3, S_4, S_5)$ or $C_5(S_1, S_2, S_3, S_4, S_5) + H$, where $S_i's$ are induced split subgraphs of G, H is nonempty and $H \sqsubseteq G$.*

It can be noted that $\mathbb{K}[C_5]$ is one of the graphs mentioned in Fig. 3 of Theorem 3.3. Also $\mathbb{K}[C_5]$ is an induced subgraph of the graphs mentioned in Theorems 3.4 and 3.5. So, we obtain the indicated coloring for the complete expansion of C_5.

Theorem 3.6. *For $1 \leq i \leq 5$, let m_i's be positive integers. Then the graph $G = \mathbb{K}[C_5](m_1, m_2, m_3, m_4, m_5)$ is k-indicated colorable for all $k \geq \chi(G)$.*

Proof of Theorem 3.6 follows by considering the cases when $\omega(G) \geq \left\lceil \frac{|V(G)|}{2} \right\rceil$ and $\omega(G) < \left\lceil \frac{|V(G)|}{2} \right\rceil$. In both the cases, we have obtained an indicated coloring for G using k colors, for all $k \geq \chi(G)$.
Corollaries 3.7, 3.8, 3.9, 3.10 and Theorem 3.11 are some of the consequences of Theorem 3.6.

Corollary 3.7. *For $1 \leq i \leq 5$, let m_i's be positive integers. Then for the graph $G = \mathbb{K}[C_5](m_1, m_2, m_3, m_4, m_5)$, $\chi(G) = \max\left\{\omega(G), \left\lceil \frac{|V(G)|}{2} \right\rceil\right\}$.*

Corollary 3.8. *If G is a $\{P_5, C_4\}$-free graph, then G is k-indicated colorable for all $k \geq \chi(G)$.*

Corollary 3.9. *Let S_1, S_2, S_3, S_4, S_5 be the split graphs. The graph $G = C_5(S_1, S_2, S_3, S_4, S_5)$ is k-indicated colorable for all $k \geq \chi(G)$.*

Corollary 3.10. *If G is connected $\{P_5, (\overline{P_2 \cup P_3}), \overline{P_5}, Dart\}$-free graph that contains an induced C_5, then G is k-indicated colorable for all $k \geq \chi(G)$.*

Theorem 3.11. *If G is a connected $\{P_2 \cup P_3, C_4\}$-free graph, then G is k-indicated colorable for all $k \geq \chi(G)$.*

Acknowledgments. For the first author, this research was supported by the Council of Scientific and Industrial Research, Government of India, File no: 09/559(0096)/2012-EMR-I. For the second author, this research was supported by the SERB DST Project, Government of India, File no: EMR/2016/007339. Also, for the third author, this research was supported by the UGC-Basic Scientific Research, Government of India.

References

1. Aravind, N.R., Karthick, T., Subramanian, C.R.: Bounding χ in terms of ω and Δ for some classes of graphs. Discrete Math. **311**, 911–920 (2011)
2. Brandstädt, A., Mosca, R.: On the structure and stability number of P_5- and co-chair-free graphs. Discrete Appl. Math. **132**, 47–65 (2004)
3. Choudum, S.A., Karthick, T.: Maximal cliques in $\{P_2 \cup P_3, C_4\}$-free graphs. Discrete Math. **310**, 3398–3403 (2010)
4. Fouquet, J.L., Giakoumakis, V., Maire, F., Thuillier, H.: On graphs without P_5 and $\overline{P_5}$. Discrete Math. **146**, 33–44 (1995)
5. Francis Raj, S., Pandiya Raj, R., Patil, H.P.: On indicated chromatic number of graphs. Graphs Comb. **33**, 203–219 (2017)
6. Grzesik, A.: Indicated coloring of graphs. Discrete Math. **312**, 3467–3472 (2012)
7. Hof, P.V., Paulusma, D.: A new characterization of P_6-free graphs. Discrete Appl. Math. **158**, 731–740 (2010)
8. Lasoń, M.: Indicated coloring of matroids. Discrete Appl. Math. **179**, 241–243 (2014)
9. Liu, J., Peng, Y., Zhao, C.: Characterization of P_6-free graphs. Discrete Appl. Math. **155**, 1038–1043 (2007)
10. Pandiya Raj, R., Francis Raj, S., Patil, H.P.: On indicated coloring of graphs. Graphs Comb. **31**, 2357–2367 (2015)
11. Sekiguchi, Y.: The game coloring number of planar graphs with a given girth. Discrete Math. **330**, 11–16 (2014)
12. Wu, J., Zhu, X.: Lower bounds for the game colouring number of partial k-trees and planar graphs. Discrete Math. **308**, 2637–2642 (2008)
13. Zhu, X.: The game coloring number of planar graphs. J. Combin. Theory Ser-B **75**, 245–258 (1999)

Line Segment Disk Cover

Manjanna Basappa[✉]

School of Computer Science and Engineering,
VIT University, Vellore 632014, India
manjanna@alumni.iitg.ernet.in

Abstract. In this paper, we consider the following variations of *Line Segment Disk Cover* (LSDC) problem.

LSDC-H: In this version of LSDC problem, we are given a set $\mathcal{S} = \{s_1, s_2, \ldots, s_n\}$ of n horizontal line segments of arbitrary length and an integer $k(\geq 1)$. Our aim is to cover all segments in $\mathcal{S}$ with k disks of minimum radius centered at arbitrary points in the plane.

LSDC-A: In this version of LSDC problem, we are given a set $\mathcal{S} = \{s_1, s_2, \ldots, s_n\}$ of n line segments of arbitrary length with arbitrary orientation and an integer $k(\geq 1)$. Our aim is to cover all segments in $\mathcal{S}$ with k disks of minimum radius centered at arbitrary points in the plane.

LSDC-D: In the discrete version of LSDC problem, we are given a set $\mathcal{S} = \{s_1, s_2, \ldots, s_n\}$ of n line segments of arbitrary length with arbitrary orientation and a set $\mathcal{D} = \{d_1, d_2, \ldots, d_m\}$ of m disks of unit radius. Our aim is to cover all segments in $\mathcal{S}$ with minimum number of disks in $\mathcal{D}$ i.e. $\mathcal{S} \subset \bigcup_{d \in D'} d$, where $D' \subseteq \mathcal{D}$ is of minimum cardinality.

For LSDC-H and LSDC-A problems, we propose $(1 + \epsilon)$-factor approximation algorithms, which run in $O((\lceil \frac{\pi}{\delta^2} \rceil)^k n(|\log r_{opt}| + \log \lceil \frac{1}{\rho} \rceil))$ time and $O((\lceil \frac{\pi}{\delta^2} \rceil)^k n \log n(|\log r_{opt}| + \log \lceil \frac{1}{\rho} \rceil))$ time respectively, where r_{opt} is the minimum radius of k disks which cover all segments in $\mathcal{S}$, and $\delta > 0$, $\rho > 0$ and $\epsilon > 0$ are fixed constants such that $\epsilon \geq (\delta + \delta\rho + \rho)$. For LSDC-D problem, we propose a $(1 + \epsilon)$-factor approximation algorithm (PTAS), which runs in $O(m^{2(\frac{8\sqrt{2}}{\epsilon})^2 + 3} + m^2 n)$ time, and a $(9 + \epsilon)$-factor approximation algorithm, which runs in $O(m^{(5 + \frac{18}{\epsilon})} \log m + m^2 n)$ time, where a constant $\epsilon > 0$.

Keywords: Approximation scheme · Fixed-parameter-tractable
Line segment disk cover

1 Introduction

In the wireless sensor network such as mobile network, deployment of base stations is a critical step in the establishment of the network. In the mobile network the two major components are mobile devices which are constantly moving and base stations which are usually fixed. Due to the constant movement of mobile

© Springer International Publishing AG 2018
B. S. Panda and P. P. Goswami (Eds.): CALDAM 2018, LNCS 10743, pp. 81–92, 2018.
https://doi.org/10.1007/978-3-319-74180-2_7

users or devices, providing end-end communication between mobile users though the base stations is a challenging task. Mobile network service providers sometimes establish additional base stations to handle the occasional heavy gathering of mobile users in certain places due to events such as sports etc. In order to find proper locations for mobile base stations to be established, we can collect the history of mobile users' movement patterns. We can then model these patterns as line segments, which a mobile user has traversed often in the past. We can now consider the placement of base stations or sensors to cover all these target line segments. We may consider several variations of positioning base stations such as placing exactly k base stations with smallest possible range and placing minimum number of base stations with fixed range. Motivated by these scenarios of base stations placement, we study the line segment disk cover problem and its variations in this paper.

In Subsect. 1.1, we discuss various covering problems involving line segments. In Sect. 2, we present an $(1 + \epsilon)$-approximation algorithm for LSDC-H problem, where $\epsilon > 0$ is any constant. In Sect. 3, we present an $(1 + \epsilon)$-approximation algorithm for LSDC-A problem, where $\epsilon > 0$ is any constant. Next, we propose a PTAS and a $(9 + \epsilon)$-approximation algorithm for LSDC-D problem, where $0 < \epsilon \leq 6$. in Sect. 4. Finally, in Sect. 5, we conclude the paper and mention some future work.

1.1   Related Work

Several variations of geometric covering problems involving line segments have already been studied in the literature. Agnetis et al. [2] studied the disk covering problem on a line. In this problem the objective is to center disks of variable radii so that a given line segment is fully covered with least total cost incurred by these disks. When only one type of disk is available, they developed a simple polynomial time algorithm, which solves the problem. When there are different types of disks, they developed a branch and bound algorithm as well as an efficient heuristic algorithm for the special case of this variation of the problem. Dash et al. [11] studied the following variant of coverage problem involving line segments. Here, a line segment ℓ is said to be k-covered if it intersects with atleast k disks. Similarly, a line segment ℓ is said to be k-uncovered if it intersects with atmost $k - 1$ disks. Given a set of n disks in the plane, Dash et al. [11] proposed algorithms to compute the smallest k-covered line segment and longest k-uncovered line segment. The time complexity of their algorithms is $O((X + n) \log n)$ for both smallest k-covered line segment and longest k-uncovered line segment, where the line segments can be of the following types: (i) line segments are axis-parallel, (ii) line segments whose one end point is fixed and are of arbitrary orientation, and X is the number of intersections among n disks. For the case that the line segments are arbitrary, their algorithm takes $(X^2 \log n + n^{\frac{11}{3}+\epsilon})$ time to determine the smallest k-covered line segment and $(X^2 \log n + n^{2+\beta+\epsilon})$ time to determine the longest k-uncovered line segment, where $\beta = \log_2(1 + \sqrt{5}) - 1$ and ϵ is a small value greater than or equal to 0. On the other side, given a set of line segments, Dash et al. [9] considered the problem of finding minimum number of disks of uniform radius to 1-cover each of these line segments. This variation of line segment disk cover problem is known to be NP-hard. Hence, they proposed constant factor approximation algorithms and PTAS for this problem. Acharyya et al. [3] studied several variations of line segment square cover problem. In this problem a set S of n line segments is given, the objective is to find the minimum number of axis-parallel unit squares, which cover at least one end point of each segment. They considered several variations depending on the orientation and the length of the line segments. They showed that some of these variations are NP-complete, and developed constant factor approximation algorithms for these problems. For some variations they proposed exact algorithms, which run in polynomial time. In the similar line we consider different orientations of line segments, and thus study some variations of line segment disk cover problem in this paper.

In the context of geometric covering with disks, we find that numerous varieties of problems have been considered, and extensively studied in the literature. In the similar line, one of the well known and well studied geometric disk covering problems is known as Discrete Unit Disk Cover (DUDC) problem. Here, we are given a set $\mathcal{P}$ of n points and a set $\mathcal{D}$ of m unit disks, the objective is to cover all points in $\mathcal{P}$ with smallest number of disks in $\mathcal{D}$. The DUDC problem is known to be NP-hard [13]. Mustafa and Ray [14] proposed a PTAS for the DUDC problem. Unfortunately, their PTAS's running time is very huge even for the largest possible approximation bound. However, several authors [1,4–8,10,12,15] proposed constant factor approximation algorithms with reasonable running times. We use the result of Mustafa and Ray [14] to develop a PTAS for one of our variations of line segment disk cover problem considered in this paper. Similarly, we use the result of Basappa et al. [4] to propose constant factor approximation algorithm with reasonable running time for the same variation of the LSDC problem.

2   LSDC-H Problem

In the LSDC-H problem, we are given a set $S = \{s_1, s_2, \ldots, s_n\}$ of n horizontal line segments of arbitrary length and a positive integer k. The objective is to compute a set $\mathcal{D}^{opt} = \{d_1^{opt}, d_2^{opt}, \ldots, d_k^{opt}\}$ of k disks of minimum radius r_{opt} such that $S \subset (\bigcup_{i=1}^{k} d_i^{opt})$. The disks in $\mathcal{D}^{opt}$ can be centered anywhere in the plane.

2.1   Preliminaries

Let $S = \{s_1, s_2, \ldots, s_n\}$ be a set of n line segments. Through out the paper, we use the following terminologies. Let p_i^L and p_i^R denote the left-end point and the right-end point of a line segment $s_i \in S$. Let p_{i*}^L be the left-most left-end point of segment s_{i*} among all segments in S. Let d_{i*}^{opt} be the disk of radius r_{opt} in an optimal solution, which covers p_{i*}^L. Let $\alpha(d)$ denote the center of disk d. Let ℓ_i be the vertical line passing through the left end-point p_i^L of segment s_i. Let ℓ_i^+ and

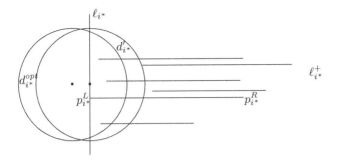

Fig. 1. Proof of Observation 1

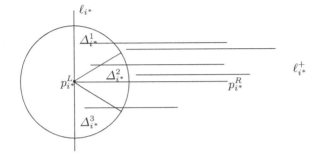

Fig. 2. Partition of the disk $\Delta(p_{i^*}^L, r) \cap \ell_{i^*}^+$ into three equal sectors $\Delta_{i^*}^1, \Delta_{i^*}^2, \Delta_{i^*}^3$

ℓ_i^- denote the right half-plane and the left half-plane respectively. The region of disk with radius r and centered at any point p is denoted by $\Delta(p, r)$.

Observation 1. *There always exists an optimal solution $\mathcal{D}^{opt}$ such that the point $\alpha(d_{i^*}^{opt})$ lies in $\ell_{i^*}^+$.*

Proof. Assume that the disk $d_{i^*}^{opt}$ has its center $\alpha(d_{i^*}^{opt})$ lying to the left of the line ℓ_{i^*}. Then, we can center a disk d_{i^*}' of radius r_{opt} on the vertical line ℓ_{i^*} such that the points $\alpha(d_{i^*}')$ and $\alpha(d_{i^*}^{opt})$ are collinear and $(d_{i^*}^{opt} \cap \mathcal{S}) \subseteq (d_{i^*}' \cap \mathcal{S})$ (See Fig. 1). $\qquad\square$

2.2 $(2 + 2\rho)$-Factor Approximation Algorithm

To develop a $(2 + 2\rho)$-factor approximation algorithm for LSDC-H problem for $\rho > 0$, we first consider the decision version of LSDC-H problem as follows:

- For a given radius r, can we center k disks of radius $2r$ at arbitrary points such that union of these k disks covers all segments in $\mathcal{S}$?

For the above decision problem, we develop an algorithm which returns a set D' of k disks of radius $2r$ if the answer is positive. Otherwise, the algorithm returns a set D' of $k+1$ arbitrary disks. We then apply doubling technique and bisection method to find the minimum radius r for which the answer to the above decision problem is positive.

Consider the left-most left-end point p_{i*}^L. Partition the half-disk $\Delta(p_{i*}^L, r) \cap \ell_{i*}^+$ into three sectors such that each pair of its consecutive radii makes an angle $\frac{\pi}{3}$ at the point p_{i*}^L. Thus, the half-disk $\Delta(p_{i*}^L, r) \cap \ell_{i*}^+$ is split into three equal sectors Δ_{i*}^1, Δ_{i*}^2 and Δ_{i*}^3 as shown in Fig. 2.

Lemma 1. *If $r \geq r_{opt}$ and $\alpha(d_{i*}^{opt}) \in \Delta_{i*}^j$, $j \in \{1, 2, 3\}$, then any disk d_{i*} of radius $2r$ centered at arbitrary point in the region Δ_{i*}^j covers at least the same portion of S that is covered by the disk d_{i*}^{opt}.*

Proof. Let us first observe that the distance between any two arbitrary points lying within any sector Δ_{i*}^j is at most r. Given the facts that (i) $r \geq r_{opt}$, and (ii) the radius of disk d_{i*} is $2r$, by the triangle inequality, the lemma follows. $\square$

Our algorithm for the decision problem is based on line-sweep technique and bounded search tree method. We know that the bounded search tree method attempts to do exhaustive search on the problem space. Hence, our algorithm which solves the decision problem for a given radius r works as follows. We start off with set $S' = S$ and set $D' = \emptyset$. We sweep the plane containing a given set S of segments from left to right by a vertical line ℓ_{i*}. At each left most left-end point p_{i*}^L, we partition the half-disk $\Delta(p_{i*}^L, r) \cap \ell_{i*}^+$ into three sectors Δ_{i*}^1, Δ_{i*}^2 and Δ_{i*}^3. We then consider each sector Δ_{i*}^j, $j = \{1, 2, 3\}$ separately. We place the disk d_{i*} of radius $2r$ centered at arbitrary point from Δ_{i*}^j. We then update the set D' by $D' \cup \{d_{i*}\}$. Let $R \subseteq S'$ be the portion of S lying inside d_{i*}. Update $S' = S' \setminus R$. We also update the left-most left-end point p_{i*}^L with respect to the updated set S' of segments. Unless $S' = \emptyset$, we repeat the above process recursively by considering the left-most left-end point p_{i*}^L. Thus, we proceed our search in a depth-first manner on a 3-way search tree where the depth of the tree is at most k. The pseudocode of the above procedure is available in Algorithm 1.

Lemma 2. *If Algorithm 1 is invoked with radius $r \geq r_{opt}$, then it always produces a positive reply with a set D' of at most k disks of radius $2r$ whose union covers all segments in S, i.e. $S \subset \cup_{d \in D'} d$.*

Proof. From Lemma 1, it is clear that $\alpha(d_{i*}^{opt})$ lies in one of three sectors Δ_{i*}^1, Δ_{i*}^2 and Δ_{i*}^3. Since we explore all possible paths in the 3-way search tree, there exists atleast one path, along which our recursive procedure ultimately results in the set $S' = \emptyset$ while the depth of recursion is atmost k. Thus, the lemma follows. $\square$

Lemma 3. *The running time of Algorithm 1 is $O(3^k n)$.*

Proof. Each segment $s_i \in S$ can intersect at most twice on the boundary of any disk. Every time we consider a new left-most left-end point p_{i*}^L, we spend at most $O(n)$ time to update the set S'. Therefore, since the degree of each internal

---

Algorithm 1. Two-Cover(S, k, r)

1: **Input:** The set S of uncovered horizontal line segments of arbitrary length, a positive integer k and a radius r.
2: **Output:** *true* if a set D' of at most k disks of radius $2r$ covers the set S; *false* otherwise and an arbitrary set D'.
3: **if** $(S = \emptyset)$ **then**
4:     **return** $(true, \emptyset)$
5: **else if** $(k = 0)$ **then**
6:     **return** $(false, \emptyset)$
7: **else**
8:     Consider the disk $\Delta(p^L, r)$ centered at left-most left-end point p^L of segment s among all segments in S, and let ℓ be a vertical line through p^L and ℓ^+ be the region right side of ℓ.
9:     Partition the right half-disk $\Delta(p^L, r) \cap \ell^+$ into three equal sectors $\Delta^1, \Delta^2, \Delta^3$ as shown in Figure 2.
10:     Set $i \leftarrow 1$; $flag \leftarrow false$; $D' \leftarrow \emptyset$
11:     **while** $(i \leq 3$ and $flag=false)$ **do**
12:         Place a disk d of radius $2r$ centered at arbitrary point in Δ^i
13:         $S' = S \setminus (S \cap d)$
14:         $(flag, D') = $ **Two-Cover**$(S', k - 1, r)$
15:         **if** $(flag=true)$ **then**
16:             $D' = D' \cup \{d\}$
17:         **end if**
18:         Set $i = i + 1$;
19:     **end while**
20: **end if**
21: **return** $(flag, D')$

---

node of the 3-way search tree the algorithm explores is 3, the running time of Algorithm 1 is given by the recurrence $T_i(S') = 3(T_{i+1}(S' \setminus (d_{i+1}^{opt} \cap S)) + n)$, where the base case $T_k(S') = O(n)$. The lemma follows from the solution of this recurrence i.e. $T_0(S) = O(3^k n)$. $\qquad\qquad\square$

Now, we describe the procedure (Algorithm 2) to find a radius r for which Algorithm 1 produces a positive reply such that $r \leq (1+\rho)r_{opt}$, where a constant $\rho > 0$. In Algorithm 2, we first use the doubling technique as follows: we initially check whether $r_{opt} > 1$ or $r_{opt} \leq 1$ by invoking Algorithm 1 with radius $r = 1$. We then repeatedly invoke Algorithm 1 with radius $r = 2^j$ for every $j = 1, 2, \ldots, j^*$ until the corresponding set D' covers S if the radius $r_{opt} > 1$, where j^* is the smallest positive integer. Similarly, if $r_{opt} \leq 1$, Algorithm 1 is invoked with radius $r = 2^{-j}$ for every $j = 1, 2, \ldots, j^*$ until the set S is covered by the union of disks in D', where j^* is the largest positive integer. Therefore, r_{opt} belongs to one of the intervals $[2^{j^*-1}, 2^{j^*}]$ and $[2^{-j^*-1}, 2^{-j^*}]$ on the real line. Note that the length of this interval is at most r_{opt} because the radius is doubled every time. Therefore, $|2^{j^*} - 2^{j^*-1}| \leq r_{opt}$ or $|2^{-j^*} - 2^{-j^*-1}| \leq r_{opt}$. We further reduce the length of this interval by bisecting it $\log \frac{1}{\rho}$ times as long as Algorithm 1 returns a positive reply. Let $[\mu, \nu]$ denote this interval before the bisection applied on it.

Algorithm 2. Rho-Cover($\mathcal{S}, k, \rho$)

1: **Input:** A set $\mathcal{S}$ of n horizontal line segments of arbitrary length, a positive integer k and a real number $\rho > 0$.
2: **Output:** a set $\mathcal{D}$ of at most k disks of radius $r' \leq (2 + 2\rho)r_{opt}$ such that union of disks in $\mathcal{D}$ covers all segments in $\mathcal{S}$.
3: $(flag, \mathcal{D},) = $ **Two-Cover**($\mathcal{S}, k, 1$) //Run Algorithm 1
4: **if** ($flag = true$) **then**
5:    Set $j = 1$
6:    **while** ($flag = true$) **do**
7:       Set $j = j - 1$
8:       $(flag, \mathcal{D}) = $ **Two-Cover**($\mathcal{S}, k, 2^{j-1}$) //Run Algorithm 1
9:    **end while**
10: **else**
11:    Set $j = 0$
12:    **while** ($flag = false$) **do**
13:       Set $j = j + 1$
14:       $(flag, \mathcal{D}) = $ **Two-Cover**($\mathcal{S}, k, 2^{j}$) //Run Algorithm 1
15:    **end while**
16: **end if**
17: Set $\mu = 2^{j-1}$, $\nu = 2^{j}$
18: **for** ($i = 1, 2, \ldots, \log\lceil\frac{1}{\rho}\rceil$) **do**
19:    Set $\gamma = \frac{\mu + \nu}{2}$
20:    $(flag, \mathcal{D}) = $ **Two-Cover**($\mathcal{S}, k, \gamma$) //Run Algorithm 1
21:    **if** ($flag = true$) **then**
22:       Set $\nu = \gamma$
23:    **else**
24:       Set $\mu = \gamma$
25:    **end if**
26: **end for**
27: $(flag, \mathcal{D}) = $ **Two-Cover**($\mathcal{S}, k, \nu$) //Run Algorithm 1
28: **Return** ($\mathcal{D}, \nu$)

Lemma 4. $(\nu - \mu) \leq \rho r_{opt}$ *after bisection applied on interval* $[\mu, \nu]$ $\log\frac{1}{\rho}$ *times.*

Proof. After bisection is applied $\lceil\log\frac{1}{\rho}\rceil$ times, $(\nu - \mu) \leq \frac{r_{opt}}{2^{\log\frac{1}{\rho}}} \leq \rho r_{opt}$ as interval is halved every time. Thus, the lemma follows. $\qquad\square$

Theorem 1. *Algorithm 2 is* $(2 + 2\rho)$-*factor approximation algorithm with running time* $O(3^k n(|\log r_{opt}| + \log\lceil\frac{1}{\rho}\rceil))$ *for LSDC-H problem, where* $\rho > 0$.

Proof. Since $r_{opt} \in [\mu, \nu]$, $\mu \leq r_{opt}$ and Algorithm 1 is invoked finally with radius $r = \nu$, $r = \nu = \mu + (\nu - \mu) \leq r_{opt} + \rho r_{opt} = (1 + \rho)r_{opt}$. From Lemma 2, for any radius $r \geq r_{opt}$ Algorithm 1 produces 2-factor approximation result and runs in $O(3^k n)$. Algorithm 2 invokes Algorithm 1 $O(|\log r_{opt}| + \log\lceil\frac{1}{\rho}\rceil)$ times. Thus, the theorem follows. $\qquad\square$

Algorithm 3. Delta-Cover(S, k, r, δ)

1: **Input:** The set S of uncovered horizontal line segments of arbitrary length, a positive integer k, a radius r and a real number $\delta > 0$.

2: **Output:** *true* if a set D' of at most k disks of radius $(1 + \delta)r$ covers the set S; *false* otherwise and an arbitrary set D'.

3: **if** $(S = \emptyset)$ **then**

4:    **return** $(true, \emptyset)$

5: **else if** $(k = 0)$ **then**

6:    **return** $(false, \emptyset)$

7: **else**

8:    Consider the disk $\Delta(p^L, r)$ centered at left-most left-end point p^L of segment s among all segments in S, and let ℓ be a vertical line through p^L and ℓ^+ be the region right side of ℓ.

9:    Partition the right half-disk $\Delta(p^L, r) \cap \ell^+$ into at most $\lceil \frac{\pi}{\delta^2} \rceil$ grid cells $(\Delta^1, \Delta^2, \ldots, \Delta^{\lceil \frac{\pi}{\delta^2} \rceil})$, each of size $\frac{\delta r}{\sqrt{2}} \times \frac{\delta r}{\sqrt{2}}$.

10:    Set $i \leftarrow 1$; $flag \leftarrow false$; $D' \leftarrow \emptyset$

11:    **while** $(i \leq \lceil \frac{\pi}{\delta^2} \rceil$ and $flag = false)$ **do**

12:       Place a disk d of radius $(1 + \delta)r$ centered at arbitrary point in Δ^i

13:       $S' = S \setminus (S \cap d)$

14:       $(flag, D') = $ **Delta-Cover**$(S', k - 1, r, \delta)$

15:       **if** $(flag = true)$ **then**

16:          $D' = D' \cup \{d\}$

17:       **end if**

18:       Set $i = i + 1$;

19:    **end while**

20: **end if**

21: **return** $(flag, D')$

2.3 $(1 + \epsilon)$-Factor Approximation Algorithm

In this subsection we develop an $(1 + \epsilon)$-factor approximation algorithm for LSDC-H problem for any $\epsilon > 0$. Here, we develop a procedure, which is a modified version of Algorithm 1. For a given radius $r \geq r_{opt}$ and a real number $\delta > 0$, this procedure computes a set D' of at most k disks of radius $(1 + \delta)r$ such that union of disks in D' covers all segments in $\mathcal{S}$. Unlike in Algorithm 1, here the half-disk $\Delta(p_{i*}^L, r) \cap \ell_{i*}^+$ is partitioned into $O(\frac{1}{\delta^2})$ sectors by placing a 2D grid of size $\frac{\delta r}{\sqrt{2}} \times \frac{\delta r}{\sqrt{2}}$ over this half-disk. This results in solution space being $\lceil \frac{\pi}{\delta^2} \rceil$-way search tree instead of 3-way search tree. The pseudocode of the procedure is formalized in Algorithm 3.

Lemma 5. *For any constant $\epsilon > 0$, there exist constants $\delta > 0$ and $\rho > 0$ such that $\epsilon \geq (\delta + \delta\rho + \rho)$.*

Proof. The inequality $\epsilon \geq (\delta + \delta\rho + \rho)$ can be rewritten as $\frac{\epsilon - \rho}{1 + \rho} \geq \delta$. Now, for a given positive real number ϵ, choose another positive real number ρ such that $\rho < \epsilon$. Then, we can always a choose a positive real number δ such that $\delta \leq \frac{\epsilon - \rho}{1 + \rho}$. Thus, the lemma follows. □

Theorem 2. *We have an $(1 + \epsilon)$-factor approximation algorithm with running time $O((\lceil \frac{\pi}{\delta^2} \rceil)^k n(|\log r_{opt}| + \log\lceil\frac{1}{\rho}\rceil))$ for LSDC-H problem, where $\delta > 0$, $\rho > 0$ and $\epsilon > 0$.*

Proof. Observe that the diameter of each grid cell of size $\frac{\delta r}{\sqrt{2}} \times \frac{\delta r}{\sqrt{2}}$ is δr. If the subroutine invoked in Algorithm 2 is Algorithm 3 instead of Algorithm 1, then the radius r' of disks in D' returned by Algorithm 3 satisfies the inequality $r' \leq (1 + \delta)r$, where $r \leq (1 + \rho)r_{opt}$ by Lemma 4. Hence, by Lemma 5 $r' \leq (1 + \delta)(1 + \rho)r_{opt} \leq (1 + \epsilon)r_{opt}$ for any $\epsilon > 0$. Thus, the theorem follows.   □

Corollary 1. *We have a polynomial time approximation scheme (PTAS) for LSDC-H problem when $\delta = O(\frac{1}{polynomial(n)})$, $k = O(\frac{1}{\delta(1+\rho)+\rho})$, $r_{opt} = O(2^{polynomial(n)})$.*

Proof. Follows as Algorithm 3 produces $(1 + \epsilon)$-approximation result in time $O(\texttt{polynomial}'(n)^{O(\frac{1}{\epsilon})}\texttt{polynomial}(n, \frac{1}{\epsilon}))$ for a given integer k and a set of n line segments such that $k = O(\frac{1}{\epsilon})$, $r_{opt} = O(2^{polynomial(n)})$ and $\delta = O(\frac{1}{polynomial(n)})$. □

Corollary 2. *We have a fully polynomial time approximation scheme (FPTAS) for LSDC-H problem when $r_{opt} = O(2^{polynomial(n)})$, and k is fixed.*

Proof. Follows as Algorithm 3 produces $(1 + \epsilon)$-approximation result in time $O(\texttt{polynomial}(n, \frac{1}{\epsilon}))$ when k is fixed and $r_{opt} = O(2^{polynomial(n)})$.   □

3   LSDC-A Problem

In the LSDC-A problem, we are given a set $\mathcal{S} = \{s_1, s_2, \ldots, s_n\}$ of n line segments of arbitrary length with arbitrary orientation and a positive integer k. The objective is to compute a set $\mathcal{D}^{opt} = \{d_1^{opt}, d_2^{opt}, \ldots, d_k^{opt}\}$ of k disks of minimum radius r_{opt} such that $\mathcal{S} \subset (\bigcup_{i=1}^{k} d_i^{opt})$. The disks in $\mathcal{D}^{opt}$ can be centered anywhere in the plane.

To solve the LSDC-A problem, as in Sect. 2 we first consider the decision problem as follows:

– For a given radius r and a real number $\delta > 0$, can we center k disks of radius $(1+\delta)r$ at arbitrary points such that union of these k disks covers all segments in $\mathcal{S}$?

In order to solve the decision problem we make the following changes in Algorithm 3 described as follows. We first set $S = \mathcal{S}$. Let us denote the convex hull of end points $\{p_1^L, p_1^R, p_2^L, p_2^R, \ldots, p_n^L, p_n^R\}$ of segments in S by $CH(S)$. We then compute the convex hull $CH(S)$ of end points of segments in S. At every node of $\lceil \frac{\pi}{\delta^2} \rceil$-way search tree explored by Algorithm 3 we can consider any vertex of the convex hull CH and rotate the axes so as this vertex becomes left-most left-end point p_{i*}^L. We then repeat the remaining steps of Algorithm 3. We update the convex hull $CH(S)$ with respect to the updated set S of line segments. We repeat the above procedure until either the depth of the tree is k or the updated set S becomes empty.

Theorem 3. *We have an $(1 + \epsilon)$-factor approximation algorithm with running time $O((\lceil \frac{\pi}{\delta^2} \rceil)^k n \log n(|\log r_{opt}| + \log \lceil \frac{1}{\rho} \rceil))$ for LSDC-A problem, where $\delta > 0$, $\rho > 0$ and $\epsilon > 0$.*

Proof. Follows as in the case of LSDC-H problem (Theorem 2) due to the facts that the search space is $\lceil \frac{\pi}{\delta^2} \rceil$-way search tree, and we spend $O(n \log n)$ time to update the convex hull $CH(S)$ whenever the set S is updated. □

Remark 1. It is further to remark that the space required in our algorithms for both LSDC-H and LSDC-A problems is only (kn).

4   LSDC-D Problem

In the LSDC-D problem, we are given a set $S = \{s_1, s_2, \ldots, s_n\}$ of n line segments of arbitrary length with arbitrary orientation and a set $D = \{d_1, d_2, \ldots, d_m\}$ of m disks of unit radius. Our aim is to cover all segments in S with minimum number of disks in D i.e. $S \subset \bigcup_{d \in D'} d$, where $D' \subseteq D$ is of minimum cardinality.

In this section, using the PTAS of Mustafa and Ray [14] for DUDC problem, we propose a PTAS for LSDC-D problem, and using the $(9+\epsilon)$-approximation result of Basappa et al. [4] for DUDC problem, we propose a $(9 + \epsilon)$-approximation algorithm for LSDC-D problem.

Given a set D of m unit disks in the plane, let us define the sector as a maximal region such that every point within that region is covered by the same set of disks in D. Thus, the region covered by the union of m disks in D is divided into many number of sectors. Let f denote the number of those sectors. Let e be the number of smallest boundary segments of disks such that each such segment appears on the boundary of at most two sectors.

Lemma 6. *For a set D of m unit disks, the number of sectors $f = O(m^2)$.*

Proof. Let $\beta(d)$ denote the boundary line of disk d. Consider some disk d in the plane containing m unit disks. The largest number of boundary segments, denoted by $T(m)$, is contributed by $\beta(d)$ when the disk d intersects with $m - 1$ other disks in the following manner. Without loss of generality, the ith disk bisects and splits two boundary segments among the segments contributed by the first $i - 1$ disks for $i = 3, 4, \ldots, m$. Thus, the largest number of boundary segments for every disk d is given by the recurrence $T(m) = T(m - 1) + 2$ for $m > 2$ with the base case $T(2) = 2$.

The solution to the above recurrence is $T(m) = 2(m - 1)$. Since there are m disks and each $\beta(d)$ contributes at most $2(m - 1)$ boundary segments, the total number of boundary segments $e = 2(m - 1)m$. Observe that each sector is bounded by at least two boundary segments of disks. Therefore, $2f \leq 2e$ as each boundary segment appears on the boundary of at most two sectors. This implies that $f \leq 2(m^2 - m)$. Thus, the lemma follows. □

To develop approximation algorithms for LSDC-D problem, we transform every instance of LSDC-D problem into an instance of DUDC problem as follows. An instance of LSDC-D problem consists of a set $\mathcal{D}$ of m disks of unit radius and a set $\mathcal{S}$ of n line segments of arbitrary length with arbitrary orientation. Then, split each line segment $s \in \mathcal{S}$ such that each of the splitted slices $sl_1, sl_2, \ldots, sl_k$ of s lies within a sector, where $k \leq f$. Now, for each set of all slices lying within one sector, we add one point into the same sector and remove all these slices. Hence, from Lemma 6 we have the number of points $n' = O(m^2)$ and number of disks $m' = m$, which is an instance of DUDC problem. Therefore, we have the following results for LSDC-D problem.

Theorem 4. *We have an $(1 + \epsilon)$-factor approximation algorithm with running time $O(m^{2(\frac{8\sqrt{2}}{\epsilon})^2+3} + m^2 n)$ for LSDC-D problem, where $\epsilon > 0$ is a real number.*

Proof. Follows by setting the number of points $n = O(m^2)$ in the PTAS of Mustafa and Ray [14] and the fact that we spend $O(m^2 n)$ time to construct this new set of $n = O(m^2)$ points. $\square$

Theorem 5. *We have a $(9 + \epsilon)$-factor approximation algorithm with running time $O(m^{(5+\frac{18}{\epsilon})} \log m + m^2 n)$ for LSDC-D problem, where $0 < \epsilon \leq 6$ is a real number.*

Proof. Follows by setting the number of points $n = O(m^2)$ in the $(9 + \epsilon)$-approximation algorithm of Basappa et al. [4], and the fact that we spend $O(m^2 n)$ time to construct this new set of $n = O(m^2)$ points. $\square$

5   Conclusion

For LSDC-H and LSDC-A problems our algorithms will turn to be fixed-parameter tractable approximation algorithms for fixed δ, where the parameter is k. For LSDC-D problem we have proposed a PTAS and a constant factor approximation algorithm with reasonable running time. In the future work, we want to investigate the complexities of LSDC-H and LSDC-A problems.

References

1. Ambühl, C., Erlebach, T., Mihalák, M., Nunkesser, M.: Constant-factor approximation for minimum-weight (connected) dominating sets in unit disk graphs. In: Díaz, J., Jansen, K., Rolim, J.D.P., Zwick, U. (eds.) APPROX/RANDOM-2006. LNCS, vol. 4110, pp. 3–14. Springer, Heidelberg (2006). https://doi.org/10.1007/11830924_3
2. Agnetis, A., Grande, E., Mirchandani, P.B., Pacifici, A.: Covering a line segment with variable radius discs. Comput. Oper. Res. **36**(5), 1423–1436 (2009)
3. Acharyya, A., Nandy, S.C., Pandit, S., Roy, S.: Covering segments with unit squares. In: Workshop on Algorithms and Data Structures, pp. 1–12 (2017)
4. Basappa, M., Acharyya, R., Das, G.K.: Unit disk cover problem in 2D. J. Discrete Algorithms **33**, 193–201 (2015)

5. Brönnimann, H., Goodrich, M.: Almost optimal set covers in finite VC-dimension. Disc. Comput. Geom. **14**, 463–479 (1995)
6. Claude, F., Das, G.K., Dorrigiv, R., Durocher, S., Fraser, R., López-Ortiz, A., Nickerson, B.G., Salinger, A.: An improved line-separable algorithm for discrete unit disk cover. Discrete Math. Algorithms Appl. **2**(1), 77–87 (2010)
7. Carmi, P., Katz, M.J., Lev-Tov, N.: Covering points by unit disks of fixed location. In: Tokuyama, T. (ed.) ISAAC 2007. LNCS, vol. 4835, pp. 644–655. Springer, Heidelberg (2007). https://doi.org/10.1007/978-3-540-77120-3_56
8. Călinescu, G., Măndoiu, I.I., Wan, P.J., Zelikovsky, A.Z.: Selecting forwarding neighbors in wireless ad hoc networks. Mob. Netw. Appl. **9**(2), 101–111 (2004)
9. Dash, D., Bishnu, A., Gupta, A., Nandy, S.C.: Approximation algorithms for deployment of sensors for line segment coverage in wireless sensor networks. Wirel. Netw. **19**(5), 857–870 (2013)
10. Das, G.K., Fraser, R., López-Ortiz, A., Nickerson, B.G.: On the discrete unit disk cover problem. Int. J. Comput. Geom. Appl. **22**(5), 407–420 (2012)
11. Dash, D., Gupta, A., Bishnu, A., Nandy, S.C.: Line coverage measures in wireless sensor networks. J. Parallel Distrib. Comput. **74**(7), 2596–2614 (2014)
12. Fraser, R., López-Ortiz, A.: The within-strip discrete unit disk cover problem. In: Proceedings of Canadian Conference on Computational Geometry, pp. 61–66 (2012)
13. Garey, M.R., Johnson, D.S.: Computers and Intractability: A Guide to the Theory of NP-Completeness. W.H. Freeman and Company, New York (1979)
14. Mustafa, N.H., Ray, S.: Improved results on geometric hitting set problems. Discrete Comput. Geom. **44**(4), 883–895 (2010)
15. Narayanappa, S., Vojtechovskỳ P.: An improved approximation factor for the unit disk covering problem. In: Proceedings of Canadian Conference on Computational Geometry, pp. 15–18 (2006)

Fixed-Parameter Tractable Algorithms for Tracking Set Problems

Aritra Banik and Pratibha Choudhary[(⊠)]

Indian Institute of Technology, Jodhpur, India
aritrabanik@gmail.com, pratibhac247@gmail.com

Abstract. We consider parameterized complexity of the recently introduced problem of tracking paths in graphs, motivated by applications in security and wireless networks. Given an undirected and unweighted graph with a specified source s and a terminal t, the goal is to find a k-sized subset of vertices that intersect with each s-t path (or s-t shortest) path in a distinct sequence (or set).

We first generalize this problem to a problem on set systems with a universe of size n and a m sized family of subsets of the universe. Using a correspondence with the well-studied TEST COVER Problem, we give a lower bound of $\lg m$ for the solution size and show the problem fixed-parameter tractable. We also show that when k is the parameter, then for such a set system

- finding a Tracking Set for such a set system of size at most $\lg m + k$ is hard for parameterized complexity class $W[2]$;
- finding a Tracking Set of size at most $m - k$ is fixed parameter tractable;
- finding a Tracking Set of size at most $n - k$ is complete for parameterized complexity class W[1].

Using the solution for the set system generalization, we show the main result of the paper that finding a Tracking Set of size at most k for shortest paths is fixed-parameter tractable. We first give an $O^*(2^{k2^k})$ algorithm using the set system solution, which we later improve to $O^*(2^{k^2})$.

1 Introduction and Motivation

In this paper, we consider parameterized complexity of the recently introduced problem of tracking shortest paths in an undirected graph with a source and a terminal. Given an undirected graph with a specified source s and a terminal t, the goal is to find the smallest subset of vertices whose intersection with every s-t path (or every shortest s-t path) is unique.

We start with motivation for the problem. Consider the security system at a large airport. Suppose there are multiple points of entry based on nationality, destination, reason for travel and some other factors. A set of carefully chosen security scan points can be selected as identification points. Every time a passenger passes through an identification point, a seal is stamped on their ticket. Hence when a traveler reaches his/her flight, by looking at the stamps on their

© Springer International Publishing AG 2018
B. S. Panda and P. P. Goswami (Eds.): CALDAM 2018, LNCS 10743, pp. 93–104, 2018.
https://doi.org/10.1007/978-3-319-74180-2_8

ticket a security inspector can identify the exact sequence of security points passed by the traveler.

Another major application scenario is tracking of moving objects in telecommunication networks and road networks. The goal can be efficient and optimized tracking of an object, for the purpose of surveillance, monitoring, intruder detection, and operations management. The problem solution can be used for reconstruction of path traced by an object in order to detect potential network flaws, to study traffic patterns of moving objects, to optimize network resources based on such patterns, and for other such network analysis based tasks.

Tracking of moving objects has been studied in the field of wireless sensor networks. See [2] for a survey of target tracking protocols using wireless sensor networks. Some researchers have studied it with respect to power management of sensors [9]. Despite being as active area of research, a major part of this research so far is heuristic-based. In [1], the authors formalized the problem of tracking in networks as a graph theoretic problem and did a systematic study. Among some other problems, they introduced following two optimization problems. $V(P)$ is used to denote the set of vertices in path P. We use $\Pi_P(V')$ to denote the sequence in which the vertices from $V(P) \cap V'$ appear in path P. A graph with a unique source s and unique destination t is called an s-t graph, and in such a graph, a path from s to t is called an s-t path. We consider the graph to be undirected and unweighted.

Problem 1. TRACKING SET FOR PATHS *Problem (TPP): Given an s-t graph $G = (V, E)$, find a minimum cardinality Tracking Set $\mathbf{T} \subseteq V$ for G, such that for any distinct two s-t paths P_1 and P_2, $\Pi_{P_1}(\mathbf{T}) \neq \Pi_{P_2}(\mathbf{T})$.*

Problem 2. TRACKING SET FOR SHORTEST PATHS *Problem (TSPP): Given an s-t graph $G = (V, E)$, find a minimum cardinality Tracking Set $\mathbf{T} \subseteq V$ for G, such that for any two distinct shortest s-t paths P_1 and P_2, $\mathbf{T} \cap V(P_1) \neq \mathbf{T} \cap V(P_2)$.*

The authors showed the problems to be NP-hard and provided a 2-approximation algorithm for tracking shortest paths in planar graphs. Our goal in this paper is to address the parameterized complexity of the problems (see Sect. 2.1 for definitions). Towards that, we first look at a more general version of the problem in terms of set systems.

A set system is a pair $\mathcal{P} = \{X, \mathcal{S}\}$, where X is a finite set and $\mathcal{S}$ is a family of subsets of X.

A Tracking Set for Set System for a set system, $\mathcal{P} = \{X, \mathcal{S}\}$ is a set of elements that has a unique intersection with each of the subsets in the family. The TRACKING SET SYSTEM Problem is defined as follows.

Problem 3. TRACKING SET SYSTEM *Problem (TSSP): Given a set system $\mathcal{P} = \{X, \mathcal{S}\}$, find a minimum cardinality Tracking Set $\mathbf{T} \subseteq X$ for $\mathcal{P}$, such that for any two distinct $S_i, S_j \in S$, $S_i \cap T \neq S_j \cap T$.*

We denote each vertex present in Tracking Set by tracker. We show a correspondence between TRACKING SET SYSTEM Problem and the TEST COVER

Problem [6]. Using this result we show that the size of a Tracking Set for Set System with n elements and m sets is at least $\lceil \lg m \rceil$[1] and using this we show that determining whether a given set system has a Tracking Set of size at most k has a $O^*(2^{k2^k})$ fixed-parameter algorithm[2]. We then consider other natural parameterizations of the solution and we show that

- Determining whether a set system with n elements and m sets has a Tracking Set of size at most $(\lg m + k)$ is hard for the parameterized complexity class W[2].
- Determining whether a set system with n elements has a Tracking Set of size at most $(n - k)$ is complete for the parameterized complexity class W[1], and
- Determining whether a set system with n elements and m sets has a Tracking Set of size at most $(m - k)$ is fixed-parameter tractable.

In Sect. 4, we consider the parameterized complexity of the Tracking Set for Shortest Paths. By using our fixed-parameter algorithm for Tracking Set for Set System, we show that Tracking Set for Paths is fixed-parameter tractable parameterized by the solution size as long as we can count the number of s-t paths in polynomial time. It is possible to count number of shortest paths in polynomial time. Thus Tracking Set for Shortest Paths is fixed-parameter tractable parameterized by solution size. In Sect. 4.2, we give an improved fixed-parameter tractable algorithm for Tracking Set for Shortest Paths using some properties of the shortest s-t paths and by using some reduction rules. Finally Sect. 5 concludes with some open problems.

2 Preliminaries

The problem of finding the minimum Tracking Set for Shortest Paths in graphs was introduced by Banik et al. [1], and was proven to be NP-hard and APX-hard for general graphs. The authors suggested multiple versions of the TRACKING SET Problem, along with providing a 2-approximation algorithm for special instance of problem, when the graph is planar.

Throughout this paper, we assume graph $G = (V, E)$ to be an s-t graph with s and t already given to us. We preprocess the graph and delete vertices that are not reachable from s (or t). This can be accomplished by any standard graph search method (BFS or DFS).

If there exists a path P_1 between vertices u and v, and there exists another path P_2 between vertices v and w, we use $P_1 \cdot P_2$ to denote the path between u and w obtained by concatenation of paths P_1 and P_2 at v.

2.1 Fixed-Parameter Tractability

A *parameterized problem* is a language $L \subseteq \Sigma^* \times \mathbb{N}$, where Σ is a fixed, finite alphabet. For an instance $(x, k) \in \Sigma^* \times \mathbb{N}, k$ is called the *parameter*. A parameterized problem $L \subseteq \Sigma^* \times \mathbb{N}$ is called *fixed-parameter tractable* (FPT) if there

---

[1] We use lg to denote logarithm to the base 2.
[2] O^* notation ignores polynomial factors.

exists an algorithm $\mathcal{A}$ (called a *fixed-parameter algorithm*), a computable function $f : \mathbb{N} \to \mathbb{N}$, and a constant c such that, given $(x, k) \in \Sigma^* \times \mathbb{N}$, the algorithm $\mathcal{A}$ correctly decides whether $(x, k) \in L$ in time bounded by $f(k) \cdot |(x, k)|^c$. The complexity class containing all fixed-parameter tractable problems is called FPT. There is also an associated hardness hierarchy and the basic hardness classes are W[1] and W[2]. The CLIQUE problem (does the given graph have a clique of size at least k) is a canonical complete problem for W[1] while the DOMINATING SET problem (does the given graph have a dominating set of size at most k) is a canonical complete problem for W[2]. See [7] for more on parameterized complexity.

Let $A, B \subseteq \Sigma^* \times \mathbb{N}$ be two parameterized problems. A *parameterized reduction* from A to B is an algorithm that, given an instance (x, k) of A, outputs an instance (x', k') of B such that

1. (x, k) is a yes-instance of A if and only if (x', k') is a yes-instance of B,
2. $k' \leq g(k)$ for some computable function g, and
3. the running time of algorithm is $f(k) \cdot |x|^{\mathcal{O}(1)}$ for some computable function f.

A reduction rule is a rule that translates a given instance into another. The rule is said to be *safe* if the reduced instance is equivalent to the original instance in the sense that the reduced instance is an YES instance if and only if the original instance is an YES instance.

3   Tracking Set for Set System

In this section we give some FPT and hardness results for TRACKING SET SYSTEM Problem. For a given set system $\mathcal{P} = \{X, \mathcal{S}\}$, Tracking Set is a set of elements that have a unique intersection with each of the subsets in the family. Formally, $T \subseteq X$ is a Tracking Set for $\mathcal{P}$ if for any two subsets $S_i, S_j \in \mathcal{S}$ ($i \neq j$), $T \cap S_i \neq T \cap S_j$.

We first observe that the TRACKING SET SYSTEM Problem is similar to the well known TEST COVER Problem and we will establish one to one correspondence between TRACKING SET SYSTEM Problem and TEST COVER Problem. Tracking Set in this section refers to Tracking Set for Set System. We refer an instance of the Tracking Set or the Test Cover problem as an (x, y) instance if the size of the universe (element set) is x and the size of the family is y. Also, we use k-Tracking Set to refer to the problem of finding a Tracking Set of size at most k.

In the TEST COVER Problem we are given a set $M = \{1, 2, \dots n\}$ of elements called vertices and a family $\mathcal{T} = \{T_1, T_2, \dots, T_m\}$ of distinct subsets of M called tests. We say that a test T_l separates a pair i and j if $|\{i, j\} \cap T_l| = 1$. A subset $\mathcal{T}'$ of $\mathcal{T}$ is called a Test Cover if for every pair of distinct vertices $i, j \in M$, there exists a test $T_i \in \mathcal{T}'$ that separates them. The TEST COVER Problem requires finding a minimum size Test Cover if one exists.

TEST COVER Problem is a well studied problem [3,8]. It is known that the TEST COVER Problem is NP-hard and APX-hard [11]. There exists an

$\mathcal{O}(\log n)$-approximation algorithm for the problem [14] and there is no $o(\log n)$-approximation algorithm unless $P = NP$ [11].

Parameterized complexity of TEST COVER Problem has also been studied extensively [5,6,10]. Given $(M, \mathcal{T})$, and $k \in \mathbb{N} \cup \{0\}$, the parameterized version of the TEST COVER Problem asks if there exists a Test Cover of size at most k.

Observe that just as $\lg n$ is a lower bound for the size of Test Cover, n and m form upper bounds on the size of Test Cover. Given lower and upper bounds of solution size, natural questions are to ask if there exists a FPT on a parameter k which determines that whether there exists a solution k greater than the lower bound or k less than the upper bound. Parameterizations of NP- optimization problems above their guaranteed lower/upper bounds was initiated by Mahajan and Raman in [12]. They proved some above guarantee versions of MAX CUT and MAX SAT to be FPT. Further in [13], the authors gave several results on parameterizing above or below the lower/upper guaranteed bounds.

On the same line we have the following results by Crowston et al. [6] which is summarized in the following theorem.

Theorem 1 ([6]). *For an (n, m)-Test Cover instance,*

(i) *there does not exist a Test Cover of size less than $\lceil \lg n \rceil$. Hence Test Cover is fixed-parameter tractable when parameterized by solution size and Test Cover can be solved in time $O^*(2^{k2^k})$.*

(ii) *Determining whether there exists a Test Cover of size at most $(m - k)$ is complete for the parameterized complexity class W[1].*

(iii) *Determining whether there exists a Test Cover of size at most $(n - k)$ is fixed-parameter tractable.*

(iv) *Determining whether there exists a Test Cover of size at most $(\lg n + k)$ is hard for the parameterized complexity class W[2].*

Next we show a tight correspondence between TRACKING SET Problem and TEST COVER Problem. Let $\mathcal{P} = \{X, \mathcal{S}\}$ be a set system, where $X = \{x_1, x_2, ..., x_n\}$ and $\mathcal{S} = \{S_1, S_2, ..., S_m\}$. Let $\mathcal{R}(\mathcal{P}) = \{M, \mathcal{T}\}$ is defined as follows, $M = \{1, ..., m\}$, and $\mathcal{T} = \{T_1, ... T_n\}$ where $T_i = \{j \mid x_i \in S_j\}$.

Lemma 1. *There exists a Tracking Set of size at most k for $\mathcal{P}$ if and only if there exists a Test Cover of size at most k for $\mathcal{R}(\mathcal{P})$.*

Proof. Let us assume that $TS \subseteq X$ is a Tracking Set for $\mathcal{P}$. We define a set $TC \subseteq \mathcal{T}$ as $TC = \{T_i \mid x_i \in TS\}$. We claim that TC is a Test Cover for $\mathcal{R}(\mathcal{P})$. Suppose not. Then there exists $i, j \in M$ such that for any set $T_r \in TC$, $|T_r \cap \{i, j\}| \neq 1$. This leads to two possibilities, the first being that i and j do not appear in any of the tests in TC, i.e. $|T_r \cap \{i, j\}| = 0$, and the second being that i appears in all those tests in TC that contain j, i.e. $|T_r \cap \{i, j\}| = 2$. We analyze these two cases individually.

1. $\forall T_r \in TC, i, j \notin T_r$

   By definition of $\mathcal{R}(\mathcal{P})$, $i, j \notin T_r$ implies that $x_r \notin S_i, S_j$. And, by definition of TC, $\forall T_r \in TC, i, j \notin T_r$ implies that $\forall x_r \in TS, x_r \notin S_i, S_j$. Hence

$TS \cap S_i = TS \cap S_j = \emptyset$. This contradicts the assumption that TS is a Tracking Set for $\mathcal{P}$.

2. $\forall T_r \in TC$, $i \in T_r$ if and only if $j \in T_r$

By definitions of $\mathcal{R}(\mathcal{P})$ and TC, we can say that this implies, $\forall x_r \in TS$, $x_r \in S_i$ if and only if $x_r \in S_j$. Hence $TS \cap S_i = TS \cap S_j$. This contradicts the assumption that TS is a Tracking Set for $\mathcal{P}$.

Conversely, assume that $TC \subseteq \mathcal{T}$ is a Test Cover for $R(P)$. Now consider $TS = \{x_i \mid T_i \in TC\}$. We will argue that TS is a Tracking Set for $\mathcal{P}$. Suppose not. Then there exists $S_i, S_j \in \mathcal{S}$ such that $TS \cap S_i = TS \cap S_j$. This means that $\forall x_r \in TS, x_r \in S_i$ whenever $x_r \in S_j$. Hence, $\forall T_r \in TC$, $i \in T_r$ whenever $j \in T_r$. Thus there does not exist a test $T_s \in TC$, such that $|T_s \cap \{i, j\}| = 1$. This contradicts the assumption that TC is a Test Cover for $\mathcal{R}(\mathcal{P})$. $\square$

Based on Theorem 1 and Lemma 1, we have the following two corollaries.

Corollary 1. *For a (n, m)-set system the following holds.*

(i) *There does not exists a Tracking Set of size less than $\lceil \lg m \rceil$.*

(ii) *Tracking Set is fixed-parameter tractable when parameterized by solution size and Tracking Set can be solved in time $O^*(2^{k2^k})$.*

(iii) *$(n - k)$-Tracking Set is W[1]-complete.*

(iv) *$(m - k)$-Tracking Set is FPT.*

(v) *$(\lg m + k)$-Tracking Set is W[2]-hard.*

4   FPT Algorithm for TSPP

In this section, we provide two FPT algorithms for finding Tracking Set in graphs, first using Tracking Set for Set System and then without using it. While the first one works for any Tracking Set problem where the number of s-t paths can be counted efficiently, the second approach is faster and is suited for the TSPP problem.

4.1   Using Tracking Set for Set System

To cast TPP and TSPP as a Tracking Set problem for a set system $\mathcal{P} = \{X, \mathcal{S}\}$, we simply take the vertex set of the given graph as the universe X and the family $\mathcal{S}$ consists of sets of vertices in all simple s-t paths or shortest s-t paths as appropriate.

In the TSPP problem since we consider only shortest s-t paths, we find and remove those edges and vertices that do not participate in any shortest s-t path. Observe that this can be done by first finding a shortest s-t path, say of length l_s, using a standard algorithm, and then removing each edge (u, v) for which the following equality does not hold:

Distance from s to $u + 1 +$ Distance from v to $t = l_s$

Now in the remaining graph none of the edges exist between vertices equidistant from s (or t). In fact, the end points of each edge are such that the difference of their distances from s (or t) is always one. Thus the vertices of graph can be categorized into layers, such that each layer consists of the vertices equidistant from s (or t). Such a graph is called a layered s-t graph. We can also consider each edge to be directed from the vertex closer to s, towards the vertex closer to t. Thus we can perceive our graph as a directed acyclic graph while solving the TSPP problem.

We define level $L(v)$ of a vertex v as the length of the shortest path from s to v. We denote out-degree of a vertex v by $d_o(v)$ and its in-degree by $d_i(v)$. In the rest of the paper, Tracking Set is used to refer Tracking Set for Paths and Tracking Set for Shortest Paths.

We can solve TPP and TSPP by modeling it as a Tracking Set problem for a set system. Observe that in order to do so we need to bound the number of s-t paths. However, for general graph, counting the number of s-t paths is hard for the complexity class $\#P$ [15]. For some special class of graphs it can be done in polynomial time. Suppose for a graph class we can enumerate all the s-t paths in $f(n)$ time. Let the number of paths be m. From Corollary 1 we know $m > 2^k$ there does not exists a of size less than equals to k. Thus we can assume that $m \leq 2^k$. Observe that no two vertex in the graph are present in the same set of s-t paths (assuming each vertex participates in at least one s-t path). Therefore number of vertices n is bounded by 2^{2^k}. We can verify whether a subset of vertices is a tracking set or not in $2^{O(k)}$ time by checking whether it intersects every path uniquely or not. Thus we have the following theorem.

Theorem 2. *If for a graph s-t paths can be enumerated in $f(n)$ time then it is possible to determining whether G has a Tracking Set of size at most k or not in time $f(n) + 2^{2^{O(k)}}$.*

Next we show that verifying whether a given a set of vertices is a Tracking Set for Shortest Paths or not can be done in polynomial time, and hence establish that TRACKING SET FOR SHORTEST PATHS is in NP. A similar proof is given in [1], however for the sake of completeness we provide a complete proof here as well. After that we show that number of shortest paths between two vertices can be counted in polynomial time.

In Lemma 2 we show that the following condition is necessary and sufficient for a set of vertices to be a Tracking Set.

Condition 1. *A set of vertices V' is said to follow Tracking Set Condition if there exists at most one shortest path between any two vertices $u, v \in V' \cup \{s,t\}$ that does not contain any vertex from $V' \cup \{s,t\} \setminus \{u,v\}$.*

As explained earlier, we perceive the graph to be directed. Hence, the shortest path between any two vertices has to be a sub-path of some shortest s-t path.

Now we have the following lemma.

Lemma 2. *Let $G = (V, E)$ be a graph and $\mathbf{T} \subseteq V$ is a set of vertices. $\mathbf{T}$ is a Tracking Set for Shortest Paths if and only if it follows Tracking Set Condition.*

Proof. Let $\mathbf{T}$ be a Tracking Set for shortest paths in G. First we prove that $\mathbf{T}$ follows *Tracking Set Condition*. Assume that *Tracking Set Condition* does not hold for $\mathbf{T}$, i.e. there exist two vertices $u, v \in \mathbf{T}' = \mathbf{T} \cup \{s, t\}$ such that there are two or more shortest paths λ_1 and λ_2 between u and v, that do not contain any vertices from $\mathbf{T}' \setminus \{u, v\}$. Observe that both u and v are part of some shortest s-t path (not necessarily the same). Let λ_s and λ_t be any shortest path from s to u and v to t. Thus λ_s and λ_t are also shortest. Observe the following

- $\forall v_i \in \lambda_s \setminus \{u\},\ L(v_i) < L(u)$
- $\forall v_j \in \lambda_1 \cup \lambda_2 \setminus \{u, v\},\ L(u) < L(v_j) < L(v)$
- $\forall v_k \in \lambda_t \setminus \{v\},\ L(v_k) > L(v)$

Hence $\lambda_s \cdot \lambda_1 \cdot \lambda_t$ and $\lambda_s \cdot \lambda_2 \cdot \lambda_t$ are two valid shortest s-t paths containing the same set of trackers. This contradicts the fact that T is a Tracking Set.

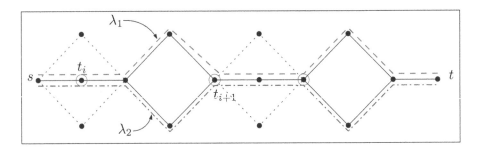

Fig. 1. Illustration of Lemma 2

Let $\mathbf{T}$ is a set of vertices which follows *Tracking Set Condition*. We prove that $\mathbf{T}$ is a Tracking Set for shortest s-t paths in G. Suppose $\mathbf{T}$ is not a Tracking Set. Then there exists at least two shortest s-t paths, say λ_1 and λ_2 that contain the same set of vertices from $\mathbf{T}$ (see Fig. 1). Observe that in a specific order O the vertices from $\mathbf{T}' = \mathbf{T} \cup \{s, t\}$ appears in λ_1 and λ_2. There are at least one pair of consecutive vertices t_i and t_{i+1} in O such that the path from t_i to t_{i+1} is different in λ_1 than in λ_2, otherwise both the paths are same. These paths are part of some shortest s-t paths. Hence there exists two shortest paths between t_i and t_{i+1}. This violates the *Tracking Set Condition*. Hence the result holds. $\square$

Verifying whether a given subset of vertices of size at most k is a Tracking Set for shortest paths can be done using Lemma 2 in $O(k^2(m + n))$ time, by checking for unique paths without a tracker between every pair of vertices in the set.

Next we have the following observation which is probably folklore, we provide details here for completeness. Level of a vertex v is the shortest distance from s to v.

Observation 1. *Let v be any vertex at level ℓ. Assume that it has k in-neighbors in level $\ell - 1$ which are $\{v_1, v_2, ..., v_k\}$. If there are $\alpha_1, \alpha_2, ..., \alpha_k$ paths from s to $v_1, v_2, ..., v_k$ respectively, then there are $\alpha_1 + \alpha_2 + ... + \alpha_k$ many paths (not necessarily disjoint) from s to v.*

The observation immediately gives a dynamic programming (on levels) algorithm to compute the number of shortest s-t paths and we have the following lemma.

Lemma 3. *The number of shortest s-t paths in a graph $G = (V, E)$ with $|V| = n, |E| = m$, can be found in $O(m + n)$ time.*

Essentially, the same algorithm works for counting the number of paths in a directed acyclic graph if we traverse the vertices in topologically sorted order to show the following (see [4]).

Lemma 4. *The number of s-t paths in a directed acyclic s-t graph $G = (V, E)$ with $|V| = n, |E| = m$, can be found in $O(m + n)$ time.*

Thus from Theorem 2 and Lemma 4, we have

Corollary 2. *When the given s-t graph is a directed acyclic graph, we can determine whether the graph has a Tracking Set for Paths of size at most k for all the paths in time $O(m + n + 2^{k2^k} k^2 (m + n))$.*

Also applying Theorem 3 in Step 1 of the algorithm of Sect. 3, we get

Theorem 3. *For a given s-t graph $G = (V, E)$ where $|V| = n$ and $|E| = m$, a Tracking Set for Shortest Paths of size at most k, if exists, can be found in $O(2^{k2^k} k^2 (m + n))$ time.*

4.2   Improved FPT Algorithm for TSPP

We obtain an improved FPT algorithm for TSPP using an additional rule which results in a larger lower bound for the number of paths, thereby giving a smaller upper bound for the size of the universe.

We start with the following reduction rules.

Reduction Rule 1. *If there is an edge from vertex x to vertex y and if both x and y have degree 2, then delete y and draw an edge between x and v_i, $\forall v_i \in N(y) \setminus \{x\}$.*

Reduction Rule 2. *If there exists a set of m degree 2 vertices, such that they are all adjacent to a pair of vertices $u, v \in V$, then arbitrarily delete $m - 1$ of these vertices and reduce k by $m - 1$.*

Proof of safeness of above two reduction rules will be provided in the full version of the paper.

Observation 2. *In a reduced instance for any pair of vertices u, v with degree at least 3, there exists at most one vertex of degree 2 that is adjacent to both u and v.*

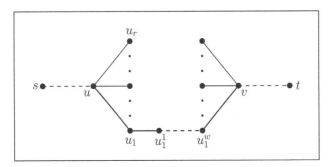

Fig. 2. Illustration of proof of Lemma 5

Lemma 5. *In a reduced instance the number of s-t paths in graph is at least $1 + \sum_{v \in V \setminus \{t\}} (d_o(v) - 1)$.*

Proof. Let $p = 1 + \sum_{v \in V \setminus \{t\}} (d_o(v) - 1)$.

We prove the claim by induction on p. Consider the base case when $p = 1$. In this case, $\sum_{v \in V \setminus \{t\}} (d_o(v) - 1) = 0$. This means that each vertex in $V \setminus \{t\}$ will have out-degree exactly 1. Clearly, the lemma holds.

For the induction hypothesis, assume that the lemma holds for the case when $p < \delta$. Now for induction, consider the case when $p = \delta$. Consider a vertex u that is closest to t among those having out-degree at least 2 (break ties arbitrarily). Assume that u has out-degree equal to r. Let $u_1, u_2, ..., u_r$ are out-neighbours of u (see Fig. 2). Consider one of the out-neighbors, say u_1. Observe that the path from u_1 to t does not contain any vertex with out-degree greater than 1. In the path from u_1 to t consider the first vertex, say v which has in-degree greater than or equal to 2 (such a vertex exists as otherwise Reduction Rule 1 would be applicable). Let the path from u_1 to v is $P_{u_1 v} = (u_1, u_1^1, u_1^2, ..., u_1^w, v)$. Delete all the vertices in the path from u_1 to v, say λ, except the vertex v. If $u_1 = v$, delete the edge between u and v. After deleting the vertices in $\lambda \setminus \{v\}$, let the new graph be $G'(V', E')$. Observe that the deletion of vertices $\{u_1^1, u_1^2, ..., u_1^w\}$ does not affect δ. However, the out-degree of u is reduced by 1 due to deletion of u_1 and consequently δ reduces by 1. Observe that the number of s-t paths in G is also reduced by at least 1. Therefore the lemma holds. $\square$

Lemma 6. *In a layered s-t graph $G = (V, E)$ on n vertices, with the property that each vertex participates in at least one s-t path, and no two vertices of degree 2 are adjacent, there exists at least $\sqrt{n}/4$ s-t paths.*

Proof. Assume that the number of vertices with degree greater than equal to 3 is n_3 and the number of vertices with degree equal to 2 is n_2. Since there can exist at most one vertex of degree 2 adjacent to a pair of vertices of degree 3 or more, n_2 is at most $\binom{n_3}{2}$. Hence n_3 is at least $\sqrt{2n_2}$.

Assume that we short-circuit all vertices of degree equal to 2 and create a new graph G'. Observe that the number of s-t paths in G is the same as the number of s-t paths in G'. Since the graph G' comprises of only vertices with degree greater than or equal to 3, the total degree of all vertices in graph is greater than equal to $3n_3$. Thus the total out-degree of all vertices in graph is greater than equal to $3n_3/2$. From Lemma 5, we know that the total number of s-t paths, say p, is at least $1 + \sum_{v \in V \setminus \{t\}} (d_o(v) - 1)$. So p is at least $3n_3/2 - n_3 = n_3/2 = n_3/4 + n_3/4 \geq (\sqrt{n_2/2})/2 + n_3/4$ which is at least $\sqrt{n}/4$ as $n_3 + n_2 = n$.

Theorem 4. *For a given s-t graph $G = (V, E)$, we can determine whether it has a size-k Tracking Set for Shortest Paths in time $O(2^{2k^2+4k}k^2(m+n))$.*

Proof. From Lemma 6, we know that m, the number of paths is at least $\sqrt{n}/4$. From Corollary 1, we know that we need at least $\lg(m) \geq \lg \sqrt{n} - 2$ trackers. Hence if $k < 0.5 \lg n - 2$, we reply that a k-sized solution is not possible. Otherwise, $k \geq \lg(\sqrt{n}/4)$, therefore $n \leq 2^{2k+4}$. Now for each subset of V of size k, we verify whether it is a Tracking Set using Lemma 2 in $O(k^2(m+n))$ time. Thus in $O(2^{2k^2+4k}k^2(m+n))$ time, we can find a Tracking Set of size at most k if one exists. □

5   Conclusions

We gave FPT algorithms for $TSPP$ and some special cases of TPP. Improving the runtime of our FPT algorithm for $TSPP$ and determining whether the general TPP is FPT are interesting open problems.

A related notion to fixed-parameter tractability is the notion of kernelization.

Gutin et al. [10] have shown that there does not exist a polynomial kernel for the TEST COVER Problem when parameterized by solution size, under standard complexity theory assumptions. Due to the correspondence with TRACKING SET Problem it will follow that the general TRACKING SET Problem also does not have a polynomial kernel under the same assumptions when parameterized by the solution size. However as $TSPP$ and TPP are special instances of Tracking Set, it is possible that they have polynomial sized kernels.

Another direction of study is further exploration of approximation algorithms for TPP and $TSPP$ in special classes of graphs.

Acknowledgments. We thank Venkatesh Raman and Saket Saurabh for fruitful discussions and valuable suggestions.

References

1. Banik, A., Katz, M.J., Packer, E., Simakov, M.: Tracking paths. In: Fotakis, D., Pagourtzis, A., Paschos, V.T. (eds.) CIAC 2017. LNCS, vol. 10236, pp. 67–79. Springer, Cham (2017). https://doi.org/10.1007/978-3-319-57586-5_7

2. Bhatti, S., Xu, J.: Survey of target tracking protocols using wireless sensor network. In: Proceedings of the 2009 Fifth International Conference on Wireless and Mobile Communications, ICWMC 2009, Washington, D.C., pp. 110–115. IEEE Computer Society (2009)

3. De Bontridder, K.M.J., Halldórsson, B.V., Halldórsson, M.M., Hurkens, C.A.J., Lenstra, J.K., Ravi, R., Stougie, L.: Approximation algorithms for the test cover problem. Math. Program. **98**(1–3), 477–491 (2003)

4. Cormen, T.H., Leiserson, C.E., Rivest, R.L., Stein, C.: Introduction to Algorithms, 3rd edn. The MIT Press, Cambridge (2009)

5. Crowston, R., Gutin, G., Jones, M., Muciaccia, G., Yeo, A.: Parameterizations of test cover with bounded test sizes. Algorithmica **74**(1), 367–384 (2016)

6. Crowston, R., Gutin, G., Jones, M., Saurabh, S., Yeo, A.: Parameterized study of the test cover problem. In: Rovan, B., Sassone, V., Widmayer, P. (eds.) MFCS 2012. LNCS, vol. 7464, pp. 283–295. Springer, Heidelberg (2012). https://doi.org/10.1007/978-3-642-32589-2_27

7. Cygan, M., Fomin, F.V., Kowalik, L., Lokshtanov, D., Marx, D., Pilipczuk, M., Pilipczuk, M., Saurabh, S.: Parameterized Algorithms, 1st edn. Springer, Cham (2015). https://doi.org/10.1007/978-3-319-21275-3

8. De Bontridder, K.M.J., Lageweg, B.J., Lenstra, J.K., Orlin, J.B., Stougie, L.: Branch-and-bound algorithms for the test cover problem. In: Möhring, R., Raman, R. (eds.) ESA 2002. LNCS, vol. 2461, pp. 223–233. Springer, Heidelberg (2002). https://doi.org/10.1007/3-540-45749-6_23

9. Ganesan, D., Cristescu, R., Beferull-Lozano, B.: Power-efficient sensor placement and transmission structure for data gathering under distortion constraints. ACM Trans. Sens. Netw. **2**(2), 155–181 (2006)

10. Gutin, G., Muciaccia, G., Yeo, A.: (Non-)existence of polynomial kernels for the test cover problem. Inf. Process. Lett. **113**(4), 123–126 (2013)

11. Halldórsson, B.V., Halldórsson, M.M., Ravi, R.: On the approximability of the minimum test collection problem. In: auf der Heide, F.M. (ed.) ESA 2001. LNCS, vol. 2161, pp. 158–169. Springer, Heidelberg (2001). https://doi.org/10.1007/3-540-44676-1_13

12. Mahajan, M., Raman, V.: Parameterizing above guaranteed values: MaxSat and MaxCut. J. Algorithms **31**(2), 335–354 (1999)

13. Mahajan, M., Raman, V., Sikdar, S.: Parameterizing above or below guaranteed values. J. Comput. Syst. Sci. **75**(2), 137–153 (2009)

14. Moret, B.M.E., Shapiro, H.D.: On minimizing a set of tests. SIAM J. Sci. Stat. Comput. **6**(4), 983–1003 (1985)

15. Valiant, L.G.: The complexity of enumeration and reliability problems. SIAM J. Comput. **8**(3), 410–421 (1979)

Exact Computation of the Number
of Accepting Paths of an NTM

Subrahmanyam Kalyanasundaram[1(✉)] and Kenneth W. Regan[2]

[1] Department of Computer Science and Engineering,
IIT Hyderabad, Sangareddy, India
subruk@iith.ac.in
[2] Department of Computer Science and Engineering,
University at Buffalo, Buffalo, USA
regan@buffalo.edu

Abstract. We look at the problem of counting the exact number of accepting computation paths of a given nondeterministic Turing machine (NTM). We give a deterministic algorithm that runs in time $\widetilde{O}(\sqrt{S})$, where S is the size (number of vertices) of the configuration graph of the NTM, and prove its correctness. Our result implies a deterministic simulation of probabilistic time classes like PP, BPP, and BQP in the same running time. This is an improvement over the currently best known simulation by van Melkebeek and Santhanam [SIAM J. Comput., 35(1), 2006], which uses time $\widetilde{O}(S^{1-\delta})$. It also implies a faster deterministic simulation of the complexity classes ⊕P and $\mathsf{Mod_k P}$.

1 Introduction

For a given nondeterministic Turing machine (NTM), counting the number of accepting computation paths is a difficult problem in general. For instance, the complexity class #P captures the complexity of counting for decision problems in NP. The computational power of #P is highlighted by a celebrated result of Toda [12]. Toda showed that a polynomial time machine with a #P oracle can perform any computation in the polynomial hierarchy.

In this paper, we prove that we can deterministically count the number of accepting paths of a k-tape NTM N in time $a^{kt/2} \cdot f(\cdot)$, where a is the alphabet size and t is the running time of N. The function f grows much slower than $a^{kt/2}$ and so does not contribute significantly to the running time. Our main theorem is:

Theorem 1. *There is a deterministic algorithm that computes the number of accepting computations of any k-tape NTM N on a given input x in time*

$$a^{kt/2} H_a^{k\sqrt{t}\log t} \cdot q^2 \mathsf{poly}(\log q, k, t, a),$$

where a is the alphabet size, t is the time complexity, and q is the number of states of N and H_a is a constant that depends only on a.

© Springer International Publishing AG 2018
B. S. Panda and P. P. Goswami (Eds.): CALDAM 2018, LNCS 10743, pp. 105–117, 2018.
https://doi.org/10.1007/978-3-319-74180-2_9

The counting variants of many specific problems have been looked at. Two classic results are Kirchhoff's matrix-tree theorem, which counts the number of spanning trees in an arbitrary graph, and Kasteleyn's theorem, which counts the number of perfect matchings in a planar graph. Both yield polynomial-time algorithms. But other problems—such as counting the number of perfect matchings in an arbitrary graph or the number of satisfying assignments of a CNF formula—are known to be #P-complete. The class #P-complete denotes the hardest problems in #P and a polynomial time algorithm for any of them would imply P = NP. Interestingly, some #P-complete problems admit a Fully Polynomial Randomized Approximation Scheme (FPRAS). Examples are counting the number of satisfying assignments of a DNF formula and counting the number of perfect matchings in a bipartite graph. See the book by Jerrum [6] for a good account of techniques used for counting problems and the paper [3] for approximability results on counting problems. However, we are interested in the counting version of a language that is accepted by an arbitrary NTM, and not an NTM that accepts a specific language. Hence we cannot assume any structure specific to the language. To the best of our knowledge, ours is the first attempt in this direction.

For a given NTM N, counting the number of accepting computation paths is a harder problem than deciding if it accepts the given input—that is, to determine if N has at least one accepting computation path. One way to go through the accepting computations would be to search the configuration graph of N. The number of strings that can be written on to the k tapes of N is already a^{kt}. In [7], it was shown that the computation of a k-tape NTM N can be simulated by a deterministic Turing machine (DTM) in time roughly[1] $a^{kt/2}$, achieving a running time approximately square root of the size of the configuration graph. This was obtained by combining two techniques used to simulate the NTM: a breadth-first search (BFS) approach and a *block trace* (BT) approach.

Counting the number of accepting computations is harder insofar as the BT component cannot simply allow computation paths to "merge" because exact counts of paths need to be preserved. We follow the structure of [7] but develop more refined techniques to track these numbers. Sections 2 and 3 explain the model and the BFS and BT approaches, giving details for reference in the proof of the main theorem in Sect. 4. Neither BFS nor BT improves the a^{kt} term by itself, but we show how the balancing idea of [7] survives the need to keep exact track of counts.

The ability to count the number of accepting computations immediately implies a faster simulation of the "counting complexity classes". The reader is referred to [2,10,13] for a detailed study on various complexity classes where membership in a language is determined as a function of the number of accepting paths in a nondeterministic TM. These include probabilistic classes such as PP, BPP and BQP. In [14], van Melkebeek and Santhanam had shown a simulation of probabilistic time t machines in deterministic time $o(2^t)$. Their improvement

---

[1] There are other multiplicative factors in the running time of the simulation, but $a^{kt/2}$ is the fastest growing factor.

comes from using techniques that save on randomness and observing that the number of random bits required can be reduced from t to $(1 - \delta)t$. However, their model assumes that there are only 2 nondeterministic choices available at each step. Our model is more general and considers all the choices available, i.e., the choices in tape movement, written alphabet and next state. Our result reduces the exponent of the running time by half, whereas the reduction in [14] is incremental in nature. Our result also implies a faster simulation of other counting classes such as $\oplus P$ (also known as ParityP) and $Mod_k P$. The connections to probabilistic and parity classes are explored in Sect. 5. See [8] and [11] for other related work.

2   Preliminaries and the BFS Approach

Given a nondeterministic Turing machine (NTM) N and an input x of length n, $t = t(n)$ is defined to be the maximum number of time steps that N takes to halt on all the computation paths on inputs of length n. At any given instant, the NTM N can have up to $a^k 3^k q$ different options for its next step, where a is the alphabet size, k is the number of tapes and q is the number of states. This is because, each of the k tape heads have a different choices for the characters to be written on the tape, and 3 different choices for the head movements – left, stay or right. We assume that $t(n)$ is time-constructible and space-constructible (see a standard textbook such as [1] or [9] for a definition).

A *configuration* of the NTM N is a snapshot of the present situation of the machine. The configuration contains the current state, the tape contents of all the k tape heads, and current position of all the heads. The total number of configurations possible is $q \cdot (a^t t)^k = qa^{kt}t^k$. The above expression follows from the fact that in a computation that uses at most t steps, the NTM traverses at most t locations in each tape. The computation starts from the standard starting configuration: start state of N, first tape containing the input x, and the remaining $k - 1$ tapes being empty with all heads at the left-most positions. Each transition follows the rules of the NTM N, and since N is nondeterministic, a configuration may be followed by 0, 1 or multiple successor configurations. The transitions naturally define a directed graph with the configurations as vertices.

The computation halts when the NTM reaches the accept or reject state. We use $\#acc_N(x)$ to denote the number of accepting computation paths of the NTM N on a given input x. We look at the problem of deterministically computing $\#acc_N(x)$ in the most efficient manner.

The naive approach would be to try a Depth First Search on the configuration graph, simulating each computation path till it halts. This is inefficient because of the high out-degree of the graph. However, a modified version of Breadth First Search works better, because each vertex in the graph is visited only once. Thus we have the following:

Theorem 2 (BFS Approach). *Given a k-tape nondeterministic Turing machine N and an input x, there is a deterministic algorithm that computes $\#acc_N(x)$, in time $q^2(3at)^k a^{kt} \mathsf{poly}(\log q, k, t, a)$.*

Proof. We consider the following modified configuration graph $\widetilde{G}$: the nodes are pairs (ρ, p), where ρ is a configuration of the NTM N and p is an integer $0 \leq p \leq t$. By the above bound, this graph has at most $qa^{kt}t^k \cdot (t+1)$ nodes. There is a directed edge from (ρ, p) to (ρ', p') if and only if ρ' is a valid successor configuration for ρ in the NTM N and $p' = p + 1$.

Notice that $\widetilde{G}$ is a directed acyclic graph, and that for any two nodes (ρ, p), $(\rho', p') \in V(\widetilde{G})$ all paths from (ρ, p) to (ρ', p') are of the same length. Let ρ_x denote the starting configuration when N takes the string x as input. We modify BFS by maintaining a count variable at each node of $\widetilde{G}$ that stores the number of shortest paths from the starting node $(\rho_x, 0)$ to that node. By the construction of $\widetilde{G}$, each path is a shortest path. Thus the BFS variant actually gives the number of paths from $(\rho_x, 0)$ to each node. We use a look up table for simulating the transition rules of N to figure out adjacencies of $\widetilde{G}$ on the fly.

Finally, we go through all the nodes, and sum up the number of paths to all the nodes corresponding to accepting configurations of N.

The dominant term in the running time comes from the number of edges of the graph $\widetilde{G}$. We have $e = |E(\widetilde{G})| \leq qa^{kt}t^k \cdot (t+1) \cdot qa^k 3^k = q^2(3at)^k a^{kt} \cdot (t+1)$. We would also need to keep track of the different adjacencies, and other bookkeeping. This brings in an additional[2] $(\log e)^{O(1)}$ factor. Thus the total time required for computing the number of accepting computations is:

$$O(e \cdot \log(e)^{O(1)}) = q^2(3at)^k a^{kt} \mathsf{poly}(\log q, k, t, a).$$

$\square$

3   Block Trace Computation

Block traces are succinct witnesses of computation paths that periodically take stock of the computation path of the NTM N. They capture information of the NTM once every d steps. The general idea of block traces has been used in the past, but we restate from [7] the particular form used here.

Definition 3 (Block trace [7]). A *segment of size d* for a k-tape NTM N is a sequence of 4-tuples

$$\tau = [(r_1, f_1, \ell_1, u_1), \ldots, (r_k, f_k, \ell_k, u_k)]$$

where for each tape j, $1 \leq j \leq k$:

- $r_j \in \{0, \ldots, d\}$ is the maximum number of cells to the right of its starting position the tape head will ever be over the next d steps,
- $f_j \in \{0, \ldots, d - r_j\}$ is the number of cells left of the position of r_j that the tape head ends up after the d-th step,

---

[2] We require $\log(qa^{kt}t^k)$ time to even read a configuration.

- $\ell_j \in \{1, \ldots, d\}$ is the number of distinct cells that shall be accessed over the next d steps on the tape j, and
- u_j is a string of length ℓ_j, which represents the final contents of those cells.

A *block trace of block-size* d is a sequence of segments of size d.

An accepting computation path is *compatible* with a block trace if the latter has $\lceil t/d \rceil$ blocks where t is the total number of steps in the path, and in every block each 3-tuple (r_j, f_j, ℓ_j) correctly describes the head locations after the corresponding d steps of the path, *and* every character in u_j is the correct final content of its cell after the d steps.

Our plan is to enumerate the number of accepting paths of N compatible to each of the block trace witnesses. We first show the simple, but critical Lemma 4 that allows us to use block traces without losing track of the count of accepting paths. We then state the key Lemma 5 that shows that we can deterministically enumerate the number of accepting computation paths corresponding to each of the witnesses. Combining Lemma 5 with Lemma 6 that bounds the number of block trace witnesses, we get a bound on the running time of the deterministic algorithm that computes the number of accepting paths of N, $\#acc_N(x)$. This bound is proved in Theorem 7.

Lemma 4. *Two different block trace witnesses give rise to a disjoint set of computation paths.*

Proof. Every computation path has a corresponding block trace witness with which it is compatible. So it is enough to show that this witness is unique. For a given fixed computation path, at each time instance and for each tape j, the values r_j, f_j, ℓ_j, u_j are fixed as in Definition 3. So the lemma follows. $\square$

Lemma 5. *The number of accepting computations of N on a given input x, that are compatible with a given block trace witness can be calculated by an algorithm in time $q^2 a^{3kd} \mathrm{poly}(\log q, k, t, a, d)$.*

Proof. We are given a block trace witness. Let $1 \le i \le \lceil t/d \rceil$ be a segment in the block trace. For each value of i, we maintain a set R_i that contains pairs (r, c). A pair $(r, c) \in R_i$ if the state r of NTM N can be reached after the ith segment (*id* time steps), and c is the number of distinct compatible computation paths that lead up to r. Initially, we have $R_0 = \{(q_0, 1)\}$, where q_0 is the start state of N.

To assist us in maintaining the sets R_i, we create a lookup table which can be thought of as a d step equivalent of the NTM transition function, along with additional information. Given an input state p and a segment, we want to know the set of states reachable from p after d steps, and the number of distinct compatible (with the given segment) computation paths for each of these states. The lookup table is indexed with (i) the state p, (ii) the segment, and (iii) for each of the k tapes, the contents of $(2d-1)$ tape cells surrounding the tape head. For each of the index values, the table returns a set of pairs $(p', c_{p'})$ which contains all the states p' reachable from p after d steps in a manner compatible with the given segment, and $c_{p'}$ is the number of distinct compatible computation paths.

We use this lookup table to generate R_i from R_{i-1} successively. For the given block trace, note that we can compute all the tape contents of the NTM after the $(i-1)$th segment. For each pair $(r, c) \in R_{i-1}$, we look up the entry corresponding to the state r, the ith segment in the block trace and the relevant tape contents. This entry will be a set of pairs (r', c'). For each pair (r', c'), we add the entry (r', cc') to R_i. This is because we had c compatible paths to reach r till the $(i - 1)$th segment and c' compatible ways to reach from r to r' during the ith segment. For the moment, we will add another entry corresponding to r' even if one already exists. We repeat this for all entries of R_{i-1}, and for each entry of R_{i-1}, all the pairs in the set returned by the lookup table. After doing this, we sort R_i according to the state values in the pairs. All of the pairs $(r', \alpha_{r'})$ are replaced by a single entry $(r', \sum \alpha_{r'})$. This ensures that the count is preserved, and that the set R_i's are of bounded size. We continue this for all the $\lceil t/d \rceil$ segments, and the entry corresponding to the accepting state q_A in $R_{t/d}$ gives the number of accepting computations.

The lookup table has $qa^{(3d-1)k}d^2$ rows. This follows from the fact that there are at most $a^{kd}d^2$ segments. If the lookup table is stored serially, the cost of each lookup is at most $qa^{(3d-1)k}d^2(\log q + 3kd \log a + 2 \log d) + q \log q$. The unsorted list of states could be of size at most q^2, and we require $q^2 \log q$ comparisons to sort, with each comparison costing $\log q$ time. Together with this, the total running time for the algorithm on t/d segments is

$$[qa^{3kd}d^2(\log q + 3kd \log a + 2 \log d) + q \log q + q^2 \log^2 q] \cdot \lceil t/d \rceil,$$

which is at most $q^2 a^{3kd} \mathsf{poly}(\log q, k, t, a, d)$. $\qquad\qquad$ $\square$

A few remarks are in order:

1. We would like to point out that the above computation can be carried out by a deterministic Turing machine having $k + 3$ tapes. The first k tapes can be used to simulate the k tapes of the NTM N, the next tape to store the lookup table used above and the last two for other computations.
2. Notice that the lookup table can be precomputed from the description of the NTM directly. Once the table is in place, it can be reused for different inputs x. The time required for computing the lookup table does not blow up the running time of our main result.
3. The counts stored in the lookup table do not contribute significantly to the running time. As noted before, at every step, the NTM has $(3a)^k q$ options, which means the total number of computation paths (including rejecting computations) is at most $((3a)^k q)^t$. The space required for storing each count is thus at most $t \cdot [k \log(3a) + \log q]$. This is absorbed in the $\mathsf{poly}(\log q, k, t, a, d)$ term above.

Once the r and f values in the block trace are fixed, the number of feasible values of ℓ that can be achieved by a TM head movement gets restricted.

This argument leads to a bound on the number of block trace witnesses. The following Lemma 6 provides this bound. Lemma 6 was originally proved in [7], so we omit the proof.

Lemma 6 ([7]). *The number B of valid segments is at most $(32a^d)^k$. Hence the number of potential block trace witnesses is at most $B^{t/d} = a^{kt}32^{kt/d}$.*

Theorem 7 (Block trace approach). *The number of accepting computation paths of a nondeterministic k-tape TM with q states and alphabet size a on an input x can be computed by a deterministic algorithm in time*

$$a^{kt}C_a^{k\sqrt{t}} \cdot q^2\mathsf{poly}(\log q, k, t, a),$$

where C_a is a constant that depends only on a.

Proof. As mentioned previously, Lemmas 6 and 5, respectively bound the number of block traces, and the time required to compute the number of compatible accepting paths. The algorithm generates all the valid block traces, and computes the number of compatible accepting paths. The running time is bounded by

$$q^2 a^{kt+3kd}32^{kt/d}\mathsf{poly}(\log q, k, t, a, d).$$

The algorithm keeps track of number of compatible accepting paths for each witness and then adds them up to get the total number of accepting computations. Lemma 4 ensures that there is no overcounting, i.e., each computation path is captured by exactly one block trace witness. We choose d such that the dominant factors are optimized. We need to minimize the product of a^{3kd} and $32^{kt/d}$. A straightforward calculation gives us that this happens when $d = \sqrt{5t/(3\log_2 a)}$. Substituting this value of d in the expression, we get a running time of

$$a^{kt}C_a^{k\sqrt{t}} \cdot q^2\mathsf{poly}(\log q, k, t, a),$$

where $C_a = 2^{\sqrt{60\log_2 a}}$. □

4   Main Theorem

In this section, we discuss an algorithm that reduces the exponent of the computation time by half. We have seen the BFS approach in Theorem 2 that requires a running time of $q^2(3at)^k a^{kt}\mathsf{poly}(\log q, k, t, a)$. The block trace approach in Theorem 7 requires a running time of $a^{kt}C_a^{k\sqrt{t}} \cdot q^2\mathsf{poly}(\log q, k, t, a)$. In both these approaches, the dominating factor is a^{kt}. On first sight, the a^{kt} factor seems unavoidable this is the number of different patterns that can be present on the k tapes of the NTM. We will see how to combine the two algorithms to bring down the exponent of the a^{kt} factor.

In the BFS approach, the number of configurations is upper bounded by $qa^{kt}t^k$. This was a consequence of the fact that the maximum tape usage is kt cells over all the k tapes. If the total tape usage were $\leq kt/2$ cells over all the k tapes, then the number of configurations would be no more than $qa^{kt/2}t^k$. Then the running time of the BFS approach would be at most $q^2(3at)^k a^{kt/2} \text{poly}(\log q, k, t, a)$.

That still leaves us with the case when N uses more than $kt/2$ space across all the tapes. We note that if the tape head is not returning ever to a tape location, then the content that the tape head writes in that location is irrelevant. If the NTM N uses more than $kt/2$ space, each of these locations are visited once for a last time. When visiting a location for the last time, we can modify the block trace method so that nothing is written in that location. This can save us a factor of $a^{kt/2}$ on the running time.

We now need a mechanism to determine which of the two situations we are in: tape usage $>kt/2$ or tape usage $\leq kt/2$. We introduce directional traces for that purpose.

Definition 8 (Directional trace [7]). A *directional segment of size d* for a k-tape NTM N with alphabet size a is a segment of size d, omitting the strings u_j, that is

$$\tau = [(r_1, f_1, \ell_1), \ldots, (r_k, f_k, \ell_k)]$$

where r_j, f_j and ℓ_j are defined as in Definition 3.

A *directional trace* of block size d, is a sequence of directional segments of size d.

By considering the number of 3-tuples (r, f, ℓ), we get a bound on the number of directional segments. A calculation similar to that in the proof of Lemma 6 gives us that for a given ℓ, the number of pairs (r, f) is at most $(d+1-\ell)^2 + 5(d+1-\ell)$. Thus the number of directional segments is at most $\sum_{i=1}^{d}(i^2 + 5i) = d(d+1)(d+8)/3 \leq d^3$, for $d \geq 6$. Thus we get the following:

Lemma 9 ([7]). *The number of directional segments of block size d is upper bounded by d^{3k}. The number of potential directional trace witnesses is at most $d^{3kt/d}$.*

The following lemma ensures that we do not over-count the number of accepting computations and is immediate from the definition of directional traces.

Lemma 10. *Two different directional trace witnesses give rise to a disjoint set of computation paths. In other words, every computation path corresponds to a unique directional trace witness.*

Now we are ready to prove the main theorem.

Proof (of Theorem 1). We assume that we know the time complexity $t = t(n)$ of the NTM N. Even if $t(n)$ is not known, we could perform a linear search by setting $t = 1, 2, 3, \cdots$, and the overhead that this will introduce can be absorbed into the $\text{poly}(t)$ component. The algorithm that computes $\#acc_N(x)$ works in three stages:

– **Preliminary Stage:** In this stage, the directional traces are used to decide which of the two approaches to use. The directional trace carries the information about the extent of tape usage for each block of size d. For each directional trace, the algorithm decides if the tape usage is $>kt/2$ or $\leq kt/2$. Corresponding to the tape usage, the suitable method is used for computation. By Lemma 9, there are $d^{3kt/d}$ directional traces. We use the block length $d = \sqrt{5t/(3\log_2 a)}$ as optimized in Theorem 7. Thus we get the upper bound on number of directional traces as

$$2^{\left\lceil \frac{\sqrt{27}}{\sqrt{20}} \cdot \sqrt{\log a} \cdot k \sqrt{t} \cdot (\log t + \log(5/(3\log_2 a))) \right\rceil},$$

which is at most $2^{\alpha_a k \sqrt{t} \log t}$, where α_a is a constant that depends[3] only on a.

From the directional traces, the algorithm computes the tape usage, and determine whether to use the BFS approach or the block trace approach. Moreover, for each tape location, the algorithm determines the block when that location was visited last. This information is stored as a lookup table for easy reference. Given a directional trace, the computation of tape usage and last visits can be done in time polynomial in k and t.

– **Block Trace Algorithm:** For the directional traces where the total tape usage is $>kt/2$, the block trace approach is used. The algorithm generates all the block traces by adding an appropriate u of length ℓ, to each 3-tuple (r, f, ℓ) of the directional trace. Recall that the time instances where the location is being visited for the last time have been identified in the preliminary stage. For the spots in u corresponding to the last visits of the respective tape locations, no symbol from the tape alphabet is assigned. Only a wildcard symbol * is assigned.

Since the tape usage is $>kt/2$, we will assign at least $>kt/2$ wildcard symbols *. This will result in a reduction in the number of block traces by a factor of at least $a^{kt/2}$. So the number of block traces that we will consider for block trace algorithm will be at most $2^{\alpha_a k \sqrt{t} \log t} a^{kt/2}$. We can modify Lemma 5 by including block trace segments with the * symbol in the lookup table. The lookup table should return all pairs (r, c) such that r is a state reachable in a manner compatible with the block trace in c different ways. This can be precomputed along with the lookup table, by combining the entries where different tape alphabets are considered in place each wildcard character.

The wildcard overhead is insignificant when compared with the running time of Lemma 5. The total running time of this is obtained by multiplying the number of block traces under consideration with the running time of Lemma 5, thus we get a running time of

$$2^{\alpha_a k \sqrt{t} \log t} a^{kt/2} \cdot q^2 a^{3kd} \mathsf{poly}(\log q, k, t, a).$$

---

[3] In fact, for values of $a \geq 4$, we can set $\alpha_a = \sqrt{(27/20)\log a}$ and for $a \in \{2,3\}$, setting α_a to be slightly greater than $\sqrt{(27/20)\log a}$ works.

Substituting the optimal value of d, we get that the total running time of this stage is at most $H_a^{k\sqrt{t}\log t}a^{kt/2} \cdot q^2\mathsf{poly}(\log q, k, t, a)$, where H_a is obtained by absorbing the a^{3kd} into the $2^{\alpha_a k\sqrt{t}\log t}$ factor. Notice that H_a depends only on a.

- **BFS Algorithm:** We do not require the directional traces when the total tape usage is $\leq kt/2$. We run the BFS algorithm as in Theorem 2 just once, but without pursuing any computations where tape usage exceeds $kt/2$. Thus the running time is at most $q^2(3at)^k a^{kt/2}\mathsf{poly}(\log q, k, t, a)$.

Notice that each computation path of N is considered in exactly one of the block trace or BFS algorithms. The output $\#acc_N(x)$ is obtained by summing the results yielded by the two algorithms. The total running time is upper bounded by

$$H_a^{k\sqrt{t}\log t}a^{kt/2}q^2\mathsf{poly}(\log q, k, t, a).$$

$\square$

Remark: We observe that we could convert the computation in the above proof into a uniform computation performed by a deterministic universal Turing machine. This DTM would take the description of the NTM N, along with its input x as arguments. This is possible because the description of N is used only in computing the lookup tables. An application of the Hennie-Stearns construction [5] on the universal machine would yield a 2-tape machine which computes $\#acc_N(x)$, but the running time incurs a small blowup. Most of the additional factors in the blowup can be accommodated by increasing H_a slightly, except for an additional q^2 factor. Thus the running time of the 2-tape DTM that computes $\#acc_N(x)$ is

$$a^{kt/2}H_a^{k\sqrt{t}\log t} \cdot q^4\mathsf{poly}(\log q, k, t, a).$$

5   Implications and Possible Extensions

We have shown techniques by which we can deterministically search the computation tree and count the number of accepting computations of an NTM in time square root of the size of the configuration graph. It would be interesting to see if one could use this approach along with additional techniques to push the running time even lower. Also, it would be interesting to see any lower bounds for the problem.

We note that our result is general enough to give a faster deterministic simulation of any language in which membership is determined as a function of the number of accepting paths.

Definition 11. The complexity class $\oplus\mathsf{P}$ (also known as $\mathsf{Parity\ P}$) is the set of all languages L, such that there exists an NTM N that runs in time polynomial in the length of the input, and

$$\forall x, \quad x \in L \iff \#acc_N(x) \equiv 1 \pmod{2}.$$

Analogously, the complexity classes $\mathsf{Mod}_k\mathsf{P}$ are defined by having $\bmod\, k$ instead of $\bmod\, 2$ in the above definition. Thus our simulation also implies a bound on the running time of the deterministic simulation of $\oplus\mathsf{P}$ and $\mathsf{Mod}_k\mathsf{P}$.

5.1   Simulating Probabilistic Classes

Another consequence of being able to count the number of accepting computations exactly is that we can deterministically simulate some randomized complexity classes. We use the following definition of a probabilistic Turing machine and prove the succeeding theorem, almost immediately.

Definition 12 ([4]). A *probabilistic Turing machine* is a TM which makes choices, possibly at each step, based on probabilities assigned to each of the choices. We say that a probabilistic TM P accepts a string x, if it accepts x with probability at least $1/2$.

A language L is said to be in the class *Probabilistic Polynomial Time* (denoted by PP) if it can be decided by a probabilistic Turing machine that runs in polynomial time.

An alternative characterization of PP is that a language L is in PP if there is a nondeterministic polynomial-time Turing machine N such that x is in L if and only if $M(x)$ has more accepting than rejecting paths.

Theorem 13. *Consider a language L, that is decided by a k-tape probabilistic TM with q states, alphabet size a and time complexity $t(n)$. Then there is a deterministic algorithm for L with time complexity*

$$a^{kt/2} H_a^{k\sqrt{t}\log t} \cdot q^2 \mathsf{poly}(\log q, k, t, a),$$

where H_a is a constant depending only on a.

Proof. Given a probabilistic machine P, which generates random coins for its computation, consider the corresponding nondeterministic Turing machine N, which makes nondeterministic choices in place of the random coins. For a given input x, P would decide on acceptance based on the number of random choices which lead to acceptance. In terms of N, this translates to the number of different nondeterministic choices which lead to acceptance.

As seen in Theorem 1, we can compute $\#acc_N(x)$ in the stated running time. $\square$

If we set $t(n)$ to be a polynomial of n, the above theorem gives us a bound on the running time of deterministically simulating of PP. van Melkebeek and Santhanam [14] gave an unconditional simulation of time-$t(n)$ probabilistic Turing machines by Turing machines operating in deterministic time $o(2^t)$. They showed that the exponent in the simulation of probabilistic TM can be reduced by a multiplicative factor smaller than 1 (as compared to our factor of $1/2$). Moreover, the class PP contains the probabilistic classes such as BPP, ZPP and BQP. Hence our simulations imply a faster simulation of these classes also.

5.2 Polynomial Hierarchy and Alternating TMs

By Toda's theorem [12], we have that the entire polynomial hierarchy (PH) is contained in $P^{\#P}$. But we cannot conclude that we have an $\tilde{O}(a^{kt/2})$ time simulation for classes in PH. This is because Toda's theorem involves a blow-up of the running time when converting a problem in say, Σ_2 to $\#P$. This negates the advantage that we gain by halving the exponent.

This leads us to a further open question. It would be interesting to see if we can simulate any of the classes in PH by $\#P$ in the same time bound. This, combined with our counting algorithm, would lead to a faster simulation of the classes in PH. Alternatively, we could try to simulate a time-$t(n)$ alternating TM, for instance a Σ_2-machine A, directly by iterating our uniform simulation for NTM's. This seems to work if the two (the existential and universal) phases of A are divided neatly into $t(n)/2$ steps each, but encounters a problem if A is existential for $t(n)(1 - \varepsilon)$ steps in some computation paths and existential for only $\varepsilon t(n)$ steps in others.

Acknowledgement. We thank Richard Lipton for helpful discussions, and the referees for comments that improved the presentation.

References

1. Arora, S., Barak, B.: Computational Complexity: A Modern Approach, 1st edn. Cambridge University Press, New York (2009)
2. Beigel, R., Gill, J., Hertramp, U.: Counting classes: thresholds, parity, mods, and fewness. In: Choffrut, C., Lengauer, T. (eds.) STACS 1990. LNCS, vol. 415, pp. 49–57. Springer, Heidelberg (1990). https://doi.org/10.1007/3-540-52282-4_31
3. Dyer, M., Goldberg, L.A., Greenhill, C., Jerrum, M.: On the relative complexity of approximate counting problems. In: Jansen, K., Khuller, S. (eds.) APPROX 2000. LNCS, vol. 1913, pp. 108–119. Springer, Heidelberg (2000). https://doi.org/10.1007/3-540-44436-X_12
4. Gill III, J.T.: Computational complexity of probabilistic Turing machines. In: Proceedings of the Sixth Annual ACM Symposium on Theory of Computing, STOC 1974, New York, pp. 91–95. ACM (1974)
5. Hennie, F.C., Stearns, R.E.: Two-tape simulation of multitape Turing machines. J. ACM **13**(4), 533–546 (1966)
6. Jerrum, M.: Counting, Sampling and Integrating: Algorithms and Complexity. Lectures in Mathematics. ETH Zürich. Birkhäuser, Basel (2003)
7. Kalyanasundaram, S., Lipton, R.J., Regan, K.W., Shokrieh, F.: Improved simulation of nondeterministic Turing machines. Theor. Comput. Sci. **417**, 66–73 (2012). Earlier version in Proceedings of the 35th International Symposium on Mathematical Foundations of Computer Science, 2010
8. Pippenger, N.: Probabilistic simulations (preliminary version). In: Proceedings of the Fourteenth Annual ACM Symposium on Theory of Computing, STOC 1982, New York, pp. 17–26. ACM (1982)
9. Papadimitriou, C.H.: Computational Complexity. Addison-Wesley Publishing Company, Reading (1994)

10. Schöning, U.: The power of counting. In: Selman, A.L. (ed.) Complexity Theory Retrospective: In Honor of Juris Hartmanison the Occasion of His Sixtieth Birthday, July 5, 1988, pp. 204–223. Springer, New York (1990)
11. Simon, J.: On some central problems in computational complexity. Technical report, Cornell University, Ithaca (1975)
12. Toda, S.: On the computational power of PP and ⊕P. In: Proceedings of the 30th Annual Symposium on Foundations of Computer Science, FOCS 1989, pp. 514–519. IEEE (1989)
13. Torán, J.: Counting the number of solutions. In: Rovan, B. (ed.) MFCS 1990. LNCS, vol. 452, pp. 121–134. Springer, Heidelberg (1990). https://doi.org/10.1007/BFb0029600
14. van Melkebeek, D., Santhanam, R.: Holographic proofs and derandomization. SIAM J. Comput. **35**(1), 59–90 (2005). Earlier version in Proceedings of the 18th Annual IEEEConference on Computational Complexity, 2003

Determining Minimal Degree Polynomials
of a Cyclic Code of Length 2^k over $\mathbb{Z}_8$

Arpana Garg and Sucheta Dutt$^{(\boxtimes)}$

Department of Applied Sciences, PEC University of Technology, Chandigarh, India
arpanapujara@gmail.com, suchetapec@yahoo.co.in

Abstract. The rank of a cyclic code of length $n = 2^k$ over $\mathbb{Z}_8$ is $n - v$ where v is the degree of a minimal degree polynomial in the code. In this paper, minimal degree polynomials in a cyclic code C of length $n = 2^k$ (where k is a natural number) over $\mathbb{Z}_8$ are determined. Further, using these minimal degree polynomials, all 95 (46 principally generated and 49 non principally generated) cyclic codes of length 4 over $\mathbb{Z}_8$ are calculated in terms of their distinguished sets of generators.

Keywords: Cyclic codes · Minimal degree polynomial · Rank

1 Introduction

A linear code C of length n over a finite commutative ring R is defined as an R-submodule of R^n. A cyclic code C over R of length n is a linear code such that whenever $(c_0, c_1, c_2, \ldots, c_{n-1})$ is in C, $(c_{n-1}, c_0, c_1, \ldots, c_{n-2})$ also belongs to C. The rank of a code C (denoted by $rank(C)$) over a ring R is defined as the minimum number of generators of C as an $R-$ module [1].

Cyclic codes over finite rings have been recently studied by many authors. For reference see ([2–5,8–10]). The structure of a cyclic code of length $n = 2^k$ over $\mathbb{Z}_8$ as an ideal of the ring $\mathbb{Z}_8[x]/\langle x^n - 1 \rangle$ is given in [6]. Using this structure of a cyclic code of length $n = 2^k$ over $\mathbb{Z}_8$, a distinguished set of generators and rank of such a cyclic code is obtained in [7] and it is proved that the rank of a cyclic code of length $n = 2^k$ over $\mathbb{Z}_8$ is $n - v$, where v is the degree of a minimal degree polynomial in the code. Moreover, the minimal degree polynomials play an important role in enumeration of cyclic codes of length 2^k over $\mathbb{Z}_8$. However, for a given code with distinguished form of generators, the value of v is not always obvious. Abualrub et al. has found the degree of minimal degree polynomial in a cyclic code of length 2^k over $\mathbb{Z}_4$ in [1].

In this paper, minimal degree polynomials with leading coefficient $2, 4$ or 6 in a cyclic code C of length $n = 2^k$ over $\mathbb{Z}_8$ are determined. Further, using these minimal degree polynomials, all 95 (46 principally generated and 49 non principally generated) cyclic codes of length 4 over $\mathbb{Z}_8$ are calculated in terms of their distinguished sets of generators.

© Springer International Publishing AG 2018
B. S. Panda and P. P. Goswami (Eds.): CALDAM 2018, LNCS 10743, pp. 118–130, 2018.
https://doi.org/10.1007/978-3-319-74180-2_10

2   Preliminaries

Let C be cyclic code of length $n = 2^k$ over $\mathbb{Z}_8$. Let $f(x) = q_0(x)$ be a minimal degree polynomial among all monic polynomials in C, $g(x) = 2q_1(x)$ be a minimal degree polynomial among all polynomials in C with leading coefficient 2 or 6 and $h(x) = 4q_2(x)$ be a minimal degree polynomial among all polynomials in C with leading coefficient 4. Let $deg(f(x)) = r$, $deg(g(x)) = s$ and $deg(h(x)) = v$. It is obvious that $r \geq s \geq v$. A unique form for the polynomials $f(x), g(x)$ and $h(x)$ is determined in Theorem 1 below. Note that the cyclic code C may or may not contain any monic polynomial or any polynomial with leading coefficient 2 or 6.

For ready reference, we recall the following results.

Corollary 1 [7]. *The following identities hold in* $\mathbb{Z}_8[x]/\langle x^n - 1\rangle$,

1. $(x + 1)^n = 2(x + 1)^{\frac{n}{2}} + 4(x + 1)^{\frac{3n}{4}} + 4(x + 1)^{\frac{n}{4}}$ *for all* $n = 2^k$, $k \geq 2$.
2. $2(x + 1)^n = 4(x + 1)^{\frac{n}{2}}$ *for all* $n = 2^k$, $k \geq 1$.

Theorem 1 [7]. *Let* C *be a cyclic code of length* $n = 2^k$ *over* $\mathbb{Z}_8$. *Let* $f(x), g(x)$ *(if they exist) and* $h(x)$ *be polynomials in* C *as defined above. Then*

1. $h(x) = 4q_2(x) = 4(x + 1)^v$.
2. *If* C *contains polynomials with leading coefficient 2 or 6, then there exists a unique polynomial in* C *of the type* $2(x + 1)^s + 4\sum_{i=0}^{v-1} \alpha_i(x + 1)^i$ *of degree* s, *where* $\alpha_i \in \mathbb{Z}_2$. *Therefore,* $g(x) = 2q_1(x)$ *can be chosen as* $2(x + 1)^s + 4\sum_{i=0}^{v-1} \alpha_i(x + 1)^i$, *where* $\alpha_i \in \mathbb{Z}_2$.
3. *If* C *contains monic polynomials, then there exists a unique polynomial in* C *of the type* $(x + 1)^r + 2\sum_{i=0}^{s-1} \beta_i(x + 1)^i + 4\sum_{i=0}^{v-1} \gamma_i(x + 1)^i$ *of degree* r, *where* $\beta_i, \gamma_i \in \mathbb{Z}_2$. *Therefore,* $f(x) = q_0(x)$ *can be chosen as* $(x + 1)^r + 2\sum_{i=0}^{s-1} \beta_i(x + 1)^i + 4\sum_{i=0}^{v-1} \gamma_i(x + 1)^i$, *where* $\beta_i, \gamma_i \in \mathbb{Z}_2$.

Theorem 2 [7]. *A cyclic code* C *of length* 2^k *over* $\mathbb{Z}_8$ *is generated by one or more polynomials from the set* $\{f(x), g(x), h(x)\}$ *where*

1. $f(x) = q_0(x) = (x + 1)^r + 2\sum_{i=0}^{s-1} \beta_i(x + 1)^i + 4\sum_{i=0}^{v-1} \gamma_i(x + 1)^i$ *with* $\beta_i, \gamma_i \in \mathbb{Z}_2$
2. $g(x) = 2q_1(x) = 2(x + 1)^s + 4\sum_{i=0}^{v-1} \alpha_i(x + 1)^i$ *with* $\alpha_i \in \mathbb{Z}_2$
3. $h(x) = 4q_2(x) = 4(x + 1)^v$

Remark 1. Referring back to Theorem 2, the generators of a cyclic code of length 2^k over $\mathbb{Z}_8$ can further be written as

1. $f(x) = q_0(x) = (x + 1)^r + 2(x + 1)^c \beta(x) + 4(x + 1)^d \gamma(x)$ where $\beta(x)$ and $\gamma(x)$ belong to $\mathbb{Z}_2[x]/\langle x^n - 1\rangle$.
2. $g(x) = 2q_1(x) = 2(x + 1)^s + 4(x + 1)^e \alpha(x)$ where $\alpha(x) \in \mathbb{Z}_2[x]/\langle x^n - 1\rangle$.

3 The Main Results

As stated in Theorem 2 above, a cyclic code of length 2^k over $\mathbb{Z}_8$ is generated by one or more polynomials from the set $\{f(x), g(x), h(x)\}$. Clearly, the minimal degree polynomial in any cyclic code of length 2^k over $\mathbb{Z}_8$ is $h(x) = 4(x+1)^v$. In case $h(x) = 4(x+1)^v$ is not a generator of the cyclic code of length 2^k over $\mathbb{Z}_8$, then the minimal degree polynomial $h(x) = 4(x+1)^v$ in the code is not obviously known. Therefore the minimal degree polynomials in the cyclic codes $\langle g(x) \rangle, \langle f(x), g(x) \rangle, \langle f(x) \rangle$ need to be determined.

Theorem 3. *Let $C = \langle g(x) \rangle$ be a cyclic code of length 2^k over $\mathbb{Z}_8$, where $g(x)$ is a polynomial in C as defined above. Then the minimal degree polynomial in C is $h(x) = 4(x+1)^v$ where v is given by the following*

1. *If $4\sum_{i=0}^{v-1} \alpha_i(x+1)^i \neq 0$, let e be the least positive integer such that $\alpha_e = 1$, then*
    (a) *If $\frac{n}{2} \neq n-s+e$ then $v = min(\frac{n}{2}, s, n-s+e)$.*
    (b) *If $\frac{n}{2} = n-s+e$ then $v = min(s, n-s+e_1)$, where e_1 is the least positive integer greater than e such that $\alpha_{e_1} = 1$. If no such e_1 exists then $v = s$.*
2. *If $4\sum_{i=0}^{v-1} \alpha_i(x+1)^i = 0$, then $v = min(s, \frac{n}{2})$.*

Proof. 1. Consider an element $l(x) = \sum_{j=0}^{n-1} c_j(x+1)^j + 2\sum_{j=0}^{n-1} d_j(x+1)^j + 4\sum_{j=0}^{n-1} f_j(x+1)^j$ in $\mathbb{Z}_8[x]/\langle x^n - 1 \rangle$, where $c_j, d_j, f_j \in \mathbb{Z}_2$ such that $g(x)l(x) = 4(x+1)^v$. Note that $g(x)l(x) = 2\sum_{j=0}^{n-s-1} c_j(x+1)^{j+s} + 4(x+1)^{\frac{n}{2}} \sum_{j=n-s}^{n-1} c_j(x+1)^{j-n+s} + 4\sum_{j=0}^{n-1} d_j(x+1)^{j+s} + 4\sum_{i=e}^{v-1} \alpha_i(x+1)^i \sum_{j=0}^{n-1} c_j(x+1)^j \equiv 0(mod 4)$ if and only if $c_j = 0$ for $j = 0, 1, 2, ..., n-s-1$. Thus

$$g(x)l(x) = 4(x+1)^{\frac{n}{2}} \sum_{j=n-s}^{n-1} c_j(x+1)^{j-n+s} + 4(x+1)^s \sum_{j=0}^{n-1} d_j(x+1)^j$$

$$+4(x+1)^{n-s+e} \left[\sum_{i=e}^{v-1} \alpha_i(x+1)^{i-e} \right] \left[\sum_{j=n-s}^{n-1} c_j(x+1)^{j-n+s} \right] \tag{1}$$

$$= 4(x+1)^{min(\frac{n}{2}, n-s+e, s)}[unit] \in C$$

This implies that $v \geq min(\frac{n}{2}, n-s+e, s)$.
Now, $2(g(x)) = 4(x+1)^s \in C$ and $(x+1)^{n-s}[g(x)] = 4(x+1)^{\frac{n}{2}} + 4(x+1)^{n-s+e}\sum_{i=e}^{v-1} \alpha_i(x+1)^i = 4(x+1)^{min(\frac{n}{2}, n-s+e)}[unit] \in C$.
(a) If $\frac{n}{2} \neq n-s+e$ then $v \leq min(s, \frac{n}{2}, n-s+e)$. Hence $v = min(s, \frac{n}{2}, n-s+e)$.
(b) If $\frac{n}{2} = n-s+e$ then $v \leq min(s, n-s+e_1)$, where e_1 is the least positive integer greater than e such that $\alpha_{e_1} = 1$. Hence $v = min(s, n-s+e_1)$. If no such e_1 exists then $v \leq s$. Hence $v = s$.

2. If $4\sum_{i=0}^{v-1}\alpha_i(x+1)^i = 0$, then

$$2(x+1)^s l(x) = 4(x+1)^{\frac{n}{2}}\sum_{j=n-s}^{n-1} c_j(x+1)^{j-n+s} + 4(x+1)^s\sum_{j=0}^{n-1} d_j(x+1)^j$$

$$= 4(x+1)^{min(s,\frac{n}{2})}[unit] \in C \tag{2}$$

This implies that $v \geq min(\frac{n}{2}, s)$.

Note that $2(g(x)) = 4(x+1)^s \in C$ and $(x+1)^{n-s}[g(x)] = 4(x+1)^{\frac{n}{2}} \in C$. This implies that $v \leq min(s, \frac{n}{2})$. Hence $v = min(\frac{n}{2}, s)$. This proves the result.

In case, a cyclic code of length 2^k over $\mathbb{Z}_8$ is generated by $f(x)$ only, the degree of a minimal degree polynomial $g(x)$ with leading coefficient 2 or 6 in the code is not obvious. The following result which can be proved on the similar lines as Theorem 3, gives the degree of $g(x)$ in the code.

Lemma 1. *Let* $C = \langle f(x)\rangle$ *be a cyclic code of length* 2^k *over* $\mathbb{Z}_8$, *where* $f(x)$ *is as defined above. Then the minimal degree polynomial in* C *with leading coefficient 2 or 6 is* $g(x) = 2(x+1)^s + 4\sum_{i=0}^{v-1}\alpha_i(x+1)^i$, *where* s *is given by the following*

1. *If* $2\sum_{i=0}^{s-1}\beta_i(x+1)^i \neq 0$ *and* c *is the least positive integer such that* $\beta_c = 1$, *then*
   (a) $s = min(r, \frac{n}{2}, n+c-r)$, *in case* $\frac{n}{2} \neq n+c-r$.
   (b) $s = min(r, n-r+c_1)$, *where* c_1 *is the least positive integer greater than* c *such that* $\beta_{c_1} = 1$, *in case* $\frac{n}{2} = n+c-r$. *If no such* c_1 *exists, then* $s = r$.
2. *If* $2\sum_{i=0}^{s-1}\beta_i(x+1)^i = 0$, *then* $s = min(r, \frac{n}{2})$.

Theorem 4. *Let* $C = \langle (x+1)^r, g(x)\rangle$ *be a cyclic code of length* 2^k *over* $\mathbb{Z}_8$, *where* $g(x)$ *is as defined above. Then the minimal degree polynomial in* C *is* $h(x) = 4(x+1)^v$ *where* v *is given by the following*

1. *If* $4\sum_{i=0}^{v-1}\alpha_i(x+1)^i \neq 0$, *let* e *be the least positive integer such that* $\alpha_e = 1$, *then*
   (a) $v = min(\frac{n}{4}, s, r-s+e)$, *in case* $r < \frac{n}{2}$.
   (b) $v = min(\frac{n}{4}, s, \frac{n}{2} - s + e)$, *in case* $r \geq \frac{n}{2}$ *provided* $\frac{n}{2} - s + e \neq \frac{n}{4}$. *If* $\frac{n}{2} - s + e = \frac{n}{4}$ *and* e_1 *is the least positive integer greater then* e *such that* $\alpha_{e_1} = 1$ *then* $v = min(s, \frac{n}{2} - s + e_1)$. *If no such* e_1 *exists then* $v = s$.
2. *If* $4\sum_{i=0}^{v-1}\alpha_i(x+1)^i = 0$, *then* $v = min(\frac{n}{4}, s)$.

Proof. A minimal degree polynomial $h(x) = 4(x+1)^v$ in C is either a multiple of $(x+1)^r$ or a multiple of $g(x)$ or a linear combination of $(x+1)^r$ and $g(x)$. First, consider an element $g_1(x) = \sum_{i=0}^{n-1} a_i(x+1)^i + 2\sum_{i=0}^{n-1} b_i(x+1)^i + 4\sum_{i=0}^{n-1} c_i(x+1)^i$ in $\mathbb{Z}_8[x]/\langle x^n - 1\rangle$ such that $4(x+1)^v = (x+1)^r g_1(x)$. Using Corollary 1, we have $4(x+1)^v = \sum_{i=0}^{n-r-1} a_i(x+1)^i + [2(x+1)^{\frac{n}{2}} + 4(x+1)^{\frac{n}{4}} + 4(x+1)^{\frac{3n}{4}}]\sum_{i=n-r}^{n-1} a_i(x+1)^{i-n+r} + 4(x+1)^{\frac{n}{2}}\sum_{i=n-r}^{n-1} b_i(x+1)^{i-n+r} + 2\sum_{i=0}^{n-r-1} b_i(x+1)^{i+r} + 4\sum_{i=0}^{n-r-1} c_i(x+1)^{i+r}$. This is possible if and only if

$$\sum_{i=0}^{n-r-1} a_i(x+1)^i + 2(x+1)^{\frac{n}{2}}\sum_{i=n-r}^{n-1} a_i(x+1)^{i-n+r} + 2\sum_{i=0}^{n-r-1} b_i(x+1)^{i+r} \equiv 0(mod 4)$$

$$\tag{3}$$

(a) If $r < \frac{n}{2}$, then Eq. (3) is possible if and only if $a_i = 0$ for $i = 0, 1, 2, \ldots, n-r-1$, $a_{n-r} = 1$ and $b_{\frac{n}{2}-r} = 1$. It follows that $(x+1)^r g_1(x) = 4(x+1)^{\frac{n}{2}} + 4(x+1)^{\frac{n}{4}} + 4(x+1)^{\frac{3n}{4}} + 4(x+1)^{\frac{n}{2}} \sum_{i=n-r}^{n-1} b_i (x+1)^{i-n+r} + 4(x+1)^r \sum_{i=0}^{n-r-1} \beta_i (x+1)^i$. This implies that $v \geq min(r, \frac{n}{4})$.

Now $4[(x+1)^r] \in C$ and $(x+1)^r \left[(x+1)^{n-r} + 2(x+1)^{\frac{n}{2}-r}\right] = 4(x+1)^{\frac{n}{2}} + 4(x+1)^{\frac{n}{4}} + 4(x+1)^{\frac{3n}{4}} \in C$. Therefore $v \leq min(\frac{n}{4}, r)$. Hence v is equal to $min(r, \frac{n}{4})$.

(b) If $r \geq \frac{n}{2}$ then Eq. (3) is possible if and only if $a_i = 0$ for $i = 0, 1, 2, \ldots, n-r-1$ and $a_{\frac{n}{2}} = 1$, $b_0 = 1$. This implies that $(x+1)^r g_1(x) = 4(x+1)^r + 4(x+1)^{r-\frac{n}{4}} + 4(x+1)^{r+\frac{n}{4}} + 4(x+1)^{\frac{n}{2}} \sum_{i=n-r}^{n-1} b_i (x+1)^{i-n+r} + 4(x+1)^r \sum_{i=0}^{n-r-1} \beta_i (x+1)^i$. This further implies that $v \geq min(r - \frac{n}{4}, \frac{n}{2})$.

Now $2(x+1)^{n-r}\left[(x+1)^r\right] = 4(x+1)^{\frac{n}{2}} \in C$ and $(x+1)^r \left[(x+1)^{\frac{n}{2}} + 2\right] = 4(x+1)^r + 4(x+1)^{r-\frac{n}{4}} + 4(x+1)^{r+\frac{n}{4}} \in C$. Therefore $v \leq min(\frac{n}{2}, r - \frac{n}{4})$. Hence v is equal to $min(\frac{n}{2}, r - \frac{n}{4})$.

Next, we consider the case when $4(x+1)^v$ is multiple of $g(x)$ only. As proved in Lemma 1, we have $deg(g(x)) = min(\frac{n}{2}, r)$. This implies that $s < \frac{n}{2}$ and therefore $n - s + e > \frac{n}{2}$. Further, if $4 \sum_{i=0}^{v-1} \alpha_i (x+1)^i \neq 0$ then as proved in Theorem 3, $v = min(\frac{n}{2}, s, n - s + e) = min(\frac{n}{2}, s)$. Also, if $4 \sum_{i=0}^{v-1} \alpha_i (x+1)^i = 0$ then as proved in Theorem 3, $v = min(s, \frac{n}{2})$.

Finally, consider elements $g_3(x)$ and $g_4(x)$ of $\mathbb{Z}_8[x]/\langle x^n - 1 \rangle$ such that $4(x+1)^v = (x+1)^r g_3(x) + g(x)g_4(x) = \sum_{i=0}^{n-1} A_i (x+1)^{i+r} + 2 \sum_{i=0}^{n-1} B_i (x+1)^{i+r} + 4 \sum_{i=0}^{n-1} C_i (x+1)^{i+r} + 2 \sum_{i=0}^{n-1} D_i (x+1)^{i+s} + 4 \sum_{i=0}^{n-1} E_i (x+1)^{i+s} + 4(x+1)^e \alpha(x) \sum_{i=0}^{n-1} D_i (x+1)^i = \sum_{i=0}^{n-r-1} A_i (x+1)^i + [2(x+1)^{\frac{n}{2}} + 4(x+1)^{\frac{n}{4}} + 4(x+1)^{\frac{3n}{4}}] \sum_{i=n-r}^{n-1} A_i (x+1)^{i-n+r} + 4(x+1)^{\frac{n}{2}} \sum_{i=n-r}^{n-1} B_i (x+1)^{i-n+r} + 2 \sum_{i=0}^{n-r-1} B_i (x+1)^{i+r} + 4 \sum_{i=0}^{n-r-1} C_i (x+1)^{i+r} + 4(x+1)^{\frac{n}{2}} \sum_{i=n-s}^{n-1} D_i (x+1)^{i-n+s} + 4 \sum_{i=0}^{n-s-1} E_i (x+1)^{i+s} + 4(x+1)^e \alpha(x) \sum_{i=0}^{n-1} D_i (x+1)^i$.

This implies that $\sum_{i=0}^{n-r-1} A_i (x+1)^i + 2(x+1)^{\frac{n}{2}} \sum_{i=n-r}^{n-1} A_i (x+1)^{i-n+r} + 2 \sum_{i=0}^{n-r-1} B_i (x+1)^{i+r} + 2 \sum_{i=0}^{n-s-1} D_i (x+1)^{i+s} \equiv 0 (mod 4)$.

This is possible if and only if $A_i = 0$ for $i = 0, 1, 2, \ldots, n-r-1$, $A_{n-r} = 1$, $B_0 = 1$, $D_{\frac{n}{2}-s} = 1$, $D_{r-s} = 1$. Thus $(x+1)^r g_3(x) + g(x)g_4(x) = 4(x+1)^{\frac{n}{2}} + 4(x+1)^{\frac{n}{4}} + 4(x+1)^{\frac{3n}{4}} + 4(x+1)^{\frac{n}{2}} \sum_{i=n-r}^{n-1} B_i (x+1)^{i-n+r} + 4(x+1)^r \sum_{i=0}^{n-r-1} C_i (x+1)^i + 4(x+1)^{\frac{n}{2}} \sum_{i=n-s}^{n-1} D_i (x+1)^{i-n+s} + 4 \sum_{i=0}^{n-s-1} E_i (x+1)^{i+s} + 4(x+1)^{\frac{n}{2}-s+e} \alpha(x) + 4(x+1)^{r-s+e} \alpha(x)$.

1. If $4 \sum_{i=0}^{v-1} \alpha_i (x+1)^i \neq 0$, then $v \geq min(\frac{n}{4}, r - s + e, \frac{n}{2} - s + e)$. Note that $(x+1)^r \left[(x+1)^{n-r}\right] + [2(x+1)^s + 4(x+1)^e \alpha(x)][(x+1)^{\frac{n}{2}-s}] = 4(x+1)^{\frac{n}{2}} + 4(x+1)^{\frac{n}{4}} + 4(x+1)^{\frac{3n}{4}} + 4(x+1)^{\frac{n}{2}-s+e} \alpha(x) = 4(x+1)^{min(\frac{n}{4}, \frac{n}{2}-s+e)} unit \in C$. $2(x+1)^r + [2(x+1)^s + 4(x+1)^e \alpha(x)][(x+1)^{r-s}] = 4(x+1)^r + 4(x+1)^{r-s+e} \alpha(x) = 4(x+1)^{r-s+e} \in C$. Therefore $v \leq min(\frac{n}{4}, \frac{n}{2} - s + e, r - s + e)$. Hence $v = min(\frac{n}{4}, \frac{n}{2} - s + e, r - s + e)$.

2. If $4\sum_{i=0}^{v-1}\alpha_i(x+1)^i = 0$, $(x+1)^r g_3(x) + g(x)g_4(x) = 4(x+1)^{\frac{n}{2}} + 4(x+1)^{\frac{n}{4}} + 4(x+1)^{\frac{3n}{4}} + 4(x+1)^{\frac{n}{2}}\sum_{i=n-r}^{n-1} B_i(x+1)^{i-n+r} + 4(x+1)^r + 4(x+1)^r\sum_{i=0}^{n-r-1} C_i(x+1)^i + 4(x+1)^{\frac{n}{2}}\sum_{i=n-s}^{n-1} D_i(x+1)^{i-n+s} + 4\sum_{i=0}^{n-s-1} E_i(x+1)^{i+s}$. This implies that $v \geq min(\frac{n}{4}, s)$. Note that $(x+1)^r(x+1)^{n-r} + 2(x+1)^s(x+1)^{\frac{n}{2}-s} = 4(x+1)^{\frac{n}{2}} + 4(x+1)^{\frac{n}{4}} + 4(x+1)^{\frac{3n}{4}} = 4(x+1)^{\frac{n}{4}}[unit] \in C$. As $4(x+1)^s$ belongs to C, we get that $v \leq min(\frac{n}{4}, s)$. Hence $v = min(\frac{n}{4}, s)$.

Taking the minimum value of v from the above cases, we get that

1. If $4\sum_{i=0}^{v-1}\alpha_i(x+1)^i \neq 0$ and let e be the least positive integer such that $\alpha_e = 1$ and $4\sum_{i=0}^{v-1}\alpha_i(x+1)^i = 4(x+1)^e\alpha(x)$ then
   (a) $v = min(\frac{n}{4}, s, r-s+e)$, in case $r < \frac{n}{2}$.
   (b) $v = min(\frac{n}{4}, s, \frac{n}{2} - s + e)$, in case $r \geq \frac{n}{2}$ provided that $\frac{n}{2} - s + e \neq \frac{n}{4}$. If $\frac{n}{2} - s + e = \frac{n}{4}$ and let e_1 is the least positive integer greater then e such that $\alpha_{e_1} = 1$ then $v = min(s, \frac{n}{2} - s + e_1)$. If e_1 does not exists then $v = s$.
2. If $4\sum_{i=0}^{v-1}\alpha_i(x+1)^i = 0$ then $v = min(\frac{n}{4}, s)$.

This proves the result.

Theorems 5, 6 and 7 can be proved in a similar fashion as Theorem 4.

Theorem 5. *Let $C = \langle (x+1)^r + 4(x+1)^d\gamma(x), g(x)\rangle$ be a cyclic code of length 2^k over $\mathbb{Z}_8$, where $g(x)$ is as defined above. Then the degree v of $h(x)$ is given by the following*

1. *If $4\sum_{i=0}^{v-1}\alpha_i(x+1)^i \neq 0$, let e be the least positive integer such that $\alpha_e = 1$, then*
   (a) $v = min(\frac{n}{4}, s, r-s+e)$, *in case $r < \frac{n}{2}$.*
   (b) $v = min(\frac{n}{4}, s, \frac{n}{2} - s + e)$, *in case $r \geq \frac{n}{2}$ provided that $\frac{n}{2} - s + e \neq \frac{n}{4}$. If $\frac{n}{2} - s + e = \frac{n}{4}$ and $\alpha_{e_1} = 1$ then $v = min(s, \frac{n}{2} - s + e_1)$. If no such e_1 exists then $v = s$.*
2. *If $4\sum_{i=0}^{v-1}\alpha_i(x+1)^i = 0$, then $v = min(\frac{n}{4}, s)$.*

Theorem 6. *Let $C = \langle (x+1)^r + 2(x+1)^c\beta(x), g(x)\rangle$ be a cyclic code of length 2^k over $\mathbb{Z}_8$, where $g(x)$ is as defined above. Then degree v of $h(x)$ in C is given by the following*

1. *If $4\sum_{i=0}^{v-1}\alpha_i(x+1)^i \neq 0$, let e be the least positive integer such that $\alpha_e = 1$, then*
   (a) $v = min(\frac{n}{4}, c, r-s+e)$, *in case $r \leq \frac{n}{2}$ provided that minimum values are not equal. Otherwise value of v depends upon the values of $\beta(x)$ and $\alpha(x)$.*
   (b) $v = min(c, \frac{n}{4}, \frac{n}{2} - s + e)$, *in case $\frac{n}{2} < r < \frac{n}{2} + c$ provided that minimum values are not equal. Otherwise value of v depends upon the values of $\beta(x)$ and $\alpha(x)$.*
   (c) *In case $r = \frac{n}{2} + c$ and $\beta(x) = 1 + (x+1)^{c_1-c}\beta_1(x)$, where c_1 is the least positive integer greater than c such that $\alpha_1(x)$ is a unit, we have the following*

    i. If $2c \leq c_1$, then $v = min(\frac{n}{4}, c, r - s + e)$ provided that minimum values are not equal. Otherwise it depends upon the value of $\beta(x)$ and $\alpha(x)$.

    ii. If $2c > c_1$, then $v = min(\frac{n}{4}, c, n - r + c_1 - s + e)$ provided that minimum values are not equal. Otherwise it depends upon the value of $\beta(x)$ and $\alpha(x)$.

    If c_1 does not exists that is $\beta(x) = 1$ then $v = min(\frac{n}{4}, c, r - s + e)$

(d) $v = min(c, \frac{n}{4}, n - r + c - s + e)$, in case $r > \frac{n}{2} + c$ provided that minimum values are not equal. Otherwise it depends upon the value of $\beta(x)$ and $\alpha(x)$ and $v \leq c$.

2. If $4 \sum_{i=0}^{v-1} \alpha_i (x + 1)^i = 0$, then $v = min(\frac{n}{4}, c)$.

Theorem 7. Let $C = \langle f(x), g(x) \rangle$ be a cyclic code of length $n = 2^k$ over $\mathbb{Z}_8$, where $f(x)$ and $g(x)$ are as defined above. Then degree v of $h(x)$ is given by the following

1. If $4 \sum_{i=0}^{v-1} \alpha_i (x + 1)^i \neq 0$ and e is the least positive integer such that $\alpha_e = 1$, then

    (a) $v = min(\frac{n}{4}, c, r - s + e)$, in case $r \leq \frac{n}{2}$ provided that minimum values are not equal. Otherwise value of v depends upon the values of $\beta(x), \gamma(x)$ and $\alpha(x)$ and $v \leq min(\frac{n}{4}, \frac{n}{2} - r + c, s)$

    (b) $v = min(c, \frac{n}{4}, s, n - r + d, \frac{n}{2} - s + e)$, in case $\frac{n}{2} < r < \frac{n}{2} + c$ provided that minimum values are not equal. Otherwise value of v depends upon the values of $\beta(x), \gamma(x)$ and $\alpha(x)$ and $v \leq min(c, r - \frac{n}{4}, s)$.

    (c) In case $r = \frac{n}{2} + c$ and $\beta(x) = 1 + (x + 1)^{c_1 - c} \beta_1(x)$, where c_1 is the least positive integer greater than c such that $\beta_1(x)$ is a unit, we have the following

        i. If $2c \leq c_1$ then $v = min(\frac{n}{4}, n - r + d, c, r - s + e)$ provided that minimum values are not equal. Otherwise it depends upon the values of $\beta(x), \gamma(x)$ and $\alpha(x)$.

        ii. If $2c > c_1$ then $v = min(\frac{n}{4}, c, n - r + d, n - r + c_1 - s + e)$ provided that minimum values are not equal. Otherwise it depends upon the values of $\beta(x), \gamma(x)$ and $\alpha(x)$.

        If c_1 does not exists that is $\beta(x) = 1$ then $v = min(\frac{n}{4}, c, n - r + d, r - s + e)$.

    (d) $v = min(c, \frac{n}{4}, s, n - r + d, n - r + c - s + e)$, in case $r > \frac{n}{2} + c$ provided that minimum values are not equal. Otherwise it depends upon the values of $\beta(x), \gamma(x)$ and $\alpha(x)$ and $v \leq c$.

2. If $4 \sum_{i=0}^{v-1} \alpha_i (x + 1)^i = 0$, then $v = min(\frac{n}{4}, c, n - r + d)$ provided that minimum values are not equal. Otherwise it depends upon the values of $\beta(x)$ and $\gamma(x)$.

Theorem 8. Let $C = \langle (x + 1)^r \rangle$ be a cyclic code of length $n = 2^k$ over $\mathbb{Z}_8$, then degree v of $h(x)$ in C is given by the following

1. $v = min(\frac{n}{4}, r)$, if $r \leq \frac{n}{2}$
2. $v = min(\frac{n}{2}, r - \frac{n}{4})$, if $r > \frac{n}{2}$.

Proof. Let $g(x)$ be a minimal degree polynomial in C with leading coefficient 2 or 6 which is a multiple of $(x+1)^r$. A minimal degree polynomial $h(x) = 4(x+1)^v$ in C is either a multiple of $(x+1)^r$ or a multiple of $g(x)$ or a linear combination of $(x+1)^r$ and $g(x)$. First, consider the case when $4(x+1)^v = (x+1)^r g_1(x) = (x+1)^r [\sum_{i=0}^{n-1} a_i(x+1)^i + 2\sum_{i=0}^{n-1} b_i(x+1)^i + 4\sum_{i=0}^{n-1} c_i(x+1)^i] = \sum_{i=0}^{n-r-1} a_i(x+1)^i + [2(x+1)^{\frac{n}{2}} + 4(x+1)^{\frac{n}{4}} + 4(x+1)^{\frac{3n}{4}}] \sum_{i=n-r}^{n-1} a_i(x+1)^{i-n+r} + 4(x+1)^{\frac{n}{2}} \sum_{i=n-r}^{n-1} b_i(x+1)^{i-n+r} + 2\sum_{i=0}^{n-r-1} b_i(x+1)^{i+r} + 4\sum_{i=0}^{n-r-1} c_i(x+1)^{i+r}$.
This is possible if and only if

$$\sum_{i=0}^{n-r-1} a_i(x+1)^i + 2(x+1)^{\frac{n}{2}} \sum_{i=n-r}^{n-1} a_i(x+1)^{i-n+r} + 2\sum_{i=0}^{n-r-1} b_i(x+1)^{i+r} \equiv 0(mod4)$$

(4)

1. If $r \le \frac{n}{2}$, then Eq. (4) holds if and only if $a_i = 0$ for $i = 0, 1, 2, \ldots, n-r-1$ and $a_{n-r} = 1$, $b_{\frac{n}{2}-r} = 1$. Thus $(x+1)^r g_1(x) = 4(x+1)^{\frac{n}{2}} + 4(x+1)^{\frac{n}{4}} + 4(x+1)^{\frac{3n}{4}} + 4(x+1)^{\frac{n}{2}} \sum_{i=n-r}^{n-1} b_i(x+1)^{i-n+r} + 4(x+1)^r \sum_{i=0}^{n-r-1} c_i(x+1)^i$. This implies that $v \ge min(r, \frac{n}{4})$.
   Note that $2[(x+1)^r] \in C$ and $(x+1)^r[(x+1)^{n-r} + 2(x+1)^{\frac{n}{2}-r}] = 4(x+1)^{\frac{n}{2}} + 4(x+1)^{\frac{n}{4}} + 4(x+1)^{\frac{3n}{4}} = 4(x+1)^{\frac{n}{4}} unit \in C$. Therefore $v \le min(r, \frac{n}{4})$. Hence $v = min(r, \frac{n}{4})$.
2. If $r > \frac{n}{2}$, then Eq. (4) holds if and only if $a_i = 0$ for $i = 0, 1, 2, \ldots, n-r-1$ and $a_{\frac{n}{2}} = 1$, $b_0 = 1$. Thus $(x+1)^r g_1(x) = 4(x+1)^r + 4(x+1)^{r-\frac{n}{4}} + 4(x+1)^{r+\frac{n}{4}} + 4(x+1)^{\frac{n}{2}} \sum_{i=n-r}^{n-1} b_i(x+1)^{i-n+r} + 4(x+1)^r \sum_{i=0}^{n-r-1} c_i(x+1)^i$. This implies that $v \ge min(r-\frac{n}{4}, \frac{n}{2})$.
   Note that $(x+1)^r 2(x+1)^{n-r} = 4(x+1)^{\frac{n}{2}} \in C$ and $(x+1)^r[(x+1)^{\frac{n}{2}} + 2] = 4(x+1)^r + 4(x+1)^{r-\frac{n}{4}} + 4(x+1)^{r+\frac{n}{4}} = 4(x+1)^{r-\frac{n}{4}} unit \in C$. Therefore $v \le min(r-\frac{n}{4}, \frac{n}{2})$. Hence $v = min(r-\frac{n}{4}, \frac{n}{2})$.

Again, in case $h(x)$ is a multiple of $g(x)$ only, we get by Lemma 1 and Theorem 3 that $v = min(r, \frac{n}{2})$, $g(x) = 2(x+1)^r$ if $r \le \frac{n}{2}$ and $v = \frac{n}{2}$, $g(x) = 2(x+1)^{\frac{n}{2}} + 4(x+1)^{\frac{n}{4}} + 4(x+1)^{\frac{3n}{4}}$ if $r > \frac{n}{2}$.

Further, in case $h(x)$ is a linear combination of $(x+1)^r$ and $g(x)$, we have the following

1. If $r \le \frac{n}{2}$ then it can easily seen that $s = r$ and $v = min(\frac{n}{4}, r)$.
2. If $r > \frac{n}{2}$ then $4(x+1)^v = (x+1)^r g_3(x) + [2(x+1)^{\frac{n}{2}} + 4(x+1)^{\frac{n}{4}} + 4(x+1)^{\frac{3n}{4}}] g_4(x) = \sum_{i=0}^{n-1} A_i(x+1)^{i+r} + 2\sum_{i=0}^{n-1} B_i(x+1)^{i+r} + 4\sum_{i=0}^{n-1} C_i(x+1)^{i+r} + 2\sum_{i=0}^{n-1} D_i(x+1)^{i+s} + 4\sum_{i=0}^{n-1} E_i(x+1)^{i+s} + 4(x+1)^{\frac{n}{4}} \sum_{i=0}^{n-1} D_i(x+1)^i + 4(x+1)^{\frac{3n}{4}} \sum_{i=0}^{n-1} D_i(x+1)^i = \sum_{i=0}^{n-r-1} A_i(x+1)^i + [2(x+1)^{\frac{n}{2}} + 4(x+1)^{\frac{n}{4}} + 4(x+1)^{\frac{3n}{4}}] \sum_{i=n-r}^{n-1} A_i(x+1)^{i-n+r} + 4(x+1)^{\frac{n}{2}} \sum_{i=n-r}^{n-1} B_i(x+1)^{i-n+r} + 2\sum_{i=0}^{n-r-1} B_i(x+1)^{i+r} + 4\sum_{i=0}^{n-r-1} C_i(x+1)^{i+r} + 4(x+1)^{\frac{n}{2}} \sum_{i=n-s}^{n-1} D_i(x+1)^{i-n+s} + 4\sum_{i=0}^{n-s-1} E_i(x+1)^{i+s} + 4(x+1)^{\frac{n}{4}} \sum_{i=0}^{n-1} D_i(x+1)^i + 4(x+1)^{\frac{3n}{4}} \sum_{i=0}^{n-1} D_i(x+1)^i$.

This implies that $\sum_{i=0}^{n-r-1} A_i(x+1)^i + 2(x+1)^{\frac{n}{2}} \sum_{i=n-r}^{n-1} A_i(x+1)^{i-n+r} + 2\sum_{i=0}^{n-r-1} B_i(x+1)^{i+r} + 2\sum_{i=0}^{n-s-1} D_i(x+1)^{i+s} \equiv 0 (mod 4)$. This is possible if and only if $A_i = 0$ for $i = 0, 1, 2, \ldots, n-r-1$, $A_{n-r} = 1$, $B_0 = 1$, $D_0 = 1$, $D_{r-s} = 1$. Thus $(x+1)^r g_3(x) + [2(x+1)^{\frac{n}{2}} + 4(x+1)^{\frac{n}{4}} + 4(x+1)^{\frac{3n}{4}}]g_4(x) = 4(x+1)^{\frac{n}{2}} + 4(x+1)^{\frac{n}{4}} + 4(x+1)^{\frac{3n}{4}} + 4(x+1)^{\frac{n}{2}} \sum_{i=n-r}^{n-1} B_i(x+1)^{i-n+r} + 4(x+1)^r + 4(x+1)^r \sum_{i=0}^{n-r-1} C_i(x+1)^i + 4(x+1)^{\frac{n}{2}} \sum_{i=n-s}^{n-1} D_i(x+1)^{i-n+s} + 4\sum_{i=0}^{n-s-1} E_i(x+1)^{i+s} + 4(x+1)^{\frac{n}{4}} + 4(x+1)^{\frac{3n}{4}} + 4(x+1)^{r-\frac{n}{4}} + 4(x+1)^{r+\frac{n}{4}}$.
This implies that $v \geq min(r - \frac{n}{4}, \frac{n}{2})$.

Note that $(x+1)^{n-r}(x+1)^r + 2(x+1)^{\frac{n}{2}} + 4(x+1)^{\frac{n}{4}} + 4(x+1)^{\frac{3n}{4}} = 4(x+1)^{\frac{n}{2}} \in C$ and $2(x+1)^r + (x+1)^{r-\frac{n}{2}}[2(x+1)^{\frac{n}{2}} + 4(x+1)^{\frac{n}{4}} + 4(x+1)^{\frac{3n}{4}}] = 4(x+1)^r + 4(x+1)^{r-\frac{n}{4}} + 4(x+1)^{r+\frac{n}{4}} = 4(x+1)^{r-\frac{n}{4}}[unit] \in C$. Therefore $v \leq min(r - n/4, n/2)$. Hence $v = min(r - \frac{n}{4}, \frac{n}{2})$

Taking the minimum value of v from the above cases, we get that

1. $v = min(\frac{n}{4}, r)$, if $r \leq \frac{n}{2}$
2. $v = min(\frac{n}{2}, r - \frac{n}{4})$, if $r > \frac{n}{2}$.

Theorems 9, 10 and 11 can be proved in a similar way as Theorem 8.

Theorem 9. *Let* $C = \langle (x+1)^r + 4(x+1)^d \gamma(x) \rangle$ *be a cyclic code of length* $n = 2^k$ *over* $\mathbb{Z}_8$, *then degree* v *of* $h(x)$ *is given by the following*

1. $v = min(\frac{n}{4}, r)$, *if* $r \leq \frac{n}{2}$
2. $v = min(\frac{n}{2}, r - \frac{n}{4})$, *if* $r > \frac{n}{2}$.

Theorem 10. *Let* $C = \langle (x+1)^r + 2(x+1)^c \beta(x) \rangle$ *be a cyclic code of length* $n = 2^k$ *over* $\mathbb{Z}_8$. *Then degree* v *of* $h(x)$ *in* C *is given by the following*

1. *If* $r < \frac{n}{2}$, *then* $v = min(\frac{n}{4}, r, \frac{n}{2} - r + c)$ *provided that* $\frac{n}{4} \neq \frac{n}{2} - r + c$. *If* $\frac{n}{4} = \frac{n}{2} - r + c$ *then* $v \leq min(\frac{n}{2} - r + c_1, r)$ *if* c_1 *exists otherwise* $v = r$
2. *If* $\frac{n}{2} \leq r < \frac{n}{2} + c$, *then* $v = min(r - \frac{n}{4}, c, \frac{3n}{4} - r + c)$ *provided that* $r - \frac{n}{4} \neq c$. *If* $r - \frac{n}{4} = c$ *and* $\beta(x) = 1 + (x+1)^{c_1-c} \beta_1(x)$ *then* $v = c_1$. *If* $\beta(x) = 1$ *then* $v = \frac{n}{2}$.
3. *If* $r = \frac{n}{2} + c$ *and* $\beta(x) = 1 + (x+1)^{c_1-c} \beta_1(x)$ *where* c_1 *is the least positive integer greater than* c *such that* $\beta_1(x)$ *is a unit, then we have the following*
   (a) $v = min(\frac{n}{4}, c_1 - c)$, *in case* $2c \leq c_1$ *and* $\frac{n}{4} \neq c_1 - c$. *Otherwise the value of* v *depends upon* $\beta_1(x)$.
   (b) $v = min(c, \frac{n}{4})$, *in case* $2c > c_1$.
   If no such c_1 *exists then* $v = \frac{n}{4}$
4. *If* $r > \frac{n}{2} + c$, *then* $v = min(c, \frac{n}{4})$.

Theorem 11. *Let* $C = \langle f(x) \rangle = \langle (x+1)^r + 2(x+1)^c \beta(x) + 4(x+1)^d \gamma(x) \rangle$ *be a cyclic code of length* $n = 2^k$ *over* $\mathbb{Z}_8$. *Then degree* v *of* $h(x)$ *in* C *is given by the following*

1. *If* $r < \frac{n}{2}$, *then* $v = min(\frac{n}{4}, r, \frac{n}{2} - r + c)$ *provided that* $\frac{n}{4} \neq \frac{n}{2} - r + c$. *If* $\frac{n}{4} = \frac{n}{2} - r + c$ *then* $v = min(\frac{n}{2} - r + c_1, r)$ *where* c_1 *is the least positive integer greater than* c *such that if* c_1 *exists otherwise* $v = r$

2. If $\frac{n}{2} \leq r < \frac{n}{2} + c$, then $v = min(r - \frac{n}{4}, c, \frac{3n}{2} - 2r + c + d, \frac{3n}{4} - r + c)$ provided that minimum values are not equal.

3. If $r = \frac{n}{2} + c$ and $\beta(x) = 1 + (x + 1)^{c_1 - c}\beta_1(x)$, where c_1 is the least positive integer greater than c such that $\beta_1(x)$ is a unit, then we have the following
   (a) $v = min(\frac{n}{4}, n - r + d, c_1 - c)$, in case $2c \leq c_1$.
   (b) $v = min(c, \frac{n}{4}, n - r + d)$ in case $2c > c_1$, provided that minimum values are not equal.

   If no such c_1 exists then $v = min(\frac{n}{4}, n - r + d)$ provided that $\frac{n}{4} \neq n - r + d$. If $\frac{n}{4} = n - r + d$ and $\gamma(x) = 1$ then $v = \frac{n}{2}$. If $\gamma(x) \neq 1$ then value of v depends upon $\gamma(x)$.

4. If $r > \frac{n}{2} + c$, then $v = min(c, \frac{n}{4}, n - r + d)$ provided that $\frac{n}{4} \neq n - r + d$. Otherwise $v = min(c, n - r + d_1)$ if d_1 exists. If no such d_1 exists then $v = c$.

4   Cyclic Codes of Length 4 over $\mathbb{Z}_8$

Using the above results, we have calculated all 95 cyclic codes of length 4 over $\mathbb{Z}_8$ in terms of their distinguished generators and listed them in the form of tables. Table 1 below lists all 46 principally generated cyclic codes of length 4 over $\mathbb{Z}_8$.

Table 1. Principally generated cyclic codes of length 4 over $\mathbb{Z}_8$

Sr. No	Cyclic code
1	$\langle 0 \rangle$
2	$\langle 4 \rangle$
3	$\langle 4(x + 1) \rangle$
4	$\langle 4(x + 1)^2 \rangle$
5	$\langle 4(x + 1)^3 \rangle$
6	$\langle 2 \rangle$
7	$\langle 2(x + 1) \rangle$
8	$\langle 2(x + 1)^2 \rangle$
9	$\langle 2(x + 1)^3 \rangle$
10	$\langle 1 \rangle$
11	$\langle (x + 1) \rangle$
12	$\langle (x + 1)^2 \rangle$
13	$\langle (x + 1)^3 \rangle$
14	$\langle 2(x + 1) + 4 \rangle$
15	$\langle 2(x + 1)^2 + 4 \rangle$
16	$\langle 2(x + 1)^2 + 4(x + 1) \rangle$
17	$\langle 2(x + 1)^2 + 4 + 4(x + 1) \rangle$
18	$\langle 2(x + 1)^3 + 4 \rangle$

(*continued*)

Table 1. (*continued*)

Sr. No	Cyclic code
19	$\langle 2(x+1)^3 + 4(x+1) + 4(x+1)^2 \rangle$
20	$\langle 2(x+1)^3 + 4(x+1) \rangle$
21	$\langle (x+1) + 4 \rangle$
22	$\langle (x+1)^2 + 4 \rangle$
23	$\langle (x+1)^3 + 4 \rangle$
24	$\langle (x+1)^3 + 4(x+1) \rangle$
25	$\langle (x+1)^3 + 4 + 4(x+1) \rangle$
26	$\langle (x+1) + 2 \rangle$
27	$\langle (x+1)^2 + 2 + 2(x+1) \rangle$
28	$\langle (x+1)^2 + 2(x+1) \rangle$
29	$\langle (x+1)^2 + 2 \rangle$
30	$\langle (x+1)^3 + 2 \rangle$
31	$\langle (x+1)^3 + 2(x+1) + 2(x+1)^2 \rangle$
32	$\langle (x+1)^3 + 2(x+1) \rangle$
33	$\langle (x+1) + 2 + 4 \rangle$
34	$\langle (x+1)^2 + 2 + 2(x+1) + 4 + 4(x+1) \rangle$
35	$\langle (x+1)^2 + 2 + 2(x+1) + 4 \rangle$
36	$\langle (x+1)^2 + 2 + 2(x+1) + 4(x+1) \rangle$
37	$\langle (x+1)^2 + 2 + 4 \rangle$
38	$\langle (x+1)^2 + 2(x+1) + 4 + 4(x+1) \rangle$
39	$\langle (x+1)^2 + 2(x+1) + 4 \rangle$
40	$\langle (x+1)^2 + 2(x+1) + 4(x+1) \rangle$
41	$\langle (x+1)^3 + 2(x+1) + 2(x+1)^2 + 4 \rangle$
42	$\langle (x+1)^3 + 2(x+1) + 2(x+1)^2 + 4(x+1) \rangle$
43	$\langle (x+1)^3 + 2(x+1) + 2(x+1)^2 + 4(x+1)^2 \rangle$
44	$\langle (x+1)^3 + 2(x+1) + 4 \rangle$
45	$\langle (x+1)^3 + 2(x+1) + 4 + 4(x+1) + 4(x+1)^2 \rangle$
46	$\langle (x+1)^3 + 2(x+1) + 4 + 4(x+1) \rangle$

Table 2 below lists all 49 non principally generated cyclic codes of length 4 over $\mathbb{Z}_8$.

Table 2. Non principally generated cyclic codes of length 4 over $\mathbb{Z}_8$

Sr. No	Cyclic code
1	$\langle 2(x+1), 4 \rangle$
2	$\langle 2(x+1)^2, 4 \rangle$
3	$\langle 2(x+1)^2, 4(x+1) \rangle$
4	$\langle 2(x+1)^3, 4 \rangle$
5	$\langle 2(x+1)^3, 4(x+1) \rangle$
6	$\langle 2(x+1)^2 + 4, 4(x+1) \rangle$
7	$\langle 2(x+1)^3 + 4(x+1), 4(x+1)^2 \rangle$
8	$\langle (x+1), 2 \rangle$
9	$\langle (x+1), 4 \rangle$
10	$\langle (x+1)^2, 2 \rangle$
11	$\langle (x+1)^2, 2(x+1) \rangle$
12	$\langle (x+1)^2, 2(x+1) + 4 \rangle$
13	$\langle (x+1)^2, 2(x+1), 4 \rangle$
14	$\langle (x+1)^2, 4 \rangle$
15	$\langle (x+1)^3, 2 \rangle$
16	$\langle (x+1)^3, 2(x+1) \rangle$
17	$\langle (x+1)^3, 4 \rangle$
18	$\langle (x+1)^3, 4(x+1) \rangle$
19	$\langle (x+1)^3, 2(x+1), 4 \rangle$
20	$\langle (x+1)^3, 2(x+1) + 4 \rangle$
21	$\langle (x+1)^3 + 4, 4(x+1) \rangle$
22	$\langle (x+1)^2 + 4, 2(x+1) \rangle$
23	$\langle (x+1)^2 + 4, 2(x+1) + 4 \rangle$
24	$\langle (x+1)^3 + 4, 2(x+1) \rangle$
25	$\langle (x+1)^3 + 4, 2(x+1) + 4 \rangle$
26	$\langle (x+1) + 2, 4 \rangle$
27	$\langle (x+1)^2 + 2 + 2(x+1), 4 \rangle$
28	$\langle (x+1)^2 + 2 + 2(x+1), 4(x+1) \rangle$
29	$\langle (x+1)^2 + 2(x+1), 4 \rangle$
30	$\langle (x+1)^2 + 2(x+1), 4(x+1) \rangle$
31	$\langle (x+1)^2 + 2, 2(x+1) \rangle$
32	$\langle (x+1)^2 + 2, 4 \rangle$
33	$\langle (x+1)^3 + 2(x+1) + 2(x+1)^2, 4 \rangle$
34	$\langle (x+1)^3 + 2(x+1) + 2(x+1)^2, 4(x+1) \rangle$
35	$\langle (x+1)^3 + 2(x+1) + 2(x+1)^2, 4(x+1)^2 \rangle$
36	$\langle (x+1)^3 + 2(x+1), 4 \rangle$
37	$\langle (x+1)^3 + 2(x+1), 2(x+1)^2 \rangle$
38	$\langle (x+1)^3 + 2(x+1), 2(x+1)^2 + 4 \rangle$
39	$\langle (x+1)^3 + 2(x+1), 2(x+1)^2, 4 \rangle$
40	$\langle (x+1)^2 + 2(x+1) + 4, 4(x+1) \rangle$
41	$\langle (x+1)^3 + 2(x+1) + 4, 2(x+1)^2 \rangle$
42	$\langle (x+1)^3 + 2(x+1) + 4, 2(x+1)^2 + 4 \rangle$
43	$\langle (x+1)^3 + 2(x+1) + 4, 2(x+1)^2 + 4 + 4(x+1) \rangle$
44	$\langle (x+1)^3 + 2(x+1) + 4, 2(x+1)^2 + 4, 4(x+1) \rangle$
45	$\langle (x+1)^3 + 2(x+1) + 4, 4(x+1) \rangle$
46	$\langle (x+1)^3 + 2(x+1) + 4 + 4(x+1), 2(x+1)^2 + 4 \rangle$
47	$\langle (x+1)^3 + 2(x+1) + 4 + 4(x+1), 2(x+1)^2 + 4 + 4(x+1) \rangle$
48	$\langle (x+1)^3 + 2(x+1) + 4 + 4(x+1), 4(x+1)^2 \rangle$
49	$\langle (x+1)^2 + 2 + 2(x+1) + 4, 4(x+1) \rangle$

References

1. Abualrub, T., Oehmke, R.: On generators of $\mathbb{Z}_4$ cyclic codes of length 2^e. IEEE Trans. Inf. Theor. **49**(9), 2126–2133 (2003)
2. Abualrub, T., Ghrayeb, A., Oehmke, R.: A mass formula and rank of $\mathbb{Z}_4$ cyclic codes of length 2^e. IEEE Trans. Inf. Theor. **50**(12), 3306–3312 (2004)
3. Dinh, H.Q.: On the linear ordering of some classes of negacyclic and cyclic codes and their distance distributions. Finite Fields Appl. **14**(1), 22–40 (2008)
4. Dinh, H.Q., Lopez Permouth, S.R.: Cyclic and negacyclic codes over finite chain rings. IEEE Trans. Inf. Theor. **50**(8), 1728–1744 (2004)
5. Dougherty, S.T., Park, Y.H.: On modular cyclic codes. Finite Fields Appl. **13**, 31–57 (2007)
6. Garg, A., Dutt, S.: Cyclic codes of length 2^k over $\mathbb{Z}_8$. Sci. Res. Open J. Appl. Sci. **2**, 104–107 (2012). October - 2012 World Congress on Engineering and Technology
7. Garg, A., Dutt, S.: On rank and MDR cyclic codes of length 2^k Over Z_8. In: Gaur, D., Narayanaswamy, N.S. (eds.) CALDAM 2017. LNCS, vol. 10156, pp. 177–186. Springer, Cham (2017). https://doi.org/10.1007/978-3-319-53007-9_16
8. Kiah, H.M., Leung, K.H., Ling, S.: Cyclic codes over $GR(p^2, m)$ of length p^k. Finite Fields Appl. **14**, 834–846 (2008)
9. Minjia, S., Shixin, Z.: Cyclic codes over the ring Z_{p^2} of length p^e. J. Electron. (China) **25**(5), 636–640 (2008)
10. Salagean, A.: Repeated-root cyclic and negacyclic codes over a finite chain ring. Discrete Appl. Math. **154**(2), 413–419 (2006)

Consistent Subset Problem with Two Labels

Kamyar Khodamoradi[1], Ramesh Krishnamurti[2], and Bodhayan Roy[3(⊠)]

[1] Department of Computing Science, University of Alberta,
Edmonton, Canada
`khodamor@cs.ualberta.ca`
[2] School of Computing Science, Simon Fraser University,
Burnaby, Canada
`ramesh@cs.sfu.ca`
[3] Faculty of Informatics, Masaryk University,
Brno, Czech Republic
`b.roy@fi.muni.cz`

Abstract. In this paper, we prove that the consistent subset problem with two labels is NP-complete.

1   Introduction

Let P be a labelled point set in the d-dimensional Euclidean space, where multiple points are allowed to have the same label. A subset C of P is said to be a *consistent subset* of P if for any point $v \in P \setminus C$, the point of C closest to v has the same label as v [5]. Consistent subsets have their applications in pattern recognizing problems [2]. The consistent subset of minimum cardinality among all consistent subsets of P is known as the *minimum consistent subset* of P. Hart [2] gave a method for computing a consistent subset, but it was not guaranteed to be an optimum solution. Ritter et al. [4] defined a related notion of *selective subsets* and gave an algorithm for computing minimum selective subsets, which has an exponential worst case running time, but an average running time of $O(n^3)$ [6].

Wilfong [5], using techniques from Masuyama et al. [3], showed that computing the minimum consistent subset with three or more labels is NP-complete. He further showed that the selective subset problem is NP-complete with two or more labels, but left the problem of computing a minimum consistent subset of a point set with only two labels open [5]. In this paper, we show that even with only two labels, computing a minimum consistent set is NP-complete.

2   NP-completeness

The consistent subset problem with any specific number of labels is in NP, as it can be verified in polynomial time whether a given subset of points is a consistent subset. It has already been proved that the consistent subset problem is NP-hard when it involves three or more labels [5]. In this section we prove that the consistent subset problem with only two labels is NP-hard. To achieve this, we reduce the *Rectilinear planar monotone 3-SAT* problem to the present problem.

© Springer International Publishing AG 2018
B. S. Panda and P. P. Goswami (Eds.): CALDAM 2018, LNCS 10743, pp. 131–142, 2018.
https://doi.org/10.1007/978-3-319-74180-2_11

2.1   Rectilinear Planar Monotone 3-SAT

Consider a 3-SAT formula θ on n variables $\{x_1, x_2, \ldots, x_n\}$ and m clauses $\{C_1, C_2, \ldots, C_m\}$. Suppose that θ is monotone, i.e. each of the clauses in θ has either all three positive or all three negative literals. These are called *positive* or *negative* clauses respectively. Then θ and a representation ξ of θ in the plane are together called a *planar rectilinear monotone 3-SAT* [1] if ξ satisfies the following properties (Fig. 1(a)):

(a) All the variables are represented by axis parallel squares of the same side a and lie on the x-axis, an equal distance b apart.
(b) All the clauses are represented by axis-parallel rectangles of height a.
(c) The rectangles representing positive clauses lie above the x-axis, and the rectangles representing negative clauses lie below the x-axis.
(d) If a variable occurs in a clause, their corresponding square and rectangle are connected by a single vertical line segment.
(e) No two line segments cross each other, and the rectangles are all pairwise disjoint.

We assume that each literal occurs in at least one clause, for otherwise we can assign 0 to it and remove the variable and all clauses containing that variable from θ. Note that the same literal can occur twice in one clause. Wlog we assume that for all pairs of variables $x_i, x_j \in \theta, i < j$, x_i lies to the left of x_j on the x-axis. For any clause C_j having variables x_i, x_j, x_k, $i < j < k$, we call x_i, x_j and x_k the *left*, *middle* and *right* literals of C_j respectively.

2.2   Overview of the Reduction

Here we give an overview of the reduction. Rectilinear planar monotone 3-SAT is an NP-complete problem [1]. The planar rectilinear drawing of θ can be modified into a bicoloured point set P for the reduction, as shown in Fig. 1. The point set P has five distinct components, namely the *clause-gadgets*, *variable-gadgets*, *CVC-paths*, *variable-line* and the *walls*. The clause-gadgets are three red points representing the three literals in a clause. The variable-gadgets are two red points representing a variable and its negation. The variable-line is a sequence of alternating green and red points running between variable-gadgets. Each CVC-path is a branching sequence of alternating red and green points connecting a point in a variable-gadget to all the clauses in which its corresponding literal occurs. All the CVC-paths have the same number of points. The *walls* are clusters of points four points thick that run alongside the CVC-paths.

We show that any consistent subset C of P must contain all points in all walls and the variable-line. C must also contain at least one among two points of each variable-gadget and at least one among three points of each clause-gadget. Furthermore, if C contains a variable-point or clause-point, then it must contain all points from their corresponding CVC-paths. Since the number points of all CVC-paths is the same, the number of points in C contained via the CVC-paths is minimized by a satisfying assignment of θ, for which the CVC-paths included

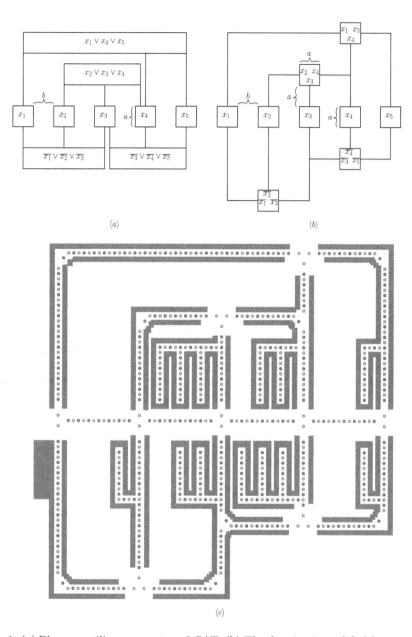

Fig. 1. (a) Planar rectilinear monotone 3-SAT. (b) The drawing is modified for our construction. (c) An approximate depiction of our final construction. Due to lack of space the positioning and number of points in the components are not precisely depicted, and the walls are depicted as continuous structures. The leftmost wall has extra point to make the total number of wall points equal W_n. The CVC-paths have extra branchings so that each of them has P_n points. (Color figure online)

in C due to the clause-gadgets are exactly the CVC-paths included in C due to the variable-gadgets.

2.3 Construction for the Reduction

From a planar rectilinear drawing ξ of θ, we now construct the point set P. We use quantities like P_n and W_n to which we will assign values in terms of m and n in the end of the construction, to prove its feasibility. We have the following steps for the construction:

(a) Stretch the line segments between the clauses and variables in ξ by a factor of m. Translate the clause rectangles vertically so that no horizontal line may intersect two clause rectangles, and two consecutive clause rectangles in their vertical ordering are exactly a distance a apart. Next shrink the clause rectangles into squares of side a and place them on the vertical segment from their middle literal. Extend the segments from the left and right literals vertically by a distance of $\frac{a}{2}$ and extend them horizontally till they hit the clause square. Now, if multiple segments originate from the same variable-square, then join them into a single segment that branches off towards different clause-squares (Fig. 1(b)).

(b) Let the bottom left vertex of the variable square for the variable x_i have coordinates (x, y). Replace this square by two red points with coordinates $(x + \frac{a}{2}, y + \frac{a}{2} + 2d)$ and $(x + \frac{a}{2}, y + \frac{a}{2} - 2d)$, representing x_i and $\overline{x_i}$ respectively. These two points constitute the variable-gadget for x_i (Fig. 3) and are called *variable-points*. Repeat the step for all variables of θ.

(c) Let the bottom left vertex of the clause square for a positive clause C_j have coordinates (x, y). Replace this square by three red points with coordinates $(x + \frac{a}{2} - d, y + \frac{a}{2} + 2d)$, $(x + \frac{a}{2}, y + \frac{a}{2} - 3d)$ and $(x + \frac{a}{2} + d, y + \frac{a}{2} + 2d)$, representing left, middle and right literals of C_j respectively (Fig. 2). These

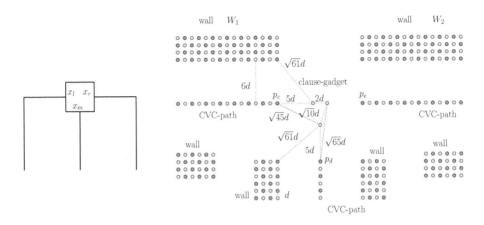

Fig. 2. A clause-gadget, its corresponding CVC-paths and walls. (Color figure online)

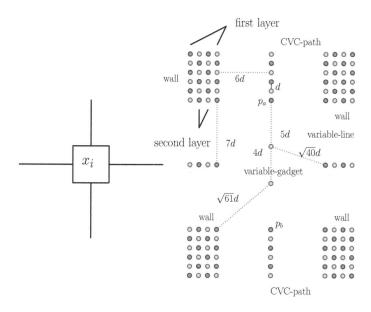

Fig. 3. A variable-gadget, its two CVC-paths, nearby points of the variable-line and walls. (Color figure online)

three points constitute the clause-gadget for C_j, and are called *clause-points*. Repeat the step for all positive clauses of θ. For negative clauses, let the top left vertex of the clause square have coordinates (x, y). Follow an analogous process, replacing the clause square by three red points with coordinates. $(x + \frac{a}{2} - d, y - \frac{a}{2} - 2d)$, $(x + \frac{a}{2}, y - \frac{a}{2} + 3d)$ and $(x + \frac{a}{2} + d, y - \frac{a}{2} - 2d)$.

(d) Let the coordinates of the top points of two consecutive variable-gadgets be (x_1, y_1) and (x_2, y_1) respectively, where $x_1 < x_2$. Then consider the line segment from point $(x_1 + 6d, y_1 - 2d)$ to $(x_2 - 6d, y_1 - 2d)$. Place a green point on $(x_1 + 6d, y_1 - 2d)$ (Fig. 3) and continue placing red and green points alternately on the line segment, with consecutive points a distance of d units apart. Repeat this process for all consecutive variable-gadgets. These points constitute the variable-line and are called *variable-line points*.

(e) Now we construct the CVC-path. We basically replace the line segments connecting the clause-gadgets and variable-gadgets with sequences of red and green alternating points. Place a green point p_a $5d$ units above the top point of each variable-gadget (Fig. 3), and continue placing red and green points alternately on the segment d units apart. Similarly, place a green point p_b $5d$ units below the bottom point of each variable-gadget (Fig. 3), and continue placing red and green points alternately on the segment.

For a clause-gadget of a positive clause, place three green points p_c, p_d and p_e, $5d$ units to the left, below and right of its left, bottom and right points respectively and place red and green points alternately d units apart in each direction (Fig. 2). For a clause-gadget of a negative clause, repeat the process

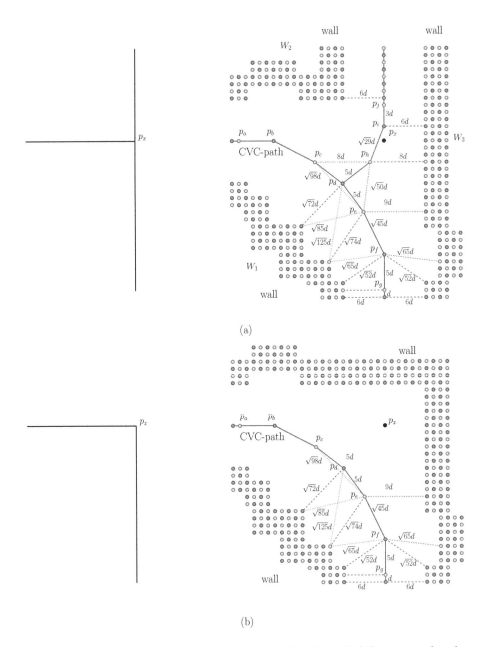

Fig. 4. (a) Points at the bifurcation of a CVC-path. The path bifurcates and makes a turn to reach a clause-gadget to the left, and also continues upwards to reach other clause-gadgets. For the CVC-path points, the distance to the nearest point of the same colour outside of the CVC-path is shown with a red dashed segment, while the distances to some other points are shown with green dotted segments. (b) A path turning without a bifurcation. (Color figure online)

by placing three green points $5d$ units to the left, right and above its left, right and top points respectively (Fig. 2).

Consider a bifurcation of the line-segments between clauses and variables. Suppose the segment bifurcates towards the left. Let the point p_x of bifurcation have coordinates (x, y). Then place red points $p_a(x - 21d, y)$, $p_c(x - 10d, y - 3d)$, $p_e(x - 3d, y - 10d)$, $p_g(x, y - 21d)$, $p_h(x - 2d, y - 3d)$ and $p_j(x, y + 5d)$ (Fig. 4(a)). Place green points $p_b(x - 16d, y)$, $p_d(x - 6d, y - 6d)$, $p_f(x, y - 16d)$ and $p_i(x, y + 2d)$ (Fig. 4(a)). Construct sequences of alternating red and green points d units apart running to the left of p_a, below p_g and above p_j (Fig. 4(a)). For path turnings without bifurcations, repeat the above process without placing p_h, p_i and p_j (Fig. 4(b)). Construct all path turnings and bifurcations analogously.

When a CVC-path has less than P_n points, within a vertical distance of a units from the variable-line, create bifurcations to the left and extend the path without reaching any other gadget or CVC-path, until the path contains P_n points (Fig. 1(c)).

(f) Now we construct the walls. The walls run along the CVC-paths. The walls are four points deep with red and green points alternating and placed d units apart. The walls are $6d$ units away from the nearest CVC-path points. At any cross section of a wall, the first and last point are said to be in the *first layer*, while the other points are said to be in the *second layer* (Figs. 3 and 2).

When a CVC-path ends at a variable-gadget, place walls to both sides of the path. The walls begin with points having the same y-coordinates as the last point of the path, the first layer points of the wall are only $6d$ units away from the nearest path points (i.e. path points with the same y-coordinates) and have the same colour as the nearest path point (Fig. 3).

When a CVC-path ends at a positive clause-gadget, if it corresponds to a middle-literal, we place the first-layer points of the wall on both sides of it, $6d$ units away from the nearest path point, and begin from the same vertical level as before. It corresponds to the left or right literals, then we place the walls above the paths, only $6d$ units away from the path and begin at the same x-coordinate as the last point of the path. However, while placing walls below the path we begin the walls (say, W_1 and W_2 only at the 10th point before the path ends (Fig. 2). Walls for CVC-paths ending at negative clauses are constructed analogously.

For a bifurcation with respect to a point $p_x(x, y)$, there are three separate walls W_1, W_2 and W_3 (Fig. 4(a)). The wall W_1 takes $90°$ turns at points $(x - 6d, y - 20d)$, $(x - 8d, y - 20d)$, $(x - 8d, y - 17d)$, $(x - 12d, y - 17d)$, $(x - 12d, y - 12d)$, $(x - 17d, y - 12d)$, $(x - 17d, y - 8d)$, $(x - 20d, y - 8d)$, $(x - 20d, y - 6d)$. The wall W_2 takes $90°$ turns at points $(x - 20d, y + 6d)$, $(x - 20d, y + 8d)$, $(x - 12d, y + 8d)$, $(x - 12d, y + 6d)$ and $(x - 6d, y + 6d)$. The wall W_3 takes $90°$ turns at points $(x + 6d, y - 20d)$, $(x + 8d, y - 20d)$, $(x + 8d, y - 12d)$ and $(x + 6d, y - 12d)$. For turnings without bifurcation the walls W_2 and W_3 join and form a single wall. Repeat an analogous process for all turnings and bifurcations.

If W_n points are not exhausted by the walls then add the rest of the points to the leftmost wall of P.

Let the total length and height of ξ be L and B respectively. Let the distance between two consecutive variable-gadgets be b units. A CVC-path can reach at most m clauses, so it can be at most $L + mB$ units long. Therefore we set $P_n = L + mB = na + (n-1)b + m(a + 2ma)$. Due to our construction, there can be two bifurcations in the same direction in the same CVC-path only $40d$ units apart. Leaving space for the turnings and walls, we see that each bifurcation can accommodate at least $a - 30$ more points in the CVC-path. Then we must have $(a - 30)(b/40) \geq P_n = na + (n-1)b + m(a + 2ma)$. Then putting $a = 160(m + 30)$ and $b = 4a + 2ma$ satisfies this relation, because by our earlier assumption each literal occurs in at least one clause, giving $3m \geq 2n$. Thus the CVC-paths can have enough bifurcations to include P_n points. Each wall runs alongside some path, and is four points deep. There are slightly more points used during turnings where the walls are thicker and the path points are separated by distances greater than d units. Since there are $2n$ CVC-paths, we set $W_n = 32P_n$. Let the total number of points in the variable-line be V_n. Then $V_n = (n-1)(b + a - 12d)$. Lastly, we simply set $a = 8d$, thus making d the unit length for our construction. Due to the relative values of a and d, the variable-line, clause-gadgets, variable-gadgets, walls and CVC-paths do not overlap with each other. Note that since a is an even multiple of d, the variable-line always has green points at both ends of each of its segments, and the CVC-paths also end in green points near clause and variable gadgets.

The above arguments and the values assigned to the constants show that P has polynomially many points in terms of the size of θ. All the operations involved in the construction can also be achieved in polynomial time. This brings us to the following lemma which we state without repeating the proof.

Lemma 1. *The construction of P from θ and ξ is feasible in polynomial time.*

2.4   Properties of the Constructed Point Set

Let C be a consistent subset of P. We start by showing that if C contains a point of a CVC-path or wall, then C must contain all points of that CVC-path, or wall. We have the following lemmas.

Lemma 2. *If C contains a point of a wall, then C must contain all points of the wall.*

Proof. By construction, the points of a wall are only d units apart. So the point closest to a wall point is another wall point. Since red and green points alternate throughout a wall, if C contains one of them then C must contain all of them. □

Lemma 3. *If C contains one point of a CVC-path, then C must contain all points of the CVC-path.*

Proof. Consecutive points of a CVC-path, except at turnings and bifurcations are only d units apart, while their nearest point outside of the CVC-path is in a wall, at a distance $6d$ units away. By construction, at turnings, the distances between two consecutive points in a CVC-path can be $3d$, $5d$, $\sqrt{29}d$ or $\sqrt{45}d$. But in each case, their nearest points outside of the CVC-path and also inside the CVC-path, are even further away (Fig. 4). Since red and green points alternate throughout a CVC-path, if one of them is in C, then it is the closest point of C to its one or two neighbours in the CVC-path, that have a different colour. So, these neighbours also must be included in C, and so on.   □

Now we explore the relationships between the different components of P in terms of their inclusion in C. We have the following lemmas.

Lemma 4. *If C contains a clause-point or variable-point corresponding to a CVC-path, then C contains all points of the CVC-path.*

Proof. Consider the point p of a CVC-path closest to a clause-gadget. The point p is green. All three clause-points are red, and by construction, the clause-point (say, q) nearest to p is only $5d$ units away while its nearest point in a wall is $6d$ away from p (Fig. 2). Suppose q is in C. By Lemma 3, if some point from the CVC-path is in C, then p is in C. Otherwise, q is the nearest point to p in C. But p and q have different colours, so p must be in C. Thus if q is in C, then p also must be in C. Then by Lemma 3 all points of the CVC-path are in C.

Now consider the point p of a CVC-path closest to its variable-gadget. By construction, the two variable-points are $5d$ and $9d$ units away from it (Fig. 3). If its corresponding variable-point i.e. the red point only $5d$ units away from it is in C, then p must also be in C because its nearest green points outside of the CVC-path are at least $6d$ units away from p. Then by Lemma 3 all points of the CVC-path are in C.   □

Lemma 5. *If C contains a point of a variable-line or variable-gadget, then C also contains all points of the variable-line and at least one point from each variable-gadget.*

Proof. Between two consecutive variable-gadgets, the points of the variable-line are only d units apart. Hence the closest point to a variable-line point is a variable-line point. Since red and green points alternate through a variable-line, if C contains one of them, C must contain all variable-line points between two consecutive variable-gadgets. By construction, the nearest point to a variable-gadget point is green (Fig. 3). Since the two endpoints of such a portion of the variable-line are green points, at least one of the two red points of the variable-gadget must also be in C. But then, the variable-point in C is only $\sqrt{40}d$ units away from its other nearest point (say, q) of the variable-line, whereas the nearest wall point to q is $7d$ units away. So, q or some point in the portion of the variable-line containing q must also be in C. Then by the first part of the proof, all points of the variable-line and at least one point from each variable-gadget must be in C.   □

Lemma 6. *If C contains a point of a CVC-path then C also contains all points of all the corresponding walls.*

Proof. If C contains a point of a CVC-path, then by Lemma 3 C must also contain all points of the CVC-path. But by construction, at anywhere other than turnings and bifurcations, the closest non wall point to a second layer wall point is a CVC-path point only $7d$ units away and of a different colour. This means that the second layer wall point must also be included in C. Then by Lemma 2 all points of the wall are in C. □

Lemma 7. *If C contains a point of a wall of some CVC-path, then C contains all points of all walls of the CVC-path.*

Proof. Other than at turnings and bifurcations, the wall is only $6d$ units away from the CVC-path. So, two consecutive walls of the path are only $12d$ units apart, and any other point closer to these walls must be from the CVC-path itself. If C contains a point of the CVC-path, then by Lemma 6 C must also contain all points of all of its walls. If C does not contain any point of the CVC-path, then consider a wall W_1 not in C that is next to a wall W_2 in C. For any point in the second layer of W_1, the closest point of C is a first layer point of W_2, which is of a different colour. So, a second layer point of W_1 must be in C. Then by Lemma 2 all points of W_1 are in C. The same argument applies for all walls of the CVC-path. □

Lemma 8. *If C contains a wall then C contains the variable-line and at least one point from each variable-gadget.*

Proof. Assume that C contains no point from the variable-line or any variable-gadget, for otherwise the claim is true by Lemma 5. If C contains a wall then by Lemma 7 C contains all walls of the same CVC-path. This means, C contains a wall-point that is only $\sqrt{61}d$ away from a variable-point (Fig. 3). This variable-point is also $5d$ away from a CVC-path point, which may or may not be in C. Both the CVC-path point and the wall point are green whereas the variable-point is red. So, one of the two variable-points must be in C. Then the claim follows from Lemma 5. □

Lemma 9. *If C contains the variable-line or a variable-point then C contains all walls.*

Proof. If C contains the variable-line or a variable-point then by Lemma 5 C contains the variable-line and at least one point from each variable-gadget. Now consider a wall point that is only $\sqrt{61}d$ away from a variable-point (Fig. 3). Call its immediate vertical neighbour on the wall q. By construction, q is a red point. Assume that C does not contain the corresponding CVC-path, for otherwise by Lemma 6 C contains the wall as well. The variable-point nearest to q is $\sqrt{72}d$ apart. The red point on the variable-line nearest to q is $\sqrt{65}d$ units away. But the green point on the variable-line nearest to q is only $8d$ units away and by our assumption it is in C. So, q must also be in C. The claim now follows from Lemma 7. □

Lemma 10. *If C contains a wall then C contains at least one point from each clause-gadget.*

Proof. If C contains a wall then by Lemma 8 C contains the variable-line and at least one point from each variable-gadget. Then by Lemma 9, C contains all walls. This means, C contains a wall-point that is only $\sqrt{61}d$ away from a clause-point (Fig. 2). This clause-point is also $5d$ away from a CVC-path point, which may or may not be in C. Both the CVC-path point and the wall point are green whereas the clause-point is red. So, one of the three clause-points must be in C. This argument applies to all clause-gadgets.   □

Now with the help of the relationships established in the above lemmas, we characterize consistent sets of P.

Lemma 11. *C contains (a) all walls of P, (b) the variable-line of P, (c) at least one variable-point from each variable-gadget, and its corresponding CVC-path, and (d) at least one clause-point clause-gadget, and its corresponding CVC-path.*

Proof. Consider any point q of C. Suppose that q is a wall point. Then by Lemma 7, C contains all wall points of the corresponding CVC-path. By Lemma 8, C contains the variable-line and at least one variable-point from each variable-gadget, and by Lemma 9, it contains all walls. By Lemma 10, C contains at least one clause-point from each clause-gadget. Then by Lemma 4, C contains the CVC-paths of the clause-points and variable-points that it contains.

Using the above part of the proof, to prove the rest of the lemma we just have to show that C contains a wall point in each case. If q is a point on the variable-line or a variable-point, then by Lemma 9 C contains a wall point. If q is a clause-point, then by Lemmas 4 and 6, C contains a wall point. Finally, if q is a point in a CVC-path, then again by Lemma 6, C contains a wall point.   □

As defined earlier, the total number of wall points in P is W_n. The number of points in each CVC-path, and the variable-line are P_n and V_n respectively.

Lemma 12. *The minimum consistent subset of P has a total of $W_n + nP_n + V_n + n + m$ points if and only if θ is satisfiable.*

Proof. First suppose that θ has a satisfying assignment. We include all walls and the variable-line in C. This accounts for $W_n + V_n$ points. Now we choose the variable-points that represent the literals that are assigned 1 in the satisfying assignment, and include them in C. This accounts for n points. For each clause, we choose one satisfied literal in the clause and include the corresponding clause-point in C. This accounts for m points. Finally, we choose the CVC-paths for all the chosen variable points. This accounts for nP_n points. For each variable-gadget, the point chosen in C is the point of C closest to the other point. The same holds for each clause-gadget. By construction, the points of the CVC-paths not included in C have wall points of the same colour as their closest points of C, since their corresponding clause-points and variable-points are not in C either. So, C is indeed a consistent subset of P.

Now suppose that a consistent subset C has only $W_n + nP_n + V_n + n + m$ points. Due to Lemma 11, and the number of points in each component, C contains exactly one clause-point from each clause-gadget and exactly one variable-point from each variable-gadget, and only the CVC-paths corresponding to these clause-points and variable-points. We build a satisfying assignment of θ from C. We assign 1 to the literals represented by the variable-points in C. So for each variable, either the variable itself or its negation is assigned 1. If this assignment does not satisfy some clause of θ, then one of the CVC-paths of its clause-gadget is in C but leads to a variable-point not in C. This means the CVC-paths contribute a total of at least $(n+1)P_n$ points, pushing the cardinality of C to at least $W_n + (n+1)P_n + V_n + n + m$, a contradiction.   □

Theorem 1. *The minimum consistent subset problem is NP-complete for point sets with two labels.*

Proof. The fact that the problem is in NP is obvious, because given a subset C of a 2-coloured point set P, it can be verified in polynomial time whether or not C is a consistent subset of P. We now prove NP-hardness of the problem.

By Lemma 12 it is possible to construct from a given planar monotone 3-SAT formula θ of n variables and m clauses, a 2-coloured point set that has a minimum consistent subset of size $W_n + nP_n + V_n + n + m$ if and only if θ is satisfiable. Furthermore, by Lemma 1, this construction is possible in time polynomial in the size of θ, thus completing the proof.   □

3   Concluding Remarks and Acknowledgments

We have proved that the consistent subset problem with only two labels is also NP-complete, a curious result given that the problem is trivial for monochrome point sets. The scope of designing parameterized algorithms and approximations better than the general case remains open for point sets with two labels.

We are grateful to Sasanka Roy for introducing the minimum consistent subset problem to us, and the discussions that followed. We thank the anonymous referees whose scrutiny of the paper has improved its presentation and clarity.

References

1. de Berg, M., Khosravi, A.: Optimal binary space partitions for segments in the plane. Int. J. Comput. Geom. Appl. **22**(3), 187–206 (2012)
2. Hart, P.E.: The condensed nearest neighbor rule (corresp.). IEEE Trans. Inf. Theor. **14**(3), 515–516 (1968)
3. Masuyama, S., Ibaraki, T., Hasegawa, T.: The computational complexity of the m-center problems on the plane. IEICE Trans. (1976–1990) **64**(2), 57–64 (1981)
4. Ritter, G.L., Woodruff, H.B., Lowry, S.R., Isenhour, T.L.: An algorithm for a selective nearest neighbor decision rule (corresp.). IEEE Trans. Inf. Theor. **21**(6), 665–669 (1975)
5. Wilfong, G.T.: Nearest neighbor problems. Int. J. Comput. Geom. Appl. **2**(4), 383–416 (1992)
6. Wilson, D.R., Martinez, T.R.: Reduction techniques for instance-based learning algorithms. Mach. Learn. **38**(3), 257–286 (2000)

The Edge Geodetic Number of Product Graphs

Bijo S. Anand[1], Manoj Changat[2(✉)], and S. V. Ullas Chandran[3]

[1] Department of Mathematics, Sree Narayana College,
Punalur 691305, Kerala, India
bijos_anand@yahoo.com
[2] Department of Futures Studies, University of Kerala,
Thiruvananthapuram 695034, Kerala, India
mchangat@gmail.com
[3] Department of Mathematics, Mahatma Gandhi College, Kesavadasapuram,
Thiruvananthapuram 695004, Kerala, India
svuc.math@gmail.com

Abstract. For a nontrivial connected graph $G = (V(G), E(G))$, a set $S \subseteq V(G)$ is called an edge geodetic set of G if every edge of G is contained in a geodesic joining some pair of vertices in S. The edge geodetic number $eg(G)$ of G is the minimum order of its edge geodetic sets. It is observed that the edge geodetic sets and numbers are interesting concepts and possess properties distinct from the vertex geodetic concepts. In this work, we determine some bounds and exact values of the edge geodetic numbers of strong and lexicographic products of graphs.

Keywords: Geodetic number · Edge geodetic number
Extreme vertex · Extreme edge · Semi-extreme vertex

AMS Subject Classification: 05C12

1 Introduction

Covering problems form one of the fundamental problems in graph theory, both vertex covering and edge covering. One of the important vertex covering problems is the geodetic covering problem, namely covering the entire vertex set of a graph using a set S of vertices with smallest cardinality such that every vertex of the graph belongs to a geodesic or shortest path between a pair of vertices in S. Harary et al. introduced the geodetic covering problem and the related graph parameter, namely the geodetic number in [5,11] followed by other authors in [4,6,8,13]. The edge version of the geodetic covering is named as the edge geodetic set, defined as the set S of vertices with smallest cardinality such that every edge of the graph belongs to a geodesic between a pair of vertices in S. The parameter, edge geodetic number of a graph was introduced and studied in [15,16]. Although the edge geodetic number is greater than or equal to the geodetic number for an arbitrary graph, the properties of the edge geodetic sets

© Springer International Publishing AG 2018
B. S. Panda and P. P. Goswami (Eds.): CALDAM 2018, LNCS 10743, pp. 143–154, 2018.
https://doi.org/10.1007/978-3-319-74180-2_12

and numbers are quite different from that of vertex geodetic concepts. There is a strong motivation to study the edge geodetic problem and the edge geodetic number, as there are many applications, for e.g., in the problem of designing the route for a shuttle transportation, as all the edges will be covered when we consider the edge geodetic sets instead of vertex geodetic sets. Recently, an application of the edge geodetic sets, the strong edge geodetic problem is introduced in [12] in the analysis of structural behavior of social networks. In particular, the edge geodetic sets are more useful than geodetic sets in regulating and routing the goods vehicles to transport the commodities to important places.

The edge geodetic number is investigated in Cartesian product graphs in [17], where we observe that the edge geodetic number and the geodetic number have significant difference. Motivated by this work, in this paper, we study the edge geodetic number in other standard graph products, namely the strong products and Lexico-graphic products. First, we fix the notation, in Sect. 2, we discuss some interesting properties of edge geodetic sets; study and estimate the upper and lower bounds on strong products and lexicographic products, respectively in Sects. 3 and 4.

By a graph $G = (V(G), E(G))$ we mean a finite undirected connected simple graph (without loops or multiple edges). The *order* and *size* of G are denoted by n and m respectively. The *distance* $d(u, v)$ between two vertices u and v in a connected graph G is the length of a shortest u - v path in G. An u - v path of length $d(u, v)$ is called an u - v *geodesic*. It is known that this distance is a metric on the vertex set $V(G)$. For a vertex v of G, the *eccentricity* $e(v)$ is the distance between v and a vertex farthest from v. The minimum eccentricity among the vertices of G is the *radius*, *rad* G, and the maximum eccentricity is its *diameter*, *diam* G of G. A vertex v of a graph G is called *simplicial* or *extreme* if its neighborhood $N(v)$ induces a clique. A *geodetic* set of G is a set $S \subseteq V(G)$ such that every vertex of G is contained in a geodesic joining some pair of vertices in S. The *geodetic number* $g(G)$ of G is the minimum order of its geodetic sets.

The *strong product* of graphs G and H, denoted by $G \boxtimes H$, has vertex set $V(G) \times V(H)$, where two distinct vertices (x_1, y_1) and (x_2, y_2) are adjacent with respect to the strong product if,

(a) $x_1 = x_2$ and $y_1 y_2 \in E(H)$, or
(b) $y_1 = y_2$ and $x_1 x_2 \in E(G)$, or
(c) $x_1 x_2 \in E(G)$ and $y_1 y_2 \in E(H)$.

The mappings $\pi_G : (x, y) \mapsto x$ and $\pi_H : (x, y) \mapsto y$ from $V(G \boxtimes H)$ onto G and H respectively are called *projections*. For a set $S \subseteq V(G \boxtimes H)$, we define the *G-projection* on G as $\pi_G(S) = \{x \in V(G) : (x, y) \in S$ for some $y \in V(H)\}$, and the *H-projection* $\pi_H(S) = \{y \in V(H) : (x, y) \in S$ for some $x \in V(G)\}$. For a walk $P : (x_1, y_1), (x_2, y_2), \ldots, (x_n, y_n)$ in $G \boxtimes H$, we define the *G-projection* $\pi_G(P)$ of P as a sequence that is obtained from $(x_1, x_2, \ldots, x_n)$ by changing each constant subsequence with its unique element. For example, if $P : (x_2, y_3), (x_2, y_4), (x_2, y_5), (x_4, y_5), (x_4, y_2), (x_3, y_2), (x_2, y_2)$, then $\pi_G(P)$ is (x_2, x_4, x_3, x_2) (it is obtained from the sequence $(x_2, x_2, x_2, x_4, x_4, x_3, x_2)$).

The H-projection $\pi_H(P)$ is defined similarly. It is clear from the definition of strong product that for any walk P in $G \boxtimes H$, both $\pi_G(P)$ and $\pi_H(P)$ are walks in the factor graphs G and H respectively.

The *lexicographic product* of graphs G and H is the graph $G \circ H$ on vertex set $V(G) \times V(H)$ in which the vertices (g_1, h_1) and (g_2, h_2) are adjacent if and only if either $g_1 g_2 \in E(G)$ or $g_1 = g_2$ and $h_1 h_2 \in E(H)$. This graph operation is also known as the graph composition and denoted by $G[H]$.

For basic graph theoretic terminology, we refer to [9]. We also refer to [5] for results on distance in graphs and to [10] for metric structures in strong product graphs. Throughout the following G denotes a connected graph with at least two vertices. The following theorem will be used in the sequel.

Theorem 1.1 [5]. *Each extreme vertex of a connected graph G belongs to every geodetic set of G.*

2   Edge Geodetic Sets

An *edge geodetic set* of a connected graph G is a set $S \subseteq V(G)$ such that every edge of G is contained in a geodesic joining some pair of vertices in S. The *edge geodetic number* $eg(G)$ of G is the minimum order of its edge geodetic sets.

For the graph G given in Fig. 1, $S = \{v_1, v_2, v_4\}$ is a minimum edge geodetic set of G so that $eg(G) = 3$. Also $S' = \{v_3, v_5\}$ is a minimum geodetic set of G so that $g(G) = 2$. Thus the geodetic number and the edge geodetic number of a graph are different.

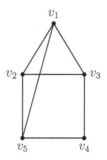

Fig. 1. G

Theorem 2.1 [16].*Every edge geodetic set of a connected graph G is a geodetic set of G.*

An edge $e = uv$ in a connected graph G is said to be an *extreme edge* if e lies on an $x - y$ geodesic in G, then $u = x$ or $v = y$. In any graph G, end edges are extreme edges but the converse need not be true. Moreover, if a graph G has diameter 2, then any edge in G is an extreme edge.

Observation 2.2. *Every edge geodetic set contains at least one end of each extreme edge. Moreover, if a connected graph G has k independent extreme edges, then $eg(G) \geq k$.*

A vertex v in G is called a *semi-extreme vertex* of G if $\Delta(\langle N(v) \rangle) = |N(v)| - 1$. That is, the induced subgraph of $N(v)$ has a full degree vertex in $N(v)$. For the graph G in Fig. 1, the vertices v_1 and v_2 are semi-extreme. The set of all semi-extreme vertices of a graph G is denoted by $sxt(G)$ and $s(G) = |Sxt(G)|$. Note that every extreme vertex is semi-extreme. A graph G is said to be *semi-complete* if all of its vertices are semi-extreme. It is clear that the complete graphs, $K_4 - e$, $G + K_2$ are semi-complete graphs. The behavior of the edge geodetic number was recently investigated in [15].

Theorem 2.3 [15]. *Let G be a connected graph. Then each semi-extreme vertex belongs to every edge geodetic set of G.*

Proof. Let T be an edge geodetic set of G. Suppose that there exists $u \in sxt(G)$ such that $u \notin T$. Since $\Delta(\langle N(u) \rangle) = |N(u)| - 1$, there exists $v \in N(u)$ such that $deg_{\langle N(u) \rangle}(v) = |N(u)| - 1$. Since T is an edge geodetic set of G, the edge $e = uv$ lies on a $x - y$ geodesic $P : x = x_0, x_1, \ldots, x_i = u, x_{i+1} = v, \ldots, x_n = y$ with $x, y \in T$. Then $v \neq x, y$. Since $deg_{\langle N(u) \rangle}(v) = |N(u)| - 1$, u is adjacent to x_{i+2}, which is a contradiction to the fact that P is a $x - y$ geodesic. Hence $sxt(G)$ is contained in every edge geodetic set of G. ∎

Theorem 2.4. *A graph G is semi-complete if and only if $eg(G) = n$.*

Proof. If G is semi-complete, then by Theorem 2.3, $eg(G) = n$. Conversely, suppose that $eg(G) = n$. If $sxt(G) \neq V(G)$, then there exists $u \in V(G)$ such that $u \notin S$. Thus $\Delta(\langle N(u) \rangle) \leq |N(u)| - 2$. It follows that $deg_{\langle N(u) \rangle}(v) \leq |N(u)| - 2$ for all $v \in N(u)$. Let $T = V(G) - \{u\}$. We claim that T is an edge geodetic set of G. Let $e = xy$ be an edge in G. If $x, y \in T$, then there is nothing to prove. So, assume that $y = u$. Then $x \in N(u) \cap T$. Let $x' \in N(u)$ be such that x and x' are non-adjacent. Then it is clear that the edge $e = xu$ lies on the geodesic $P : x, u, x'$ with $x, x' \in T$. Hence T is an edge geodetic set of G and so $eg(G) \leq n - 1$, which is a contradiction to the fact that $eg(G) = n$. Thus $sxt(G) = V(G)$. ∎

3   Edge Geodetic Sets in Strong Product Graphs

In this section, we investigate the behavior of edge geodetic sets on strong products of graphs. We establish both lower and upper bounds for the edge geodetic number and obtain the exact value of this parameters for a number of strong product graphs.

Theorem 3.1 [6,13]. *Let G and H be connected graphs with (u, v) and (x, y) arbitrary vertices of the strong product $G \boxtimes H$ of G and H. Then $d_{G \boxtimes H}((u, v), (x, y)) = max\{d_G(u, x), d_H(v, y)\}$. Moreover, if P a $(u, v) - (u', v')$ geodesic in $G \boxtimes H$ of length n and if $d_G(u, u') \geq d_H(v, v')$, then $\pi_G(P)$ is a $u - u'$ geodesic in G of length n.*

Theorem 3.2 [6,13]. *If G and H are non trivial connected graphs, then $g(G \boxtimes H) \geq 4$*

Theorem 3.3. *Let G and H be connected graphs. Then*

$$sxt(G \boxtimes H) = [sxt(G) \times V(H)] \cup [V(G) \times sxt(H)]$$

Proof. Let $g \in sxt(G)$ and $h \in V(H)$. Then there exists $g' \in N_G(g)$ such that $deg_{\langle N(g) \rangle}(g') = |N_G(g) - 1|$. Thus the vertex g' is adjacent to all the neighbors of g in G. Let (x, y) be any neighbor of (g, h) different from (g', h) in $G \boxtimes H$. Then we have that $x = g$ or x is adjacent to g in G. Also, $y = h$ or y is adjacent to h in H. This shows that the vertex (g', h) is adjacent to (x, y) in $G \boxtimes H$ and so $(g, h) \in sxt(G \boxtimes H)$. On the otherhand, let (x, y) be any semi-extreme vertex in $G \boxtimes H$. Assume the contrary that both x and y are not semi-extreme vertices in the corresponding components. Now, let (u, v) be a neighbor of (x, y) in $G \boxtimes H$ such that it is adjacent to all other neighbors of (x, y) in $G \boxtimes H$. Then we have that $u \neq x$ or $v \neq y$, say $u \neq x$. Since x is not a semi-extreme vertices in G, we can choose $u' \neq u$ such that u and u' are non-adjacent neighbors of x. Then it is clear that the vertex (u', y) is a neighbor of (x, y) in $G \boxtimes H$ such that (u, v) and (u', y) are non-adjacent in $G \boxtimes H$, which is a contradiction. Hence the result follows. ∎

Corollary 3.4. *Let G and H be connected graphs of order n and m respectively. Then*

$$eg(G \boxtimes H) \geq m.s(G) + n.s(H) - s(G)s(H)$$

Corollary 3.5. *For any connected graph G of order n, $eg(G \boxtimes K_n) = mn$.*

Remark 3.6. *Let G and H be connected graphs of order n and m respectively. Let S be an edge geodetic set of $G \boxtimes H$. If $sxt(G) \neq \emptyset$, then it follows from Theorem 3.3 that $\pi_H(S) = V(H)$. This shows that $m \leq eg(G \boxtimes H)$. Moreover, if $sxt(G) \neq \emptyset$ and $sxt(H) \neq \emptyset$, then $max\{m, n\} \leq eg(G \boxtimes H) \leq mn$.*

Theorem 3.7. *Let G and H be connected graphs and let S be an edge geodetic set in $G \boxtimes H$. Then $\pi_G(S)$ is an edge geodetic set in G and $\pi_H(S)$ is an edge geodetic set in H.*

Proof. Let $e_G = xx'$ be an edge in G and let y be any vertex in H. Then (x, y) and (x', y) are adjacent in $G \boxtimes H$, call this edge as e. Since S is an edge geodetic set in $G \boxtimes H$, there exist vertices $(g_0, h_0), (g_n, h_n) \in S$ such that e lies on a $(g_0, h_0)-(g_n, h_n)$ geodesic, say $P : (g_0, h_0), (g_1, h_1), \ldots, (g_i, h_i) = (x, y), (x', y) = (g_{i+1}, h_{i+1}), \ldots, (g_n, h_n)$. This shows that $l(\pi_H(P)) < l(P)$ and so it follows from Theorem 3.1 that $l(\pi_G(P)) = l(P)$ and $\pi_G(P)$ is a $g_0 - g_n$ geodesic in G containing the edge $e_G = xx'$. ∎

Corollary 3.8. *Let G and H be connected graphs of order n and m respectively. Then*

$$eg(G \boxtimes H) \geq max\{m.s(G) + n.s(H) - s(G)s(H), eg(G), eg(H)\}$$

Proof. This follows from Theorem 3.7 and Corollary 3.4.   ∎

Theorem 3.9. *Let G and H be connected graphs. Then $G \boxtimes H$ is semi-complete if and only if G or H is semi-complete.*

Proof. If G or H is semi-complete, say G, then $sxt(G) = V(G)$. Then it follows from Theorem 3.3 that $sxt(G \boxtimes H) = V(G) \times V(H)$ and hence $G \boxtimes H$ is semi-complete. conversely, suppose that $G \boxtimes H$ is semi-complete. Then $sxt(G \boxtimes H) = V(G) \times V(H)$. If both G and H are not semi-complete, choose vertices x in G and y in H such that both x and y are not semi-extreme vertices in G and H respectively. Now, since (x, y) is semi-extreme vertex in $G \boxtimes H$, there exists a neighbor (u, v) of (x, y) in $G \boxtimes H$ such that (u, v) is adjacent to all other neighbors of (x, y) in $G \boxtimes H$. Now, we have that $u \neq x$ or $v \neq y$, say $u \neq x$. Now, choose $u' \in N(x)$ such that u and u' are non-adjacent in G. Then it is clear that the vertex (u', y) is neighbor of (x, y) in $G \boxtimes H$ which is non-adjacent with the vertex (u, v). This is a contradiction to the fact that (x, y) is a semi-extreme vertex in $G \boxtimes H$.   ∎

Remark 3.10. *If $G \boxtimes H$ is semi-complete, then both G and H need not be semi-complete graphs. For example, the graph $C_4 \boxtimes K_2$ is semi-complete, whereas C_4 is not semi-complete.*

Theorem 3.11. *Let G and H be connected graphs of order n and m respectively. Then $eg(G \boxtimes H) \leq n.eg(H) + m.eg(G) - eg(G)eg(H)$.*

Proof. Let S be a minimum edge geodetic set in G and let T be a minimum edge geodetic set in H. We claim that the set $W = (V(G) \times T) \cup (S \times V(H))$ is an edge geodetic set in $G \boxtimes H$. For, let $e = (x_1, y_1)(x_2, y_2)$ be an edge in $G \boxtimes H$.

Case 1. $x_1 \neq x_2$. Then x_1 and x_2 are adjacent in G and $d(y_1, y_2) \leq 1$. Since S is an edge geodetic set in G, there exist vertices $g_1, g_r \in S$ such that the edge $x_1 x_2$ lies on a $g_1 - g_r$ geodesic in G, say $P : g_1, g_2, \ldots, g_i = x_1, g_{i+1} = x_2, \ldots, g_n$. Then it follows from Theorem 3.1 that the path $Q : (g_1, y_1), (g_2, y_1), \ldots, (g_i, y_1) = (x_1, y_1), (x_2, y_2) = (g_{i+1}, y_2), \ldots, (g_r, y_2)$ is a $(g_1, y_1) - (g_r, y_2)$ geodesic in $G \boxtimes H$ containing the edge e with $(g_1, y_1), (g_r, y_2) \in S \times V(H)$.

Case 2. $x_1 = x_2$. Then $y_1 y_2$ is an edge in H. Since T is an edge geodetic set in H, there exist vertices $h_1, h_s \in T$ such that the edge $y_1 y_2$ lies on a $h_1 - h_s$ geodesic in H, say $P' : h_1, h_2, \ldots, g_j = y_1, g_{j+1} = y_2, \ldots, h_s$. Then it follows from Theorem 3.1 that the path $Q' : (x, h_1), (x, h_2), \ldots, (x, h_j) = (x, y_1), (x, y_2) = (x, h_{j+1}), \ldots, (x, h_s)$ is a $(x, h_1) - (x, h_s)$ geodesic in $G \boxtimes H$ containing the edge e with $(x, h_1), (x, h_s) \in V(G) \times T$. Hence the result follows.   ∎

As a direct consequence of the above theorem, the following results are obtained.

Corollary 3.12. *Let G and H be connected graphs of order n and m respectively. Then*

$$max\{m.s(G) + n.s(H) - s(G)s(H), eg(G), eg(H), 4\} \leq eg(G \boxtimes H) \leq$$
$$n.eg(H) + m.eg(G) - eg(G)eg(H).$$

Corollary 3.13. *Let T_1 be a tree of order m with r end vertices and T_2 be a tree or order n with s end vertices. Then*

$$eg(T_1 \boxtimes T_2) = ms + nr - rs.$$

Corollary 3.14. *For integers $m, n \geq 4$,*

$$eg(W_{1,m} \boxtimes W_{1,n}) = mn + m + n.$$

4 Edge Geodetic Sets of Lexicographic Product of Graphs

There is a constant research interest for lexicographic product of graphs over the years, which overlaps several branches of graph theory. This is evident from the references, [1–3,7]. For more on lexicographic products in general we recommend [10].

A *2-edge geodetic set* of a graph G is a set S of vertices of G such that every edge on G lies on some geodesic of length at most 2 joining two vertices in S [14]. The minimum cardinality of a 2-edge geodetic set of G is its *2-edge geodetic number* $eg_2(G)$ [14]. A 2-edge geodetic set of cardinality $eg_2(G)$ is called eg_2-*set* of G. Notice that $eg(G) \leq eg_2(G)$, since every 2-edge goedetic set is an edge geodetic set. Certainly, the converse of these assertion is not far being true in general, unless the graph has diameter 2.

The following result is a direct consequence of the definition of lexicographic product.

Theorem 4.1. *Let G and H be graphs. Then*

1. *The graph $G \circ H$ is connected if and only if G is connected.*
2. *The lexicographic product is associative but not commutative.*
3. *If G is connected, then $d_{G \circ H}((g, h), (g', h')) = d_G(g, g')$ if $g \neq g'$*
   $d_{G \circ H}((g, h), (g, h')) = 2$ if $hh' \notin E(H)$
   $d_{G \circ H}((g, h), (g, h')) = 1$ if $hh' \in E(H)$

Theorem 4.2. *Let G and H be two connected graphs. Then*

$$sxt(G \circ H) = \begin{cases} (V(G) \times sxt(H)) \cup (sxt(G) \times V(H)), & \text{if } H \text{ contains full degree vertex;} \\ V(G) \times sxt(H), & \text{otherwise.} \end{cases}$$

Proof. Case 1: H contains full degree vertices.

Let $(x, y) \in sxt(G \circ H)$ and let (u, v) be a neighbour of (x, y) such that (u, v) is adjacent with all other neighbours of (x, y).

Assume the contrary that $x \notin sxt(G)$ and $y \notin sxt(H)$. First suppose that $x = u$, then $y \neq v$. Since $y \notin sxt(H)$, we can chose a neighbour v' of y but v' is not adjacent with v. This shows that (x, v') is a neighbour of (x, y) which is non-adjacent with (x, v). This is a contradiction.

Now consider the case that $x \neq u$. Then x and u should be adjacent in G and so all the vertices in the layer $^u H$ are neighbours of the vertex (x, y) in $G \circ H$. Hence it follows from the choice of the vertex (u, v) that it must be adjacent with all other vertices in the layer ^uH. Thus v must be a full degree vertex in H. Since $x \notin sxt(G)$, we can chose a neighbour u' of x such that u and u' are non-adjacent in G. But all the vertices of the layer $^{u'}H$ are neighbours of the vertex (x, y) and so (u, v) must be adjacent with all the vertices of the layer $^{u'}H$, particularly the vertex (u', v). This is possible only when u and u' are adjacent in G, a contradiction. This shows that $x \in sxt(G)$ or $y \in sxt(H)$ and so $sxt(G \circ H) \subseteq (V(G) \times sxt(H)) \cup (sxt(G) \times V(H))$.

Conversely suppose that $x \in sxt(G)$. We claim that $(x, y) \in sxt(G \circ H)$ for all $y \in V(H)$. Let v be a full degree vertex in H. Choose a neighbour u of x such that u is adjacent with all other neighbours of x. Now let (s, t) be any neighbour of (x, y) in $G \circ H$ different from (u, v). First suppose that $s = x$, then (u, v) must be adjacent with (s, t). So let $s \neq x$. Then s is a neighbour of x in G. If $s \neq u$, then it follows from the choice of u that s and u must be adjacent in G. Thus (u, v) is adjacent with (s, t) in $G \circ H$. Otherwise, if $s = u$, then $v \neq t$. since v is a full degree vertex, we have that v and t are adjacent in H. Therefore in this case also the vertex (u, v) is adjacent with (s, t). Hence $(x, y) \in sxt(G \circ H)$ and so $sxt(G) \times V(H) \subseteq sxt(G \circ H)$. Now let $y \in sxt(H)$. We show that $(x, y) \in sxt(G \circ H)$, for all $x \in V(G)$. Since $y \in sxt(H)$, we can choose a neighbour v of y such that v is adjacent with all other neighbours of y. Let (s, t) be any neighbour of (x, y) such that $(s, t) \neq (x, v)$. If $s = x$, then t must be adjacent with y and so the vertex v must be adjacent with t. Hence (s, t) is adjacent with (x, v) in $G \circ H$. Similarly if $s \neq x$, then x and s are adjacent in G and so (x, v) is adjacent with (s, t). This shows that $(x, y) \in sxt(G \circ H)$.

Case 2: H has no full degree vertices.

Let $(x, y) \in sxt(G \circ H)$. Choose a neighbour (u, v) of (x, y) such that (u, v) must be adjacent with all other neighbours of (x, y). If $u \neq x$, then all vertices in the layer ^uH are the neighbours of (x, y) and so (u, v) must be adjacent to all other vertices in the layer ^uH. This shows that v is a full degree vertex in H, which is impossible. Hence we can assume that $u = x$. Now we claim that y is a semi-extreme vertex in H. For let v' be any neighbour of y different from v. Then (x, v') is a neighbour of (x, y) and so (x, v) must be adjacent with (x, v') in $G \circ H$. This is possible only when v is adjacent to v' in H. Hence $y \in sxt(H)$.

On the other hand, suppose that $y \in sxt(H)$. we claim that $(x, y) \in sxt(G \circ H)$ for all $x \in V(G)$. For, let v be any neighbour of y such that v is adjacent to all other neighbours of y in H. Let (s, t) be any neighbour of (x, y) in $G \circ H$ such that $(s, t) \neq (x, v)$. If $s = x$, then t is a neighbour of y and so v must be adjacent with t in H. Hence the vertex (x, v) is adjacent to (s, t) in $G \circ H$. So assume that $s \neq x$. Then x and s are adjacent in G and so (s, t) is adjacent with (x, v) in $G \circ H$. This shows that $(x, y) \in sxt(G \circ H)$.

Corollary 4.3. *Let G and H be connected graphs of order n and m respectively. Then $eg(G \circ H) \geq n.sxt(H)$. Moreover, if H contains full degree vertices, then $eg(G \circ H) \geq nm - n + sxt(G)$.*

Proof. By Theorem 4.2, we have that $V(G) \times sxt(H) \subseteq sxt(G \circ H)$. Hence $eg(G \circ H) \geq sxt(G \circ H) = n.sxt(H)$. Moreover, if H contains full degree vertices, then $sxt(H) = m$ or $sxt(H) = m-1$. Hence we can assume that $sxt(H) \geq m-1$. Now the result follows from Theorem 4.2.

Corollary 4.4. *Let G and H be connected graphs such that H has no full degree vertices. Then $G \circ H$ is semi-complete graph if and only if H is a semi-complete graph.*

Remark 4.5. If H contains full degree vertices, then Corollary 4.4 need not be true. For consider $G = K_n$ and $H = K_{1,n}$. Then it follows from Theorem 4.2 that $sxt(G \circ H) = n(n+1)$ and so $G \circ H$ is semi-complete, whereas H is not semi-complete.

Proposition 4.6. *Let G and H be connected graphs such that H has a unique full degree vertex. Then $G \circ H$ is semi-complete if and only if G is semi-complete.*

Proof. Let n and m be the order of G and H respectively. Since H has a unique full degree vertex, we have that $sxt(H) = m - 1$. Hence by Theorem 4.2, we get $sxt(G \circ H) = mn - n + sxt(G)$. This shows that $G \circ H$ is semi-complete if and only if G is semi-complete.

Remark 4.7. If H contains more than one full degree vertices, then H is a semi-complete graph and hence by Theorem 4.2, $sxt(G \circ H) = V(G) \times sxt(H) = V(G \circ H)$ and so $G \circ H$ is semi-complete.

Remark 4.8. If H contains more than one full degree vertex, then the above proposition need not be true. For, consider $G = C_4$ and $H = K_3$, then $G \circ H$ is semi-complete, whenever G is not semi-complete.

Corollary 4.9. *For any positive integer $n \geq 2$, $eg(G \circ K_n) = n.|V(G)|$*

Corollary 4.10. *For integers $n \geq 2$ and $m \geq 3$, $eg(K_n \circ W_{1,m}) = n.(m+1)$*

Lemma 4.11. *Let G and H be connected graphs and let S be an edge geodetic set in $G \circ H$. Then for each $x \in G$, the set $S \cap V(^xH)$ is a $2-$edge geodetic set in the induced subgraph of $V(^xH)$ in $G \circ H$.*

Proof. Let $e = (x,y)(x,y')$ be any edge in ^xH. Since S is an edge geodetic set in $G \circ H$, we have that e lies on a $(u,v) - (u',v')$ geodesic P with $(u,v),(u',v') \in S$. Let the geodesic P be $P : (u,v) = (u_0,v_0),\dots,(u_i,v_i) = (x,y),(x,y') = (u_{i+1},v_{i+1}),\dots,(u_n,v_n) = (u',v')$. If $u_{i-1} \neq x$, then u_{i-1} must be adjacent with x and so (u_{i-1},v_{i-1}) is adjacent with $(x,y') = (x_{i+1},v_{i+1})$, which is impossible. Hence $u_{i-1} = x$. Continue this argument, we can prove that $u_0 = u_1 = \dots = u_i = \dots = u_n = x$ and hence $(u,v),(u',v') \in^x H$. By Theorem 4.1, $d((u,v),(u',v')) = 1$ or 2 and so $S \cap V(^xH)$ is a 2-edge geodetic set in the induced subgraph of ^xH.

Theorem 4.12. *Let G and H be two connected graphs. Then $neg_2(H) \leq eg(G \circ H) \leq neg_2(H) + eg(G)m - eg(G)eg_2(H)$, where n and m are the number of vertices of G and H respectively.*

Moreover, both the bounds are tight.

Proof. The lower bound is the direct consequence of Lemma 4.11. To show the sharpness of lower bound take $G = C_6$ and any graph H (see Theorem 4.14). Now, we prove the upper bound. Let S_1 be an edge geodesic set of G and S_2 be a 2-edge geodesic set of H. We claim that the set $S = (V(G) \times S_2) \cup (S_1 \times V(H)) - (S_1 \times S_2)$ is an edge geodesic set of $G \circ H$. For, let $e = (g_1, h_1)(g_2, h_2)$ be an edge in $G \circ H$.

Case 1: $g_1 = g_2 = g$ and $h_1 h_2 \in E(H)$
Since S_2 is the 2-edge geodetic set of H, there exist two vertices h and h' in S_2 such that $h_1 h_2$ is in $h - h'$ geodesic of length atmost 2. It follows from Theorem 4.1 that the corresponding path in the g-layer of H is a $(g, h_1) - (g, h_2)$ geodesic in $G \circ H$ containing the edge e.

Case 2: $g_1 g_2 \in E(G)$: Since S_1 is the edge geodetic set of G, there exist vertices g and g' in S_1 such that $g_1 g_2$ is in a $g - g'$ geodesic in G, say $P : g = u_0, u_1, \ldots, u_i = g_1, g_2 = u_{i+1}, \ldots, u_n = g'$. By Theorem 4.1, the path $Q : (g, h_1) = (u_0, h_1), (u_1, h_1), \ldots, (u_i, h_1) = (g_1, h_1), (g_2, h_2) = (u_{i+1}, h_2), \ldots, (u_n, h_2) = (g', h_2)$ is a $(g, h_1) - (g', h_2)$ geodesic in $G \circ H$ containing the edge e. This shows that S is an edge geodetic set in $G \circ H$. Hence the theorem.

To show the sharpness of this upper bound take $G = C_5$ and $H = C_4$ (see Theorem 4.14).

Lemma 4.13. *Let G and H be connected graphs and let $e = uv$ be an extreme edge in G. Then every edge geodetic set of $G \circ H$ contains all the vertices in the $u - $ layer of H or it contains all the vertices in the $v - $ layer of H.*

Proof. Let S be an edge geodetic set of $G \circ H$. Let $e = uv$ be an edge in G. Suppose that $V(^u H) \nsubseteq S$ and $V(^v H) \nsubseteq S$. Choose $(u, u') \in V(^u H)$ and $(v, v') \in V(^v H)$ such that $(u, u'), (v, v') \notin S$, Since S is an edge geodetic set, it follows that the edge in $G \circ H$ lies on a $x - y$ geodesic P with $x, y \in S$. We may assume that the geodesic P be $P : x = (x_0, y_0), (x_1, y_1), \ldots, (x_i, y_i) = (u, u'), (v, v') = (x_{i+1}, y_{i+1}), \ldots (x_n, y_n) = y$ with $1 \leq i \leq n - 2$ Since $l(P) \geq 3$, it follows from the definition of $G \circ H$ that $x_i = x_j$ for all $i \neq j$. Hence $\pi_G(P) : x_0, x_1, \ldots, x_i = u, v = x_{i+1}, \ldots x_n$ is a geodesic in G with $1 \leq i \leq n - 2$. This shows that uv is not an extreme edge in G, which is impossible. Hence $V(^u H) \subseteq S$ or $V(^v H) \subseteq S$.

Theorem 4.14. *For any positive integer $n \geq 4$,*
$$eg(C_n \circ H) = \begin{cases} 2|V(H)| + 2eg_2(H), & if \, n = 4; \\ 3|V(H)| + 2eg_2(H), & if \, n = 5; \\ n.eg_2(H), & if \, n \geq 6. \end{cases}$$

Proof. Let H be any connected graph of order m.
Case 1: $n = 4$.

Let the cycle C_4 be $C_4 = x_1, x_2, x_3, x_4$. Then $diam(G) = 2$ and so every edge in C_4 is an extreme edge. Hence it follows from Lemmas 4.11 and 4.13 that $eg(C_4 \circ H) \geq 2.|V(H)| + 2.eg_2(H)$. On the other hand, let $S = [x_1 \times V(H)] \cup [x_3 \times V(H)] \cup (x_2 \times T) \cup (x_4 \times T)$, where T is a minimum 2-edge geodetic set in H.

Then $|S| = 2.|V(H)| + 2eg_2(H)$. We claim that S is an edge geodetic set of $C_4 \circ H$. For, let $e = (x, y)(x', y')$ be any edge in $G \circ H$. First suppose that $x = x'$. Then y and y' are adjacent in H. If $x = x_1$ or $x = x_3$, then $(x, y)(x', y) \in S$. So assume that $x = x_2$. Since T is a 2-edge geodetic set, it follows that there exists vertices $v, v' \in T$ such that the edge yy' lies on a $v - v'$ geodetic set in T of length at most 2. This shows that the corresponding $(x, v), (x, v')$ path in the $x-$ layer of H in $C_4 \circ H$ is a geodesic in $C_4 \circ H$ containing the edge e with $(x, v), (x', v) \in S$.

Next suppose that $x \neq x'$. Then x and x' should be adjacent in G. Without loss of generality, we may assume that $x = x_1$ and $x' = x_2$. Let v be any vertex in H such that $v \in T$. Then the edge $e = (x_1, y)(x_2, y')$ lies in the $(x_1, y) - (x_3, v)$ geodesic $P : (x_1, y)(x_2, y')(x_3, v)$ in $G \circ H$ with $(x_1, y), (x_3, v) \in S$. This shows that S is an edge geodetic set in $C_4 \circ H$ so $eg(C_4 \circ H) = 2m + 2eg_2(H)$.

Case 2: $n = 5$. Let $C_5 : x_1, x_2, x_3, x_4, x_5, x_1$. Since $diam(C_5) = 2$, we have that every edge is an extreme edge. Then by Lemmas 4.11 and 4.13, we get $eg(C_5 \circ H) \geq 3m + 2eg_2(H)$.

Let $S = \{x_1, x_3, x_5\} \times V(H) \cup \{x_2, x_5\} \times T$, where T is a minimum 2-edge geodetic set in H. Then as in the previous case, we can prove that S is an edge geodetic set of $C_5 \circ H$ and so $eg(C_5 \circ H) = |S| = 3m + 2eg_2(H)$.

Case 3: $n \geq 6$.

In this case C_n has no extreme edges. Let $C_n = x_1, x_2, \ldots x_n, x_1$. By Lemma 4.11, we have that $eg(C_n \circ H) \geq n.eg_2(H)$. Now let $S = V(C_n) \times T$, where T is any minimum 2-edge geodetic set of H. Then $|S| = n.eg_2(H)$. We claim that S is an edge geodetic set in $C_n \circ H$. For let $(x, y)(x', y')$ be any edge in $C_n \circ H$. First suppose that $x = x'$, then y and y' are adjacent in H. Since T is a 2-edge geodetic set in H, there exists vertices $v, v' \in T$ such that the edge yy' lies on $v - v'$ geodesic of length at most 2 in H. This implies that the corresponding $(x, v), (x, v')$ geodesic in $C_n \circ H$ containing the edge e with $(x, v), (x, v') \in S$.

Next assume that $x \neq x'$. Then x and x' are adjacent in C_n. Without loss of generality, we may assume that $x = x_1$ and $x' = x_2$. Then the edge $x_1 x_2$ lies on the $x_n - x_3$ geodesic $P : x_n, x_1, x_2, x_3$ of length 3 in C_n. Now let v be any vertex in H such that $v \in T$. Then the edge $e = (x_1, y)(x_2, y')$ lies on the $(x_n, v) - (x_3, v)$ geodesic $P : (x_n, v)(x_1, y)(x_2, y)(x_3, v)$ in $C_n \circ H$ with $(x_n, v), (x_3, v) \in S$. Hence S is an edge geodetic set in $C_n \circ H$ and so $eg(C_n \circ H) = n.eg_2(H)$.

Theorem 4.15. *Let T be any tree of order n with k support vertices, then for any connected graph H, $eg(T \circ H) = k.|V(H)| + (n - k)eg_2(H)$.*

Proof. Since the tree T has k support vertices, it follows that T contains k independent end edges. Each end edge is an extreme edge. Thus T has k independent extreme edges and so it follows from Lemmas 4.11 and 4.13 that $eg(T \circ H) \geq k.|V(H)| + (n-k)eg_2(H)$. Now let $x_1, x_2, \ldots, x_k$ be the k support vertices and let $S' = \{x_1, x_2, \ldots, x_k\} \times V(H) \cup [(V(T) - \{x_1, x_2, \ldots, x_k\}) \times S']$, where S is a minimum 2-edge geodetic set in H. Then $|S'| = k.|V(H)| + (n-k)eg_2(H)$. Then as in the previous theorem, one can easily verify that S' is an edge geodetic set in $T \circ H$.

Theorem 4.16. *Let H be any connected graph. Then for any positive integer $n \neq 2$, $eg(K_n \circ H) = (n-1)|V(H)| + eg_2(H)$*

Proof. Since $diam(K_n) = 1$, we have that each edge is extreme and so it follows from Lemmas 4.11 and 4.13 that $eg(K_n \circ H) \geq (n-1)|V(H)| + eg_2(H)$. Let T be a minimum 2-edge geodetic set and let $x \in V(K_n)$. Then one can easily verify that the set $S = (V(K_n - x)) \times V(H) \cup (\{x\} \times T)$ is an edge geodetic set of $K_n \circ H$. Hence the result follows.

References

1. Anand, B.S., Changat, M., Klavžar, S., Peterin, I.: Convex sets in lexicographic products of graphs. Graphs Comb. **28**(1), 77–84 (2012)
2. Anand, B.S., Changat, M., Narasimha-Shenoi, P.G.: Helly and exchange numbers of geodesic and steiner convexities in lexicographic product of graphs. Discret. Math. Algorithms Appl. **7**(04), 1550049 (2015)
3. Anand, B.S., Changat, M., Peterin, I., Narasimha-Shenoi, P.G.: Some steiner concepts on lexicographic products of graphs. Discret. Math. Algorithms Appl. **6**(04), 1450060 (2014)
4. Brešar, B., Klavžar, S., Horvat, A.T.: On the geodetic number and related metric sets in cartesian product graphs. Discret. Math. **308**(23), 5555–5561 (2008)
5. Buckley, F., Harary, F.: Distance in Graphs. Addison-Wesley, Redwood City (1990)
6. Cáceres, J., Hernando, C., Mora, M., Pelayo, I.M., Puertas, M.L.: On the geodetic and the hull numbers in strong product graphs. Comput. Math. Appl. **60**(11), 3020–3031 (2010)
7. Cagaanan, G.B., Canoy, S.: On the geodetic covers and geodetic bases of the composition g [km]. Ars Comb. **79**, 33–45 (2006)
8. Chartrand, G., Harary, F., Zhang, P.: On the geodetic number of a graph. Networks **39**(1), 1–6 (2002)
9. Chartrand, G., Zhang, P.: Introduction to Graph Theory. McGraw-Hill, New York (2006)
10. Hammack, R.H., Imrich, W., Klavžar, S., Imrich, W., Klavžar, S.: Handbook of Product Graphs. CRC Press, Boca Raton (2011)
11. Harary, F., Loukakis, E., Tsouros, C.: The geodetic number of a graph. Math. Comput. Modell. **17**(11), 89–95 (1993)
12. Manuel, P., Klavžar, S., Xavier, A., Arokiaraj, A., Thomas, E.: Strong edge geodetic problem in networks. Open Math. **15**(1), 1225–1235 (2016)
13. Santhakumaran, A.P., Ullas Chandran, S.V.: The geodetic number of strong product graphs. Discuss. Math. Graph Theory **30**(4), 687–700 (2010)
14. Santhakumaran, A.P., Ullas Chandran, S.V.: The k-edge geodetic number of a graph. Utilitas Math. **88**, 119–137 (2012)
15. Santhakumaran, A.P., Ullas Chandran, S.V.: Comment on "Edge Geodetic Covers in Graphs". Proyecciones (Antofagasta) **34**(4), 343–350 (2015)
16. Santhakumaran, A.P., John, J.: Edge geodetic number of a graph. J. Discret. Math. Sci. Cryptogr. **10**(3), 415–432 (2007)
17. Santhakumaran, A.P., Ullas Chandran, S.V.: The edge geodetic number and Cartesian product of graphs. Discuss. Math. Graph Theory **30**(1), 55–73 (2010)

Burning Spiders

Sandip Das, Subhadeep Ranjan Dev$^{(\boxtimes)}$, Arpan Sadhukhan, Uma kant Sahoo, and Sagnik Sen

Indian Statistical Institute, Kolkata, India
info.subhadeep@gmail.com

Abstract. Graph burning is a graph process modeling the spread of social contagion. Initially all the vertices of a graph G are unburned. At each step an unburned vertex is put on fire and the fire from burned vertices of the previous step spreads to their adjacent unburned vertices. This process continues till all vertices are burned. The burning number $b(G)$ of the graph is the minimum number of steps required to burn all the vertices in the graph. The burning number conjecture by Bonato et al. states that for a connected graph G of order n, its burning number $b(G) \leq \lceil \sqrt{n} \rceil$. It is easy to observe that in order to burn a graph it is enough to burn its spanning tree. Hence it suffices to prove that for any tree T of order n, its burning number $b(T) \leq \lceil \sqrt{n} \rceil$. A spider S is a tree with one vertex of degree at least 3 and all other vertices with degree at most 2. Here we prove that for any spider S of order n, its burning number $b(S) \leq \lceil \sqrt{n} \rceil$.

1 Introduction

Graph burning is a process that captures the spread of social contagion and was introduced by Bonato et al. [1]. We first describe the process of burning a simple graph $G(V, E)$ of order n. Graph burning consists of discrete steps. Each vertex is either *burned* or *unburned*, once a vertex is burned it remains burned till the end. Initially all the vertices are unburned. In the first step we burn a vertex. At each subsequent step, first, a new unburned vertex is burned; second, the fire spreads from each burned vertex of the previous step to its neighboring unburned vertices. The process ends when all the vertices are burned. The *burning number*, denoted by $b(G)$, is the minimum number of steps taken for this process to end.

The burning problem asks, given a graph G and an integer $k \geq 2$, whether $b(G) \leq k$. Bonato et al. [2] proved that the burning problem is NP-complete even for spider graphs and path-forests.

An intuitive way to look at this process is to cover the vertices of the graph G by $b(G)$ balls of radius $0, 1, \ldots, b(G) - 1$, placed at appropriate vertices. A ball of radius r placed at a vertex v can cover vertices which are at a distance of at most r from v.

For $m, n > 1$, it is easy to see that $b(K_n) = 2$, $b(K_{m,n}) = 3$. A slightly complicated example is burning paths. For a path of n vertices, its burning number $b(P_n) = \lceil \sqrt{n} \rceil$ [1]. To see that $b(P_{n^2}) = n$, observe that the appropriately

© Springer International Publishing AG 2018
B. S. Panda and P. P. Goswami (Eds.): CALDAM 2018, LNCS 10743, pp. 155–163, 2018.
https://doi.org/10.1007/978-3-319-74180-2_13

chosen balls of radius $0, 1, \ldots, n - 1$ cover the path on n^2 vertices. Each ball of radius r covers exactly $2r + 1$ vertices, so in the path, the balls individually cover $1, 3, 5, \ldots, 2n - 1$ vertices whose sum equals n^2. Observe that each vertex here is covered by exactly one ball, so we cannot cover P_{n^2} by k balls of radius $0, 1, \ldots, k - 1$, with $k < n$; hence the bound is tight.

In order to burn a graph G it suffices to burn any of its spanning trees. Also $b(G) = \min\{b(T) \mid T \text{ is a spanning tree of } G\}$ [1]. The *burning number conjecture* [1] states that for any connected graph G with order n, its burning number $b(G) \leq \lceil \sqrt{n} \rceil$. Any tree T on n vertices is a subgraph of a tree T' on $n + 1$ vertices. By Corollary 2.8 of [1], $b(T) \leq b(T')$. So in order to settle this conjecture observe that it suffices to prove that for any natural number n and any tree of order n^2, its burning number $b(T) \leq n$. As a step in this direction, we prove that the burning number conjecture holds for spiders. A *spider S* is a tree with one vertex of degree at least 3 and the rest vertices with degree at most 2.

Theorem 1. *The burning number of a spider on n vertices is at most $\lceil \sqrt{n} \rceil$.*

Bonato et al. [1] proved that for a graph G of order n, $b(G) \leq 2\sqrt{n} - 1$. Recently Land and Lu [3] improved it to $b(G) \leq \lceil \frac{-3 + \sqrt{24n + 33}}{4} \rceil$. Amongst other works in this area, Mitsche et al. [4] gave bounds on burning number of various graph products.

In the rest of this section we give some preliminary definitions. In Sect. 2, we develop a technique to understand the burning process in certain disjoint paths. In Sect. 3, we prove the burning number conjecture for spiders using the results (Theorem 2 and Corollary 1) obtained in Sect. 2.

Preliminary Definitions. A *path-forest* is a disjoint union of a collection of paths. In a spider, the vertex with degree at least 3 is called the *head* of the spider. The disjoint paths obtained after deleting the head of the spider are called its *arms*. The *front vertex* of each arm is adjacent to the head of the spider. The length of each arm of the spider adds to $n - 1$, where n is the order of the spider. For $n \in \mathbb{N}$, let $[n] = \{1, 2, \ldots, n\}$. We follow the standard notation of West [6].

For a graph G, let $b(G) = k$, so the vertices of G can be burnt in k steps by placing fires at appropriate vertices. The first fire is represented by (k) as after k steps it burns vertices within the ball of radius $k - 1$, centered at the location of fire. So the process of burning by fires $(k), (k - 1), \ldots, (1)$ can be thought of as covering by balls of radius $\{k - 1, k - 2, \ldots, 0\}$, centered at the respective locations of fire.

2   Covering Certain Path-Forests

In this section we develop a method to understand how many steps we require to burn a specific type of path-forest which will help us in proving the burning number conjecture for spider graphs. Although Roshanbin [5] proved that for a path forest PF of order N and m components, $b(PF) \leq \sqrt{N} + m - 1$, we find exact burning number for specific type of path forests. To be more specific we need Theorem 2 and Corollary 1 presented in this section to prove the burning number conjecture for spider graphs. We need the following definitions.

Let N, m, n be positive integers such that $m \leq n$. The set $\mathcal{P} = \{a_i \mid i \in [m],$ $a_i \in \mathbb{N}\}$ is an (m, n)-partition of N if $\sum_{i \in [m]} a_i = N$ and for $j \in [m-1]$, $a_j \leq n$. But a_m can be greater than n. Let $S = \{1, 3, 5, \ldots, 2n - 1\}$. An (m, n)-partition $\mathcal{P}$ of N is solvable if for each element a_i of $\mathcal{P}$ we can assign a set $S_i \subseteq S$ such that $a_i \leq \sum_{s \in S_i} s$ and the sets $S_1, S_2, \ldots, S_m$ are all disjoint and their union equals the set S. We call set S_i as the covering set of a_i and the sets $\{S_1, S_2, \ldots, S_m\}$ together as the set of covering sets with respect to the (m, n)-partition of N.

N is said to be solvable with respect to (m, n) if every (m, n)-partition of N is solvable. In other words we can find a set of covering sets with respect to every (m, n)-partition of N. We define $I_{m,n}$ to be the maximum natural number which is solvable with respect to (m, n), for example $I_{2,4} = 15$, $I_{3,3} = 5$, $I_{3,4} = 12$ etc. In the rest of this section we find a general formula for calculating $I_{m,n}$ (refer Theorem 2). We begin with the following results.

Lemma 1. For all $N \leq I_{m,n}$, N is solvable with respect to (m, n).

Proof. For $N < m$, there is no (m, n)-partition of N, hence N is trivially solvable. Now for $m \leq N \leq I_{m,n}$, any (m, n)-partition $\mathcal{P} = \{a_i \mid \forall i \in [m], a_i \in \mathbb{N}\}$ of N can be reduced to an (m, n)-partition $\mathcal{P}' = \{a_i' \mid \forall i \in [m], a_i' \in \mathbb{N}\}$ of $I_{m,n}$ by setting $a_i' = a_i$, for all $i \in [m-1]$ and setting $a_m' = a_m + (I_{m,n} - N)$. Since, by definition, $I_{m,n}$ is solvable with respect to (m, n) therefore for each (m, n)-partition $\mathcal{P}'$ of $I_{m,n}$ there is a set of covering sets $S' = \{S_1, S_2, \ldots, S_m\}$ which covers it. From the construction of $\mathcal{P}'$ the set S' is also a set of covering sets for the partition $\mathcal{P}$. Therefore N is also solvable with respect to (m, n). □

Lemma 2. $I_{m,n} \leq n^2 - (m-1)^2$, $\forall m \leq n, n > 1, n, m \in \mathbb{N}$.

Proof. Consider an (m, n)-partition $\mathcal{P} = \{2, 2, 2, \ldots, 2, n^2 - m^2 + 2\}$ of $N = n^2 - (m-1)^2 + 1$. We claim that this partition is not solvable with respect to (m, n). We have $S = \{1, 3, 5, \ldots, 2n - 1\}$. Notice that to cover $a_i = 2$, for all $i \in [m-1]$, we need at least one number from $S \setminus \{1\}$. In the best case we have to use the numbers in $\{3, 5, 7, \ldots, 2m - 1\}$ to cover the a_i's, for all $i \in [m-1]$. Therefore total sum of the numbers used to cover all $a_i, i \in [m-1]$ is at least $m^2 - 1$. Our claim holds if a_m is greater than the sum of the remaining numbers i.e. $n^2 - m^2 + 2 > n^2 - (m^2 - 1)$. This holds trivially. So the given partition is not solvable with respect to (m, n). Hence by Lemma 1, $I_{m,n} \leq n^2 - (m-1)^2$. □

Lemma 3. Suppose for a particular (m, n)-partition $\{a_1, a_2 \ldots, a_m\}$ of $N = n^2 - (m-1)^2$, we can find a subset $S' = \{s_1, s_2, \ldots, s_{m-1}\}$ such that $S' \subseteq S = \{1, 3, 5 \ldots, 2n - 1\}$ and $\forall i \in [m-1]$, $0 \leq s_i - a_i \leq 2i - 1$, then the corresponding (m, n)-partition of $n^2 - (m-1)^2$ is solvable.

Proof. Let $\mathcal{P} = \{a_i \mid i \in [m]\}$ be any (m, n)-partition of $N = n^2 - (m - 1)^2$, and the set S' is as defined. As $m \leq n$, $S \setminus S'$ is not empty. $\mathcal{P}$ is solvable if the set $S \setminus S'$ acts as a covering set for a_m i.e. the sum of the elements in $S \setminus S'$ is at least a_m as then the set of covering sets for the (m, n) partition of N will be $\{S_1 = \{s_1\}, S_2 = \{s_2\}, \ldots, S_{m-1} = \{s_{m-1}\}, S_m = S \setminus S'\}$. Let us denote this sum of the elements in $S \setminus S'$ by Q and suppose, to the contrary, $Q < a_m$. Now,

$$s_i - a_i \leq 2i - 1, \quad \forall i \in [m - 1]$$

$$\Rightarrow \sum_{i=1}^{m-1} s_i \leq \sum_{i=1}^{m-1} a_i + \sum_{i=1}^{m-1} (2i - 1),$$

$$\Rightarrow \sum_{i=1}^{m-1} s_i \leq \sum_{i=1}^{m-1} a_i + (m - 1)^2,$$

$$\Rightarrow \sum_{i=1}^{m-1} s_i + Q < \sum_{i=1}^{m-1} a_i + (m - 1)^2 + a_m,$$

$$\Rightarrow n^2 < \sum_{i=1}^{m} a_i + (m - 1)^2,$$

$$\Rightarrow n^2 - (m - 1)^2 < \sum_{i=1}^{m} a_i.$$

This is a contradiction as $\mathcal{P} = \{a_i \mid i \in [m]\}$ is a (m, n)-partition of $N = n^2 - (m - 1)^2$. Hence the (m, n)-partition of N is solvable.   □

Theorem 2. $I_{m,n} = n^2 - (m - 1)^2$ *for* $1 \leq m \leq n$, $n > 1, n, m \in \mathbb{N}$.

Proof. For $m = 1$, the theorem trivially holds as $\sum_{i=1}^{n} (2i - 1) = n^2$; so suppose $m > 1$. Let $N = n^2 - (m - 1)^2$ and $\mathcal{P} = \{a_i \mid i \in [m], a_i \in \mathbb{N}\}$ be an arbitrary (m, n)-partition of N with $a_1 \leq a_2 \leq \ldots \leq a_{m-1} \leq n$. From definition it follows that, a_m can be greater than n and assume $a_{m-1} \leq a_m$. If we can prove that $\mathcal{P}$ is solvable with respect to (m, n), then we are done. Let $S = \{1, 3, 5, \ldots, 2n - 1\}$. Now we give an algorithm to partition set S into the covering sets S_i for each $a_i, i \in [m]$; thereby proving that $\mathcal{P}$ is solvable with respect to (m, n). We define $S'_i = S_1 \cup \ldots \cup S_{i-1}$ for all $2 \leq i \leq m$. Now we will construct the sets S_i, for all $i \in [m]$.

Choose s_1 to be the least odd number in S not less than a_1. Since $a_1 < n$ such a number always exists. Set $S_1 = \{s_1\}$. We continue this process and pick an element $s_i \in S \setminus S'_i$, such that $0 \leq (s_i - a_i) \leq 2i - 1$ and set $S_i = \{s_i\}$. Let i^* be the first index such that we cannot pick $s_{i^*} \in S \setminus S'_{i^*}$ with $0 \leq (s_{i^*} - a_{i^*}) \leq 2i^* - 1$

and let $M = \{2i - 1 \mid n \le 2i - 1 \le 2n - 1\}$. If $i^* = m$, then by Lemma 3, the corresponding (m, n)-partition of N is solvable. So we assume $i^* < m$ and this implies that $a_{i^*} \le n$.

Claim 1. $M \subseteq S'_{i^*}$ and hence $i^* > \lceil \frac{n}{2} \rceil$.

Proof of Claim 1: By definition of i^* there are no odd numbers in $[a_{i^*}, a_{i^*} + 2i^* - 1] \cap (S \setminus S'_{i^*})$. But since the interval $[a_{i^*}, a_{i^*} + 2i^* - 1]$ has i^* many odd numbers and to form $S \setminus S'_{i^*}$ we have only removed $i^* - 1$ odd numbers from S, therefore there is at least one odd number in the interval $[a_{i^*}, a_{i^*} + 2i^* - 1]$ which does not belong to S. Hence $a_{i^*} + 2i^* - 1 > 2n - 1$, for if $a_{i^*} + 2i^* - 1 \le 2n - 1$ then all odd numbers in the interval $[a_{i^*}, a_{i^*} + 2i^* - 1]$ belongs to $\{1, 3, 5, \ldots 2n - 1\} = S$. Now since $i^* < m$, $a_{i^*} \le n$, also we have $a_{i^*} + 2i^* - 1 > 2n - 1$, so $M \subseteq [a_{i^*}, a_{i^*} + 2i^* - 1]$. Hence all odd numbers in the set $M = \{2i - 1 \mid n \le 2i - 1 \le 2n - 1\}$ belongs to S'_{i^*}. Otherwise, if $c \in M$ and $c \notin S'_{i^*}$ then $c \in [a_{i^*}, a_{i^*} + 2i^* - 1]$ and $c \in S \setminus S'_{i^*}$, which is a contradiction to the definition of i^*. Hence $i^* > |M| = \lceil \frac{n}{2} \rceil$. This concludes the proof of the claim.

Now we give a strategy to find $S_j \subseteq S \setminus S'_j$ for $i^* \le j \le m$ such that $0 \le \sum_{s \in S_j} s - a_j \le 2j - 1$. To choose S_j, we first choose a subset $S_{j_1} \subseteq S \setminus S'_j$ such that $0 \le (\sum_{s \in S_{j_1}})s - a_j$ holds. If $\sum_{s \in S_{j_1}} s - a_j \le 2j - 1$ we are done, therefore assume $(\sum_{s \in S_{j_1}})s - a_j > 2j - 1$. From claim 1 it is clear that the numbers in the set $S \setminus S'_{j'}$ are all at most $n - 1$ for all $j' \ge i^*$ Hence we claim that we can delete some numbers from S_{j_1} such that the difference between sum of remaining numbers in S_{j_1} and a_j is at most $2j - 1$ and greater than or equal to 0.

Recall that $j \ge i^* > \lceil n/2 \rceil$, so $2j - 1 > n$. Since $0 \le (\sum_{s \in S_{j_1}})s - a_j$, we go on deleting numbers from S_{j_1}, till the sum of remaining numbers in S_{j_1} is less than a_j. Let s^* be the last number deleted and S_j^{left} denote the set of remaining numbers in S_{j_1}. Since $(\sum_{s \in S_j^{left}})s - a_j < 0$, $s^* \le n - 1$ and $2j - 1 > n$, $(\sum_{s \in S_j^{left}})s - a_j + s^* < n < 2j - 1$. This proves our claim.

Now fix $S_j = S_j^{left} \cup \{s^*\}$. So $0 \le (\sum_{s \in S_j} s) - a_j \le 2j - 1$. This is how we can choose S_j. If we can choose S_j for all $j \le m$, then this algorithm partitions set S into S_i's, for $i \in [m]$, such that $a_i \le \sum_{s \in S_i} s$, where $a_i \in \mathcal{P}$; thereby proving that $\mathcal{P}$ is solvable with respect to (m, n).

Otherwise, let $j = j^* \le m$ be the first index such that we cannot choose $S_{j^*} \subset S \setminus S'_{j^*}$ such that $0 \le (\sum_{s \in S_{j^*}} s) - a_{j^*} \le 2j^* - 1$. (Notice that j^* can be equal to i^* i.e. we cannot even choose subset S_{i^*}.) Let $S'' = S \setminus S'_{j^*}$ and $Q = \sum_{s \in S''} s$. Since we were unable to choose S_{j^*}, we were also unable to choose S_{j^*1}. This implies $Q < a_{j^*}$. Notice that Q might also be 0. So we have the following:

$$\sum_{j=1}^{j^*-1}\sum_{s\in S_j} s \leq \sum_{i=1}^{j^*-1} a_i + \sum_{i=1}^{j^*-1}(2i-1),$$

$$\Rightarrow \sum_{j=1}^{j^*-1}\sum_{s\in S_j} s \leq \sum_{i=1}^{j^*-1} a_i + (j^*-1)^2,$$

$$\Rightarrow \sum_{j=1}^{j^*-1}\sum_{s\in S_j} s + Q < \sum_{i=1}^{j^*-1} a_i + (j^*-1)^2 + a_{j^*},$$

$$\Rightarrow \sum_{j=1}^{j^*-1}\sum_{s\in S_j} s + Q < \sum_{i=1}^{j^*} a_i + (j^*-1)^2,$$

$$\Rightarrow n^2 < \sum_{i=1}^{j^*} a_i + (j^*-1)^2,$$

$$\Rightarrow n^2 - (j^*-1)^2 < \sum_{i=1}^{j^*} a_i.$$

So if we cannot choose S_{j^*} for some $i^* \leq j^* \leq m$, we always have the following,

$$n^2 - (j^*-1)^2 < \sum_{i=1}^{j^*} a_i.$$

Since

$$\sum_{i=1}^{j^*} a_i \leq N = n^2 - (m-1)^2,$$

we have

$$n^2 - (j^*-1)^2 < n^2 - (m-1)^2.$$

But this leads to a contradiction as $j^* \leq m$.

Therefore the algorithm will never fail. So it partitions set S into S_i's, for $i \in [m]$, such that $a_i \leq \sum_{s\in S_i} s$, where $a_i \in \mathcal{P}$, thereby proving that $\mathcal{P}$ is solvable with respect to (m, n). Since $\mathcal{P}$ was an arbitrary partition of N, we conclude that N is solvable with respect to (m, n).

Now using Lemma 2, we have $I_{m,n} = n^2 - (m-1)^2$. ☐

Let $m, n \in \mathbb{N}$ such that $m \leq n$ and $PF_{m,n}$ be the set of path-forests of order at most $n^2 - (m-1)^2$ having m components of which $m-1$ have order at most n. Observe that we can visualize a path of order k as the natural number k. Note that if we have n fires to burn a path then the first fire has the capacity of burning $2n - 1$ vertices, similarly the ith fire has the capacity to burn $2i - 1$ vertices, so the set $S = \{1, 3, 5, \ldots, 2n-1\}$ represents the burning capacities if we have n

fires to be put on a path. So if we have m disjoint paths of length $\{a_1, a_2, \ldots, a_m\}$ and n fires with $m \leq n$, it is easy to see that if $N = \sum_{i=1}^{m} a_i \leq I_{m,n}$ then we can burn it completely using at most n fires or steps. Covering an a_i by some subset S_i of S, means burning the path of length a_i with a specific sequence of burns. So if we correlate the results we have obtained into a graph theoretic form, we see that Theorem 2 implies any path-forest $PF_{m,n}$ can be burned using n steps or fires. Recall that in Lemma 2, we had given a partition of $n^2 - (m-1)^2 + 1$ that is not solvable with respect to (m, n). So we have the following corollary.

Corollary 1. *For positive integers $m \leq n$ and $PF_{m,n}$ be a path-forest of order at most $n^2 - (m-1)^2$ having m components of which $m-1$ have order at most n, then $b(PF_{m,n}) \leq n$.*

3   Proof of Theorem 1

Any spider S on $|S|$ vertices is a subgraph of a spider S' on $|S| + 1$ vertices. By Corollary 2.8 of [1], $b(S) \leq b(S')$. So in order to prove the theorem it suffices to prove that burning number of a spider of order $(n + 1)^2$ is at most $n + 1$. We use induction on n. The spider of order 4 can be burnt in 2 steps by placing fire (2) at its head and the fire (1) in the one remaining vertex. Now we assume that for some natural number $n \geq 2$ any spider of order n^2 has burning number at most n.

Consider a spider of order $(n + 1)^2$. Let v_0 be its head and $P_1, \ldots, P_m, \ldots, P_k$ be its k arms and let p_i denote the order of arm P_i. In each arm P_i the vertex adjacent to v_0 is called the front vertex of P_i, and the vertex at other end i.e. whose degree is 1, is called the end vertex of P_i. Without loss of generality $P_1, P_2, \ldots, P_m, \ldots, P_k$ are arranged such that for $i \in [m]$, $p_i > n$ and for $j \in [k] \setminus [m]$, $p_j \leq n$. Also $p_1 \leq p_2 \leq \cdots \leq p_m$. Clearly $\sum_{i=1}^{m} p_i \leq n^2 + 2n$.

If $p_m \geq 2n + 1$, then we select a sub-path of order $2n + 1$ from the end vertex of P_m and place the fire $(n + 1)$ at its center. After $n + 1$ steps it would burn this subpath completely. The rest of the spider is connected and has order $(n + 1)^2 - (2n + 1) = n^2$. By our hypothesis, this has burning number at most n. So spiders of order $(n + 1)^2$ with $p_m \geq 2n + 1$ will have burning number at most $n + 1$.

Henceforth we deal with the case when $p_m \leq 2n$. First we claim that $m \leq n$. Suppose to the contrary $m > n$; since $\forall i \in [m], p_i > n$, the number of vertices in the spider is at least $(n + 1)(n + 1) + 1 > (n + 1)^2$, which is a contradiction. This implies $m \leq n$.

Consider the case when $m < n$. We claim that burning number of such a spider is at most $n + 1$. We place the $(n + 1)$ fire on the head of the spider. It burns the head of the spider, all paths P_j for $j \in [k] \setminus [m]$, and n vertices of each P_i for $i \in [m]$. The remaining unburned vertices induce a path-forest with m paths each of which has order at most n. The order of this path-forest is at most $\min(mn, (n + 1)^2 - (mn + 1)) = \min(mn, n^2 - mn + 2n)$.

We claim that $\min(mn, n^2 - mn + 2n) \leq n^2 - (m - 1)^2$ for all $m < n$. For $n = 2, 3$ we can check the inequality by checking all pairs of m and n.

Now we consider $n \geq 4$. If $n \geq 2m - 2$, $\min(mn, n^2 - mn + 2n) = mn$. Then $mn \leq \left(\frac{n+2}{2}\right)n = \frac{n^2}{2} + n \leq \frac{n^2}{2} + \frac{n^2}{4} = n^2 - \frac{n^2}{4} \leq n^2 - (m-1)^2$. If $n < 2m - 2$, $\min(mn, n^2 - mn + 2n) = n^2 - mn + 2n$. Also $n \geq 4$ implies $m > 3$. Then $m + \frac{1}{m-2} < m + 1 \leq n$. So $n^2 - mn + 2n = n^2 - (m-2)n < n^2 - (m-2)\left(m + \frac{1}{m-2}\right) = n^2 - m(m-2) - 1 = n^2 - (m-1)^2$. So $\min(mn, n^2 - mn + 2n) \leq n^2 - (m-1)^2$ for all $m < n$.

By Corollary 1, the burning number of this path-forest is at most n. Hence the burning number of a spider of order $(n+1)^2$ with $p_m \leq 2n$ and $m < n$, is at most $n + 1$.

For the case when $m = n$, $\sum_{i=1}^{m} p_i \geq n(n+1)$. Therefore $\sum_{j=m+1}^{k} p_j \leq n + 1$ with $p_j \leq n$ for $j \in [k] \setminus [m]$. So if we place fire $(n+1)$ at the head, it alone burns at least $n^2 + 1$ vertices after $n + 1$ steps. The remaining $2n$ vertices will induce a path-forest of at most n components.

If $\sum_{j=m+1}^{k} p_j > 0$, then there are at most $2n - 1$ remaining vertices which induce a path-forest of at most n paths. Since $I_{n,n} = 2n - 1$, Corollary 1 implies that burning number of this path-forest is at most n, and hence burning number of this spider is at most $n + 1$.

So it suffices to consider the cases where $\sum_{j=n+1}^{k} p_j = 0$ i.e. $k = m = n$. Observe that $p_1 \not> n + 2$, else the spider will have more than $(n+1)^2$ vertices.

If $p_1 = n + 1$, then we place fire $(n+1)$ at the head vertex of P_1. After $n + 1$ steps this fire alone burns $n + 1 + 1 + (n-1)(n-1)$ vertices. The remaining $3n - 2$ vertices will induce a path-forest of at most $n - 1$ components, since P_1 is completely burnt. $I_{n-1,n} = 4n - 4 > 3n - 2$, for $n \geq 2$. We handle the case $n = 2$ in our base case of this induction hypothesis. Corollary 1 implies that burning number of this path-forest is at most n, and hence burning number of this spider is at most $n + 1$.

If $p_1 = n + 2$, then $p_1 = \cdots = p_m = n + 2$. We place the fire $(n+1)$ at the vertex in P_1, adjacent to its head vertex. After $n + 1$ steps, this fire alone burns $n + 2 + 1 + (n-2)(n-1)$ vertices. The remaining $4n - 4$ vertices will induce a path-forest of at most $n - 1$ paths, since P_1 is completely burnt. Since $I_{n-1,n} = 4n - 4$, Corollary 1 implies that burning number of this path-forest is at most n, and hence burning number of this spider is at most $n + 1$.

So burning number of a spider of order $(n+1)^2$ is at most $n + 1$. This completes the proof of Theorem 1.   □

4   Conclusion

In this article we prove that the burning number conjecture is true for spiders. To do so, we developed a method to cover certain disjoint paths of total order at most $I_{m,n}$, by intervals of odd lengths. We use this method several times in our proof. Another conducive factor used in the proof is the structural properties of spiders.

In case of trees once the vertices with degree at least three are burned, the remaining unburned vertices induce a path forest. Hence this path covering technique might be extendable to other classes of trees.

Acknowledgement. We thank the anonymous reviewers for their valuable comments and suggestions to improve the clarity of the paper.

References

1. Bonato, A., Janssen, J.C.M., Roshanbin, E.: How to burn a graph. Internet Math. **12**(1–2), 85–100 (2016)
2. Bonato, A., Janssen, J.C.M., Roshanbin, E.: Burning a graph is hard. ArXiv e-prints arXiv:1507.06524 (2015)
3. Land, M.R., Lu, L.: An upper bound on the burning number of graphs. In: Bonato, A., Graham, F.C., Prałat, P. (eds.) WAW 2016. LNCS, vol. 10088, pp. 1–8. Springer, Cham (2016). https://doi.org/10.1007/978-3-319-49787-7_1
4. Mitsche, D., Pralat, P., Roshanbin, E.: Burning graphs: a probabilistic perspective. Graph. Comb. **33**(2), 449–471 (2017)
5. Roshanbin, E.: Burning a graph as a model of social contagion. Ph.D. thesis, Dalhousie University (2016)
6. West, D.B.: Introduction to Graph Theory, 2nd edn. Pearson Education, London (2002)

Drawing Graphs on Few Circles
and Few Spheres

Myroslav Kryven[1](✉), Alexander Ravsky[2], and Alexander Wolff[1]🄳

[1] Universität Würzburg, Würzburg, Germany
myroslav.kryven@uni-wuerzburg.de
[2] Pidstryhach Institute for Applied Problems of Mechanics and Mathematics,
National Academy of Sciences of Ukraine, Lviv, Ukraine
alexander.ravsky@uni-wuerzburg.de

Abstract. Given a drawing of a graph, its *visual complexity* is defined as the number of geometrical entities in the drawing, for example, the number of segments in a straight-line drawing or the number of arcs in a circular-arc drawing (in 2D). Recently, Chaplick et al. [4] introduced a different measure for the visual complexity, the *affine cover number*, which is the minimum number of lines (or planes) that together cover a crossing-free straight-line drawing of a graph G in 2D (3D). In this paper, we introduce the *spherical cover number*, which is the minimum number of circles (or spheres) that together cover a crossing-free circular-arc drawing in 2D (or 3D). It turns out that spherical covers are sometimes significantly smaller than affine covers. Moreover, there are highly symmetric graphs that have symmetric optimum spherical covers but apparently no symmetric optimum affine cover. For complete, complete bipartite, and platonic graphs, we analyze their spherical cover numbers and compare them to their affine cover numbers as well as their segment and arc numbers. We also link the spherical cover number to other graph parameters such as chromatic number, treewidth, and linear arboricity.

1 Introduction

A drawing of a given graph can be evaluated by many different quality measures depending on the concrete purpose of the drawing. Classical examples are the number of crossings, the ratio between the lengths of the shortest and the longest edge, or the angular resolution. Clearly, different layouts (and layout algorithms) optimize different measures. Hoffmann et al. [12] studied ratios between optimal values of quality measures implied by different graph drawing styles. For some pairs of styles, they proved constant ratios; for others, they showed that the ratio is unbounded.

A few years ago, a new type of quality measure was introduced: the number of geometric objects that are needed to draw a graph given a certain style. Schulz [20] termed this measure the *visual complexity* of a drawing. More concretely, Dujmović et al. [6] defined the *segment number* seg(G) of a graph G to

© Springer International Publishing AG 2018
B. S. Panda and P. P. Goswami (Eds.): CALDAM 2018, LNCS 10743, pp. 164–178, 2018.
https://doi.org/10.1007/978-3-319-74180-2_14

be the minimum number of straight-line segments over all straight-line drawings of G. Similarly, Schulz [20] defined the *arc number* $\text{arc}(G)$ with respect to circular-arc drawings of G and showed that circular-arc drawings are an improvement over straight-line drawings not only in terms of visual complexity but also in terms of area consumption. Mondal et al. [17] showed how to minimize the number of segments in convex drawings of 3-connected planar graphs off and on the grid. Igamberdiev et al. [13] fixed a bug in the algorithm of Mondal et al. and compared the resulting algorithm to two other algorithms in terms of angular resolution, edge length, and face aspect ratio. Recently, Kindermann et al. [14] presented a user study showing that people without mathematical or computer science background prefer drawings that consist of few line segments, that is, drawings of low visual complexity (whereas people with such a background seem to prefer drawings that are more symmetric).

For this paper, the most important precursor is the work of Chaplick et al. [4] who introduced another measure for the visual complexity, namely the *affine cover number*. Given a graph G, they defined $\rho_d^l(G)$ to be the minimum number of l-dimensional affine subspaces that together cover a crossing-free straight-line drawing of G in d-dimensional space. It turned out that it suffices to investigate the parameters $\rho_2^1(G)$, $\rho_3^1(G)$, and $\rho_3^2(G)$. Among others, Chaplick et al. showed that the affine cover number can be asymptotically smaller than the segment number, constructing an infinite family of triangulations $(T_n)_{n>1}$ such that T_n has n vertices and $\rho_2^1(T_n) = O(\sqrt{n})$, but $\text{seg}(T_n) = \Omega(n)$. On the other hand, they showed that $\text{seg}(G) = O(\rho_2^1(G)^2)$ for any connected planar graph G.

Our contribution. Combining the approaches of Schulz and Chaplick et al., we introduce the *spherical cover number* $\sigma_d^l(G)$ of a graph G to be the minimum number of l-dimensional spheres in $\mathbb{R}^d$ such that G has a crossing-free circular-arc drawing that is contained in the union of these spheres. For $\sigma_2^1(G)$ we insist that G is planar. Note that any drawing with straight-line segments and circular arcs can be transformed into a drawing that uses circular arcs only.

Proposition 1. *Given a graph G and a drawing Γ of G that represents edges as straight-line segments or circular arcs on r l-dimensional planes or spheres in $\mathbb{R}^d$, there is a circular-arc drawing Γ' of G on r l-dimensional spheres in $\mathbb{R}^d$. In particular, $\sigma_d^l(G) \leq \rho_d^l(G)$ for any graph G and $1 \leq l < d$.*

Proof. Take an arbitrary sphere $S \subset \mathbb{R}^d$ whose center is not contained in any of the given spheres and which does not intersect any of the given planes. Invert the drawing with respect to S by the map $x \mapsto \rho x / \|x\|$, which assumes that S has radius ρ and is centered at the origin. The resulting drawing is a circular-arc drawing of G on r l-dimensional spheres in $\mathbb{R}^d$. Indeed, using basic properties of the inversion (see, for instance, [8] or [3, Chap. 5.1]), it can be proved that this inversion transforms planes into spheres of the same dimension and preserves spheres, in other words, the set of images of points on a sphere forms another sphere of the same dimension. □

Therefore, we may consider any line a "circle of infinite radius", any plane a "sphere of infinite radius", and any affine cover a spherical cover. By "line" we always mean a straight line.

Trivial bounds on $\sigma_3^1(G)$ follow from the fact that every circle is contained in a plane and that we have more flexibility for drawing in 3D than in 2D.

Proposition 2. *For any graph G, it holds that $\rho_3^2(G) \leq \sigma_3^1(G)$. If G is planar, we additionally have $\sigma_3^1(G) \leq \sigma_2^1(G)$.*

The spherical cover number $\sigma_3^2(G)$ can be considered a characteristic of a graph G that lies between its *thickness* $\theta(G)$, which is the smallest number of planar graphs whose union is G, and its *book thickness* bt(G), also called page number, which is the minimum number of pages (halfplanes) needed to draw the edges of G when the vertices lie on the *spine* of the book (the line that bounds all halfplanes).

Proposition 3. *For every graph G, it holds that $\theta(G) \leq \sigma_3^2(G) \leq \lceil \text{bt}(G)/2 \rceil$.*

Proof. Each sphere covers a planar subgraph of G, so $\sigma_3^2(G)$ is bounded from below by $\theta(G)$. On the other hand, given a book embedding of a graph G with the minimum number of pages (equal to bt(G)), we put the vertices from the spine along a circle which is the common intersection of $\lceil \text{bt}(G)/2 \rceil$ spheres. Then, for each page, we draw all its edges as arcs onto a hemisphere. Thus, we obtain a drawing witnessing $\sigma_3^2(G) \leq \lceil \text{bt}(G)/2 \rceil$. See Fig. 1 for an example.   □

We obtain bounds for the spherical cover number σ_3^2 of the complete and complete bipartite graphs which show that spherical covers can be asymptotically smaller than affine covers; see Table 1 and Sect. 2.

Then we turn to platonic graphs, that is, to 1-skeletons of platonic solids; see Sect. 3. These graphs possess several nice properties: they are regular, planar and Hamiltonian. We use them as indicators to compare the above-mentioned measures of visual complexity; we provide bounds for their segment and arc numbers (see Table 2) as well as for their affine and spherical cover numbers (see Table 3). For the lower bounds, we present straight-line drawings with (near-) optimal affine cover number ρ_2^1 and circular-arc drawings with optimal spherical cover number σ_2^1; see Figs. 3, 4 and 5. These illustrate another advantage of optimal spherical covers with respect to affine covers: potentially, the former better reflect the symmetry of the given graph.

For general graphs, we present lower bounds for the spherical cover numbers by means of many combinatorial graph characteristics, in particular, by the chromatic number, treewidth, balanced separator size, linear arboricity, and bisection width; see Sect. 4.

We decided to start with our more concrete (and partially stronger) results and postpone the structural observations to Sect. 4, although this means that we'll sometimes have to use forward references to Theorem 3, our main result in Sect. 4. Finally, we formulate an integer linear program that yields a lower bound for the segment number of a given graph; see Sect. 5. For the platonic solids, the lower bounds (see Table 4) turned out to be tight. We conclude with a few open problems.

Table 1. Lower and upper bounds on the three-dimensional line, plane, circle, and sphere cover numbers of K_n for any $n \geq 1$ and of $K_{p,q}$ for any $p, q \geq 3$.

G	$\rho_3^1(G)$	$\rho_3^2(G)$	$\sigma_3^1(G)$	$\sigma_3^2(G)$
K_n	$\binom{n}{2}$	$\frac{n^2-n}{12} \ldots \frac{n^2+5n+6}{6}$	$\lfloor \frac{n^2}{8} \rfloor \ldots \frac{n^2+5n+6}{6}$	$\lfloor \frac{(n+7)}{6} \rfloor \ldots \lceil \frac{n}{4} \rceil$
$K_{p,q}$	$pq - \lfloor \frac{p}{2} \rfloor - \lfloor \frac{q}{2} \rfloor$	$\lceil \frac{\min\{p,q\}}{2} \rceil$	$\lceil \frac{pq}{4} \rceil \ldots \lceil \frac{p}{2} \rceil \lceil \frac{q}{2} \rceil$	$\lceil \frac{pq}{2(p+q-2)} \rceil \ldots \lceil \frac{\min\{p,q\}}{2} \rceil$

Table 2. Bounds on $\mathrm{seg}(G)$ and $\mathrm{arc}(G)$. We obtained the lower bounds on $\mathrm{seg}(G)$ with the help of an integer linear program; see Table 4 in Sect. 5.

$G = (V, E)$	$|V|$	$|E|$	$|F|$	$\mathrm{seg}(G)$	Upper bd.	$\mathrm{arc}(G)$	Lower bd.	Upper bd.
Tetrahedron	4	6	4	6		3		
Octahedron	6	12	8	9	Fig. 2a	3	# int. pts	Fig. 2c
Cube	8	12	6	7	Fig. 3a	4	[6, Lemma 5]	Fig. 3d
Dodecahedron	20	30	12	13	Fig. 4a	10	[6, Lemma 5]	Fig. 4d
Icosahedron	12	30	20	15	Fig. 5a	7	Theorem 3(a)	Proposition 6

Table 3. Bounds on the affine cover numbers ρ_d^l and the spherical cover numbers σ_d^l for platonic graphs. The lower bounds on σ_2^1 and σ_3^1 stem from Theorem 3(a).

$G = (V, E)$		ρ_2^1	ρ_3^1	Lower bd.	Upper bd.	σ_2^1	σ_3^1	Upper bd.
Tetrahedron	T	6	6			3	3	
Octahedron	O	9	9	Proposition 5(a)	Fig. 2a	3	3	Fig. 2c
Cube	C	7	7	Proposition 5(b)	Fig. 3a	4	4	Fig. 3d
Dodecahedron	D	9...10	9...10	Proposition 5(c)	Fig. 4a	5	5	Fig. 4d
Icosahedron	I	13...15	13...15	Proposition 5(d)	Fig. 5a	7	7	Fig. 5c

2   Complete and Complete Bipartite Graphs

In this section we investigate the spherical cover numbers of complete graphs and complete bipartite graphs.

Theorem 1. *(a) For any $n \geq 3$, it holds that $\lfloor (n+7)/6 \rfloor \leq \sigma_3^2(K_n) \leq \lceil n/4 \rceil$.*
(b) For any $1 \leq p \leq q$, it holds that $pq/(2p + 2q - 4) \leq \sigma_3^2(K_{p,q}) \leq p$ and, if additionally $q > p(p-1)$, it holds that $\sigma_3^2(K_{p,q}) = \lceil p/2 \rceil$.

Proof. (a) By Proposition 3, $\theta(K_n) \leq \sigma_3^2(K_n) \leq \lceil \mathrm{bt}(K_n)/2 \rceil$. It remains to note that, e.g., Duncan [7] showed that $\theta(K_n) \geq \lfloor (n+7)/6 \rfloor$ and Bernhart and Kainen [2] showed that $\mathrm{bt}(K_n) = \lceil n/2 \rceil$.

(b) Again, it suffices to bound the values of the graph's thickness and book thickness. It can be easily shown that $\mathrm{bt}(K_{p,q}) \leq \min\{p, q\}$. On the other hand, Harary [11, Sect. 7, Theorem 8] showed that $\theta(K_{p,q}) \geq pq/(2p + 2q - 4)$. Due to Proposition 3, $\theta(K_{p,q}) \leq \sigma_3^2(K_{p,q}) \leq \min\{p, q\} \leq p$. In particular, if $q > p(p-1)$ then $\mathrm{bt}(K_{p,q}) = p$, due to Bernhart and Kainen [2, Theorem 3.5] and $\lceil pq/(2p + 2q - 4) \rceil = \lceil p/2 \rceil$, so in this case $\sigma_3^2(K_{p,q}) = \lceil p/2 \rceil$.   □

Theorem 1 implies that for any n-vertex graph $G = (V, E)$, $\sigma_3^2(G) \leq \lceil n/4 \rceil$. On the other hand, by Theorem 3(e), $\sigma_3^1(G) \geq \mathrm{bw}(G)$, where $\mathrm{bw}(G)$ is the *bisection width* of G, that is, the minimum number of edges between the two sets (W_1, W_2) of a *bisection* of V, that is, $|W_1| = \lceil n/2 \rceil$ and $|W_2| = \lfloor n/2 \rfloor$.

Proposition 4. *For any n, p, and q, $\mathrm{bw}(K_n) = \lfloor n^2/4 \rfloor$ and $\mathrm{bw}(K_{p,q}) = \lceil pq/2 \rceil$.*

Proof. Let (W, W') be a bisection of $V(K_n)$ with $|W| = \lfloor n/2 \rfloor$. Then the width of this bisection is $\lfloor n^2/4 \rfloor$. Now let (W, W') be a bisection of $K_{p,q}$ that contains r_1 vertices from the p-partition and r_2 vertices from the q-partition (with $r_1 + r_2 = \lfloor (p + q)/2 \rfloor$). Then the width of this bisection is $r_1(q - r_2) + r_2(p - r_1)$. The minimum of this value can be found by a routine calculation of the minimum of an integer quadratic polynomial. □

Theorem 2. *For any n, p, q, and $d \geq 3$, $\sigma_d^1(K_n) \geq \lfloor n^2/8 \rfloor$ and $\lceil pq/4 \rceil \leq \sigma_d^1(K_{p,q}) \leq \lceil p/2 \rceil \lceil q/2 \rceil$.*

Proof. The lower bounds follow from Theorem 3(e) and Proposition 4. The upper bound for $\sigma_d^1(K_{p,q})$ can be seen as follows. It suffices to consider the case $d = 3$. Let $p' = \lceil p/2 \rceil \geq p/2$ and $q' = \lceil q/2 \rceil \geq q/2$. Draw a bipartite graph $K_{2p',2q'} \supset K_{p,q}$ in 3D as follows. Let $V(K_{2p',2q'}) = P \cup Q$ be the natural bipartition of its vertices. Fix any family of p' distinct spheres with a common intersection circle. Place the $2q'$ vertices of Q on q' distinct pairs of antipodal points on the circle. Consider a line going through the center of the circle and orthogonal to its plane. Place the $2p'$ vertices of P into p' pairs of distinct intersection points of the line with the circles of the family, the points from each pair belonging to the same sphere. Now each pair of antipodal points in Q together with each pair of cospheric points in P determines a unique circle that contains all these points and provides a drawing of the four edges between them. The union of all these circles is the desired drawing of $K_{2p',2q'}$ onto $p'q'$ circles. □

By Propositions 1 and 2, $\rho_3^2(G) \leq \sigma_3^1(G) \leq \rho_3^1(G)$ for each graph G. For $\sigma_3^1(K_n)$, we can improve the upper bound $\rho_3^1(K_n) = \binom{n}{2}$ by using a combinatorial cover of K_n with copies of K_3: the proof of [4, Theorem 13] immediately implies that $(n^2 + 5n + 6)/6$ copies suffice. Now we can simply place the vertices of K_n in general position in 3D and draw each copy of K_3 as a circle.

Table 1 summarizes the known bounds for the affine cover numbers [4] and the new bounds for the spherical cover numbers of complete (bipartite) graphs.

3 Platonic Graphs

In this section we analyze the segment numbers, arc numbers, affine cover numbers, and spherical cover numbers of platonic graphs. We provide upper bounds via the corresponding drawings; see Figs. 2, 3, 4 and 5, and, for the more complicated icosahedron, in Proposition 6.

To lowerbound the spherical cover number σ_2^1 of the platonic graphs, we use a single combinatorial argument—Theorem 3(a); see Sect. 4. For the affine cover number ρ_2^1, a similar combinatorial arguments fails [4, Lemma 9(a)]. Therefore, we lowerbound ρ_3^1 (and, hence, also ρ_2^1) for each platonic graph individually; see Proposition 5. For an overview of our results, see Tables 2 and 3. The abbreviations that we use for the platonic graphs are listed in Table 3.

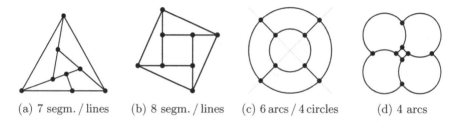

(a) 9 segm. / lines   (b) 4 arcs   (c) 3 arcs

Fig. 1. $\sigma_3^2(K_5) \leq 2$ **Fig. 2.** Drawings of the octahedron

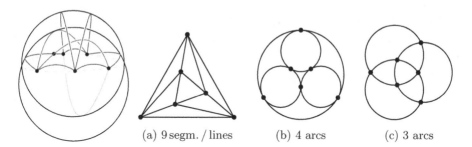

(a) 7 segm. / lines   (b) 8 segm. / lines   (c) 6 arcs / 4 circles   (d) 4 arcs

Fig. 3. Drawings of the cube

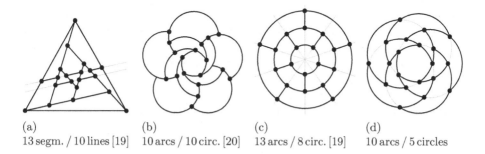

(a)   (b)   (c)   (d)
13 segm. / 10 lines [19]   10 arcs / 10 circ. [20]   13 arcs / 8 circ. [19]   10 arcs / 5 circles

Fig. 4. Drawings of the dodecahedron

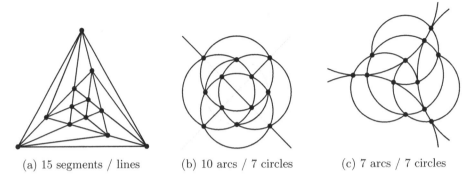

(a) 15 segments / lines   (b) 10 arcs / 7 circles   (c) 7 arcs / 7 circles

Fig. 5. Drawings of the icosahedron

Proposition 5. *(a) $\rho_3^1(O) \geq 9$; (b) $\rho_3^1(C) \geq 7$; (c) $\rho_3^1(D) \geq 9$; (d) $\rho_3^1(I) \geq 13$.*

Proof. (a) Consider a straight-line drawing of the octahedron O covered by a family $\mathcal{L}$ of ρ lines. Observe that every vertex of the octahedron is adjacent to every other except the opposite vertex. Therefore, no line in $\mathcal{L}$ can cover more than three vertices, otherwise the edges on the line would overlap. Hence, every line covers at most two edges, and these must be adjacent. Moreover, the two end vertices of these length-2 paths cannot be adjacent. Since there are only three pairs of such vertices, at most three lines cover two edges each. Since the octahedron has twelve edges, $\rho \geq 9$.

(b) Now consider a straight-line drawing of the cube C covered by a family $\mathcal{L}$ of ρ lines. We distinguish two cases.

Assume first that the drawing of the cube lies in a single plane. Each embedding of the cube contains two nested cycles, namely, the boundary of the outer face and the innermost face. We consider three cases depending on the shape of the outer face. (i) If the outer face is drawn as a convex quadrilateral, then none of the lines covering its sides can be used to cover the edges of the innermost cycle, therefore, it needs three additional lines. (ii) If the outer face is drawn as a non-convex quadrilateral, then we need three additional lines to cover the three edges going from its three convex angles to the innermost cycle. (iii) Now assume that the outer cycle is drawn as a triangle. Then none of the lines covering its sides can be used to cover the edges of the innermost cycle. If this cycle is drawn as a quadrilateral, then we need four additional lines to cover its sides. If the innermost cycle is drawn as a triangle, then we need three lines for the triangle and an additional line to cover the edge incident to the vertex of the innermost cycle which is not a vertex of the triangle. In each of the three cases (i)–(iii), we need at least seven lines to cover the cube.

Now assume that the drawing of the cube is not contained in a single plane. Then its convex hull has (at least) four extreme points. In order to cover the cube, we need at least one pair of intersecting lines of $\mathcal{L}$ for each vertex of the cube and at least three such pairs for each extreme point, that is, at least $4 + 4 \cdot 3 = 16$ pairs of intersecting straight lines in total. So, $\binom{\rho}{2} \geq 16$ and $\rho \geq 7$.

(c) Consider a straight-line drawing of the dodecahedron D covered by a family $\mathcal{L}$ of ρ lines. Again we distinguish two cases.

Assume first that the drawing of the dodecahedron lies in a single plane. To cover the edges on the outer face, we need a family $\mathcal{L}_0$ consisting of at least three lines. Again we make a case distinction depending on the shape of the outer face. (i) If the outer face is convex then none of them covers any of 15 vertices remaining in its interior. Thus each of these vertices is an intersection point of two lines of $\mathcal{L} \setminus \mathcal{L}_0$. Since $\mathcal{L} \setminus \mathcal{L}_0 \leq \rho - 3$, this family of lines can generate at most $\binom{\rho-3}{2}$ intersection points. Therefore, $\binom{\rho-3}{2} \geq 15$ and, hence, $\rho \geq 9$. (ii) Assume that the outer face is drawn as a non-convex quadrilateral. Then the drawing is contained in a convex angle opposite to the reflex angle. To cover the angle sides, we need a family $\mathcal{L}_0$ consisting of at least two lines. None of them covers any of the at least $15 + 1$ vertices remaining in the interior of the angle. Similarly to the previous paragraph, we obtain $\binom{\rho-2}{2} \geq 16$ and, hence, $\rho \geq 9$. (iii) Assume that the outer face is drawn as a pentagon P. Since the angle sum of a pentagon is 3π, P has at most two reflex angles, and therefore, at least three convex angles. Each vertex of D drawn as a vertex of a convex angle is an intersection point of (at least) three covering lines, because it has degree 3. There exists an edge e of P such that P is contained in one of the half-planes created by the line ℓ spanned by e (see, for instance, [18]). It is easy to check that ℓ can cover only edge e of the outer face of D. Then the family $\mathcal{L} \setminus \{\ell\}$ covers all edges of G but e. The angles of P incident to e are convex. Let v be a vertex of D drawn as a vertex of a convex angle not incident to e. In order to cover D, we need at least one pair of intersecting lines from $\mathcal{L} \setminus \{\ell\}$ for each vertex of D different from v and at least three such pairs for v, that is, at least $19 + 3 = 22$ pairs of intersecting lines in total. Therefore, $\binom{\rho-1}{2} \geq 22$ and, hence, $\rho \geq 9$. Note that, in each of the three cases (i)–(iii), we have $\rho \geq 9$.

Now assume that the drawing of D is not contained in a single plane. Then its convex hull has (at least) four extreme points. In order to cover D, we need at least one pair of intersecting lines of $\mathcal{L}$ for each vertex of D and at least three such pairs for each extreme point, that is, at least $16 + 4 \cdot 3 = 28$ pairs of intersecting lines in total. Therefore, $\binom{\rho}{2} \geq 28$. But if we have equality then any two lines of $\mathcal{L}$ intersect. So all of them share a common plane or a common point. In the first case the drawing is contained in a single plane; in the second case the family $\mathcal{L}$ cannot cover the drawing. Thus $\binom{\rho}{2} > 28$, and, hence, $\rho \geq 9$.

(d) If the drawing of the icosahedron I is not contained in a single plane, then we can pick four extreme points of the convex hull of the drawing. Each of these points represents a vertex of degree 5, so we need five lines to cover edges incident to this vertex, that is, 20 lines in total, but we have doublecounted the lines that go through pairs of the extreme points that we picked. Of these, there are at most $\binom{4}{2} = 6$. Thus we need at least $20 - 6 = 14$ lines to cover the drawing.

Now assume that there exists a straight-line drawing of the icosahedron in a single plane covered by a family $\mathcal{L}$ of twelve lines. Let u, v, w be the vertices of

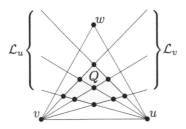

Fig. 6. The families $\mathcal{L}_v$ and $\mathcal{L}_u$

the outer face of I. Clearly, three distinct lines in $\mathcal{L}$ form the triangle uvw. For $s \in \{u, v, w\}$, we denote by $\mathcal{L}_s$ the lines in $\mathcal{L}$ that go through s and do not cover edges of the outer face. Since I is 5-regular, $|\mathcal{L}_s| = \deg(s) - 2 = 3$. Consider the set P of intersection points between the line families $\mathcal{L}_u$ and $\mathcal{L}_v$. The set P lies in the triangle uvw and is bounded by the quadrilateral Q formed by the outer pairs of lines in $\mathcal{L}_v$ and $\mathcal{L}_u$; see Fig. 6.

The quadrilateral Q is convex and eight of the nine points in P lie on the boundary of Q, hence, for any line ℓ in $\mathcal{L}_w$, we have $|\ell \cap P| \leq 3$. Observe that $|\ell \cap P| = 3$ implies that ℓ goes through the only point of P that lies in the interior of Q. Thus the lines in $\mathcal{L}_w$ can create at most seven triple intersection points with the lines in $\mathcal{L}_u$ and $\mathcal{L}_v$.

The icosahedron is 5-regular, so all vertices must be placed at the intersection of at least three lines. We need at least nine triple intersection points in order to place all $12 - 3$ inner vertices of the icosahedron—a contradiction. ☐

Proposition 6. $\mathrm{arc}(I) \leq 7$.

Proof. To construct the required drawing (see Fig. 7a), we first cover the edges of the icosahedron by seven objects, grouped into a single cycle K and two sets $L = \{L_0, L_1, L_2\}$ and $M = \{M_0, M_1, M_2\}$, where K is a cycle of length 6 and all elements of L and M are simple paths of length 4; see Fig. 7a. We identify the paths and cycles with their drawings as arcs and circles. For a set $S \in \{\{K\}, L, M\}$ and a number $i \in \{0, 1, 2\}$, let (d_S, α_{S_i}) be the polar coordinates of the center $c(S_i)$ of the circle of radius r_S that covers arc $S_i \in S$ (see Fig. 7b). We set the coordinates and radii as follows:

$$\alpha_K = 0 \qquad\qquad d_K = 0 \qquad\qquad r_K = 1$$

$$\alpha_{L_i} = i \cdot 2\pi/3 \qquad d_L = (3 + \sqrt{3})/2 \qquad r_L = \sqrt{5/2 + \sqrt{3}}$$

$$\alpha_{M_i} = \pi/2 + i \cdot 2\pi/3 \qquad d_M = (3 - \sqrt{3})/2 \qquad r_M = \sqrt{5/2 - \sqrt{3}}$$

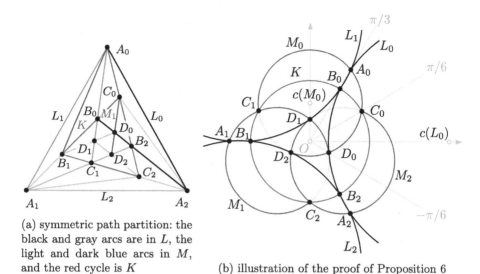

(a) symmetric path partition: the black and gray arcs are in L, the light and dark blue arcs in M, and the red cycle is K

(b) illustration of the proof of Proposition 6

Fig. 7. Bounding the arc number of the icosahedron

Using the law of cosines, it is easy to compute the intersection points:

$$\{A_i\} := L_i \cap L_{i+1} \cap M_i \qquad \Rightarrow A_i = \big(i \cdot 2\pi/3, (1+\sqrt{3})/2\big);$$
$$\{B_i\} := L_i \cap L_{i+1} \cap K \qquad \Rightarrow B_i = (i \cdot 2\pi/3, 1);$$
$$\{C_i\} := M_i \cap M_{i+2} \cap K \qquad \Rightarrow C_i = (\pi/3 + i \cdot 2\pi/3, 1);$$
$$\{D_i\} := L_i \cap M_i \cap M_{i+1} \qquad \Rightarrow D_i = \big(\pi/2 + i \cdot 2\pi/3, (\sqrt{3}-1)/2\big).$$

For $i = 0, 1, 2$, let L_i be the larger arc of the covering circle between the points A_i and B_i, let M_i be the larger arc of the covering circle between the points C_{i+1} and D_{i+2} (with indices modulo 3), and let K be the whole unit circle.   □

4   Lower Bounds for σ_d^1

Given a graph G, we obtain lower bounds for $\sigma_d^1(G)$ via standard combinatorial characteristics of G similarly to bounds for $\rho_d^1(G)$ [4]. In particular, we prove a general lower bound for $\sigma_d^1(G)$ in terms of the treewidth $\mathrm{tw}(G)$ of G, which follows from the fact that graphs with low parameter $\sigma_d^1(G)$ have small separators. This fact is interesting by itself and has yet another consequence: graphs with bounded vertex degree can have linearly large value of $\sigma_d^1(G)$ (hence, the factor of n in the trivial bound $\sigma_d^1(G) \leq m \leq n \cdot \Delta(G)/2$ is best possible).

We need the following definitions. The *linear arboricity* $\mathrm{la}(G)$ of a graph G is the minimum number of linear forests that partition the edge set of G [10]. Let $W \subseteq V(G)$. A set of vertices $S \subset V(G)$ is a *balanced W-separator* of the graph G if $|W \cap C| \leq |W|/2$ for every connected component C of $G - S$. Moreover, S is a *strongly balanced W-separator* if there is a partition $W \setminus S = W_1 \cup W_2$ such that $|W_i| \leq |W|/2$ for both $i = 1, 2$ and there is no path between W_1 and W_2 that avoids S. Let $\mathrm{sep}_W(G)$ ($\mathrm{sep}_W^*(G)$) denote the minimum k such that G has a (strongly) balanced W-separator S with $|S| = k$. Furthermore, let $\mathrm{sep}(G) = \mathrm{sep}_{V(G)}(G)$ and $\mathrm{sep}^*(G) = \mathrm{sep}_{V(G)}^*(G)$. Note that $\mathrm{sep}_W(G) \leq \mathrm{sep}_W^*(G)$ for any $W \subseteq V(G)$ and, in particular, $\mathrm{sep}(G) \leq \mathrm{sep}^*(G)$.

It is known [9, Theorem 11.17] that $\mathrm{sep}_W(G) \leq \mathrm{tw}(G) + 1$ for any $W \subseteq V(G)$. On the other hand, $\mathrm{tw}(G) \leq 3k$ if $\mathrm{sep}_W(G) \leq k$ for every W with $|W| = 2k + 1$.

Recall that the *bisection width* $\mathrm{bw}(G)$ of a graph $G = (V, E)$ is the minimum number of edges between two sets of vertices W_1 and W_2 with $|W_1| = \lceil n/2 \rceil$ and $|W_2| = \lfloor n/2 \rfloor$ partitioning V. Note that $\mathrm{sep}^*(G) \leq \mathrm{bw}(G) + 1$.

Now we use these graph parameters to lowerbound the spherical cover number. The proofs are similar to those regarding the affine cover number [4].

Theorem 3. *For any integer $d \geq 1$ and any graph G with n vertices and m edges, the following bounds hold:*

(a) $\sigma_d^1(G) \geq \frac{1}{2}\left(1 + \sqrt{1 + 2\sum_{v \in V(G)} \left\lceil \frac{\deg v}{2} \right\rceil \left(\left\lceil \frac{\deg v}{2} \right\rceil - 1\right)}\right)$,

(b) $\sigma_d^1(G) \geq \frac{1}{2}\left(1 + \sqrt{2m^2/n - 2m + 1}\right)$ *for any graph G with $m \geq n \geq 1$.*

(c) $\sigma_d^1(G) \geq \chi(G)/3$, *where $\chi(G)$ is the chromatic number of G;*

(d) $\lceil \frac{3}{2}\sigma_d^1(G) \rceil \geq \mathrm{la}(G)$;

(e) $\sigma_d^1(G) \geq \mathrm{bw}(G)/2$;

(f) $\sigma_d^1(G) > n/10$ *for almost all cubic graphs with n vertices;*

(g) $\sigma_d^1(G) \geq \mathrm{sep}_W^*(G)/2$ *for every $W \subseteq V(G)$;*

(h) $\sigma_d^1(G) \geq \mathrm{tw}(G)/6$.

Proof. The proofs for (a), (b), (c), (e), (g), and (h) are similar to the proofs of Lemma 9(a), Lemma 9(b), Theorem 8, Theorem 11(a), Theorem 11(c), and Theorem 11(d) in [4], respectively.

(d) Given the drawing of the graph G on $r = \sigma_d^1(G)$ circles, we remove an edge from each of the circles (provided such an edge exists), obtaining at (most) r linear forests. The removed edges we group into (possible, degenerated) pairs, obtaining at most $\lceil r/2 \rceil$ additional linear forests. So, $\mathrm{la}(G) \leq r + \lceil r/2 \rceil$.

(f) The proof is similar to that of Theorem 11(b) in [4]; the claim follows from (e) and from the fact that a random cubic graph on n vertices has bisection width at least $n/4.95$ with probability $1 - o(1)$ [15].   □

Corollary 1. $\sigma_d^1(G)$ *cannot be bounded from above by a function of $\mathrm{la}(G)$ or $v_{\geq 3}(G)$ or $\mathrm{tw}(G)$, where $v_{\geq 3}(G)$ is the number of vertices with degree at least 3.*

Proof. $\mathrm{la}(G)$: Akiyama et al. [1] showed that, for any cubic graph G, $\mathrm{la}(G) = 2$. On the other hand, $v_{\geq 3}(G) = n$, so $\sigma_3^1(G) > \sqrt{n}$ by Theorem 3(a). Theorem 3(f) yields an even larger gap.

$v_{\geq 3}(G)$: Let G be the disjoint union of k cycles. Then $v_{\geq 3}(G) = 0$. Clearly, an arrangement A of ℓ circles has at most ℓ^2 vertices. Each cycle of G "consumes" at least two vertices of A or a whole circle, so $\sigma_d^1(G) = \Omega(\sqrt{k})$.

tw(G): Let G be a caterpillar with linearly many vertices of degree 3. Then, tw$(G) = 1$. On the other hand, by Theorem 3(a), we have $\sigma_d^1(G) = \Omega(\sqrt{n})$.   □

Lemma 1. *A circular-arc drawing $\Gamma \subset \mathbb{R}$ of a graph G that contains k nested cycles cannot be covered by fewer than k circles.*

Proof. Fix any point inside the closed Jordan curve in Γ that corresponds to the innermost cycle of G. Let ℓ be an arbitrary line through this point. Then ℓ crosses at least twice each of the j Jordan curves that correspond to the k nested cycles in G. Hence, there are at least $2k$ points where ℓ crosses Γ.

On the other hand, consider any set of r circles whose union covers Γ. Then it is clear that ℓ crosses each of these r circles in at most two points, so there are at most $2r$ points where ℓ crosses Γ. Putting together the two inequalities, we get $r \geq k$ as desired.   □

At last we remark that $\sigma_3^1(G)$ is a lot smaller than $\sigma_2^1(G)$ for some graphs G.

Theorem 4. *For infinitely many n there is a planar graph G on n vertices with $\sigma_2^1(G) = \Omega(n)$ and $\sigma_3^1(G) = O(n^{2/3})$.*

Proof. We use the graph G of [4, Theorem 24(b)] with $\rho_2^1(G) = \Omega(n)$ and $\rho_3^1(G) = O(n^{2/3})$. The lower bound on $\sigma_2^1(G)$ follows from Lemma 1. The upper bound on $\sigma_3^1(G)$ follows from Proposition 1 for $l = 1$ and $d = 3$.   □

5   ILP Formulation for Optimal Segment Drawing

To obtain lower bounds on the segment numbers of planar graphs, we formulate an integer linear program (ILP). For the platonic graphs, it turns out that the bounds are tight; see Tables 2 and 4. Our ILP determines a locally consistent angle assignment [5] with the maximum number of π-angles between incident edges. Note that such an angle assignment is not necessarily realizable with straight-line edges in the plane. This is why the ILP yields only an upper bound for the number of π-angles—and a lower bound for the segment number.

Let $G = (V, E)$ be a 3-connected graph with fixed embedding given by a set $\mathcal{F}$ of faces and an outer face f_0. For any $v \in V$ and $f \in \mathcal{F}$, we introduce a variable $x_{v,f}$ with $0 < x_{v,f} \leq 2$ whose value is intended to be the angle of vertex v in face f divided by π. Thus $(\pi \cdot x_{v,f})_{v \in V, f \in \mathcal{F}}$ is an angle assignment. The following constraints guarantee that the assignment is locally consistent.

$$\sum_{f \sim v} x_{v,f} = 2 \qquad \text{for each } v \in V$$

$$\sum_{v \sim f} x_{v,f} = \deg f - 2 \qquad \text{for each } f \in \mathcal{F} \setminus \{f_0\}$$

$$\sum_{v \sim f_0} x_{v,f_0} = \deg f_0 + 2$$

For any vertex v, let $L_v = \langle v_1, \ldots, v_k \rangle$ be the vertices adjacent to v, in clockwise order as they appear in the embedding. Due to the 3-connectivity of G, any two consecutive vertices v_t, v_{t+1} adjacent to v uniquely define a face $f \sim v, v_t, v_{t+1}$ [16, Lemma 2]. We express the angle between two adjacent vertices v_i, v_j, $i < j$ of v as the sum of the angles assigned to the faces incident to v between v_i and v_j:

$$\angle(v_i v v_j) = \pi \cdot \sum_{t=i}^{j-1} x_{v, f \sim v, v_t, v_{t+1}}.$$

We want to maximize the number of π-angles between any two edges incident to the same vertex. To this end, we introduce a 0–1 variable s_{v, v_i, v_j} for any vertex v with two different neighbors v_i and v_j. The intended meaning of $s_{v, v_i, v_j} = 1$ is that $\angle(v_i v v_j) = \pi$. We add the following constraints to the ILP:

$$\frac{\angle(v_i v v_j)}{\pi} \leq 2 - s_{v, v_i, v_j} \qquad \text{for each } v \in V; v_i, v_j \in L_v \text{ with } i < j \qquad (1)$$

$$s_{v, v_i, v_j} \leq \frac{\angle(v_i v v_j)}{\pi} \qquad \text{for each } v \in V; v_i, v_j \in L_v \text{ with } i < j \qquad (2)$$

If $\angle(v_i v v_j) > \pi$, constraint (1) will force s_{v, v_i, v_j} to be 0 and constraint (2) will not be effective. If $\angle(v_i v v_j) < \pi$, constraint (2) will force s_{v, v_i, v_j} to be 0 and constraint (1) will not be effective. Only if $\angle(v_i v v_j) = \pi$, both constraints will allow s_{v, v_i, v_j} to be 1.

To obtain a balanced angle assignment, we introduce the following additional variables $\alpha_l, \alpha_u \in (0, 2)$ which are intended to describe the smallest and the largest angles in the angle assignment, respectively. This is achieved as follows.

$$x_{v, f} \geq \alpha_l \qquad \text{for each } v \in V \text{ and } f \in \mathcal{F}$$

$$x_{v, f} \leq \alpha_u \qquad \text{for each } v \in V \text{ and } f \in \mathcal{F}$$

Primarily, we want to maximize the number of π-angles between incident edges, that is, $\text{ang}_\pi(G) := \sum_{v \in V} \sum_{v_i, v_j \in L_v, i < j} s_{v, v_i, v_j}$. As a secondary objective, we want to maximize the angle resolution. The following linear objective function achieves both the primary and secondary objective:

$$\text{maximize} \quad \text{ang}_\pi(G) + (\alpha_l - \alpha_u)/2$$

For every π-angle between incident edges, we can use an already drawn segment to accommodate another edge; hence,

$$\text{seg}(G) = |E| - \text{ang}_\pi(G). \qquad (3)$$

The ILP gives an upper bound on $\text{ang}_\pi(G)$, thus Eq. (3) provides a lower bound for the segment number $\text{seg}(G)$. The experimental results for the platonic graphs are displayed in Table 4.

Table 4. Upper bounds on $\text{ang}_\pi(G)$ and corresponding lower bounds on $\text{seg}(G)$ obtained by the ILP and sizes of the ILP formulation. Runtimes where measured on a 64-bit machine with 7.7 GB main memory and four Intel i5 cores with 1.90 GHz, using the ILP solver IBM ILOG CPLEX Optimization Studio 12.6.2.

Graph G	Octahedron	Cube	Dodecahedron	Icosahedron
$\text{ang}_\pi(G) \leq$	3	5	17	15
$\text{seg}(G) \geq$	9	7	13	15
Variables	62	50	122	182
Constraints	185	162	387	515
Runtime [s]	0.2	0.2	0.2	2.8

6 Open Problems

What are optimal affine covers for the icosahedron and the dodecahedron? We conjecture that $\rho_3^1(I) = 15$ and $\rho_3^1(D) = 10$.

We have already seen that $\sigma_3^2(K_n)$ grows asymptotically more slowly than $\rho_3^2(K_n)$. Is there a family of planar graphs where σ_2^1 grows asymptotically more slowly than ρ_2^1?

Chaplick et al. [4] showed that the affine hierarchy "collapses" in the sense that, for any integers $1 \leq l \leq d$, $d \geq 3$, and for any graph G, it holds that $\rho_d^l(G) = \rho_3^l(G)$. Does the spherical hierarchy collapse, too?

References

1. Akiyama, J., Exoo, G., Harary, F.: Covering and packing ingraphs III: cyclic and acyclic invariants. Math. Slovaca **30**, 405–417 (1980)
2. Bernhart, F., Kainen, P.C.: The book thickness of a graph. J. Combin. Theory Ser. B **27**(3), 320–331 (1979). http://www.sciencedirect.com/science/article/pii/0095895679900212
3. Brannan, D.A.: Geometry. Cambridge University Press, Cambridge (1999)
4. Chaplick, S., Fleszar, K., Lipp, F., Ravsky, A., Verbitsky, O., Wolff, A.: Drawing graphs on few lines and few planes. In: Hu, Y., Nöllenburg, M. (eds.) GD 2016. LNCS, vol. 9801, pp. 166–180. Springer, Cham (2016). https://doi.org/10.1007/978-3-319-50106-2_14
5. Di Battista, G., Eades, P., Tamassia, R., Tollis, I.G.: Graph Drawing: Algorithms for the Visualization of Graphs. Prentice Hall, Upper Saddle River (1999)
6. Dujmović, V., Eppstein, D., Suderman, M., Wood, D.: Drawings of planar graphs with few slopes and segments. Comput. Geom. Theory Appl. **38**, 194–212 (2007)
7. Duncan, C.A.: On graph thickness, geometric thickness, and separator theorems. Comput. Geom. Theory Appl. **44**(2), 95–99 (2011). http://www.sciencedirect.com/science/article/pii/S0925772110000738
8. Edelsbrunner, H.: Lecture notes for Computational Topology (CPS296.1) (2006). http://www.cs.duke.edu/courses/fall06/cps296.1/Lectures/sec-III-3.pdf
9. Flum, J., Grohe, M.: Parametrized Complexity Theory. Springer, Heidelberg (2006). https://doi.org/10.1007/3-540-29953-X

10. Harary, F.: Covering and packing in graphs I. Ann. N.Y. Acad. Sci. **175**, 198–205 (1970)
11. Harary, F.: A Seminar on Graph Theory. Dover Publications, New York (2015)
12. Hoffmann, M., van Kreveld, M., Kusters, V., Rote, G.: Quality ratios of measures for graph drawing styles. In: CCCG 2014, pp. 33–39 (2014). http://www.cccg.ca/proceedings/2014/papers/paper05.pdf
13. Igamberdiev, A., Meulemans, W., Schulz, A.: Drawing planar cubic 3-connected graphs with few segments: algorithms and experiments. In: Di Giacomo, E., Lubiw, A. (eds.) GD 2015. LNCS, vol. 9411, pp. 113–124. Springer, Cham (2015). https://doi.org/10.1007/978-3-319-27261-0_10
14. Kindermann, P., Meulemans, W., Schulz, A.: Experimental analysis of the accessibility of drawings with few segments. In: Frati, F., Ma, K.L. (eds.) GD 2017. LNCS, vol. 10692. Springer (to appear, 2018). arxiv.org/abs/1708.09815
15. Kostochka, A.V., Melnikov, L.S.: On a lower bound for the isoperimetric number of cubic graphs. In: Proceedings of the 3rd International Petrozavodsk Conference on Probabilistic Methods in Discrete Mathematics, pp. 251–265. TVP, VSP, Moskva, Utrecht (1993)
16. Kryven, M., Ravsky, A., Wolff, A.: Drawing graphs on few circles and few spheres. ArXiv e-print arxiv.org/abs/1709.06965 (2017)
17. Mondal, D., Nishat, R.I., Biswas, S., Rahman, M.S.: Minimum-segment convex drawings of 3-connected cubic plane graphs. J. Comb. Opt. **25**(3), 460–480 (2013). https://doi.org/10.1007/s10878-011-9390-6
18. Moscow Mathematical Olympiad, problem no. 78223 (1960). http://www.problems.ru/view_problem_details_new.php?id=78223 (in Russian)
19. Scherm, U.: Minimale Überdeckung von Knoten und Kanten in Graphen durch Geraden. Bachelor's Thesis, Institut für Informatik, Universität Würzburg (2016)
20. Schulz, A.: Drawing graphs with few arcs. J. Graph Alg. Appl. **19**(1), 393–412 (2015)

On a Lower Bound for the Eccentric Connectivity Index of Graphs

Devsi Bantva$^{(\boxtimes)}$

Lukhdhirji Engineering College, Morvi 363 642, Gujarat, India
devsi.bantva@gmail.com

Abstract. The eccentric connectivity index of a graph G, denoted by $\xi^c(G)$, defined as $\xi^c(G) = \sum_{v \in V(G)} \epsilon(v) \cdot d(v)$, where $\epsilon(v)$ and $d(v)$ denotes the eccentricity and degree of a vertex v in a graph G, respectively. The volcano graph $V_{n,d}$ is a graph obtained from a path P_{d+1} and a set S of $n - d - 1$ vertices, by joining each vertex in S to a central vertex/vertices of P_{d+1}. In [4], Morgan *et al.* proved that $\xi^c(G) \geq \xi^c(V_{n,d})$ for any graph of order n and diameter $d \geq 3$. In this paper, we present a short and simple proof of this result by considering the adjacency of vertices in graphs.

Keywords: Eccentricity (in graph) · Eccentric connectivity index
Volcano graph

1 Introduction

Let G be a finite, connected and undirected graph without loops and multiple edges. We denote the *vertex set* of G by $V(G)$. The *distance* between two vertices u and v of G, denoted by $d(u, v)$, is the least length of a $u, v-$path in G. The *eccentricity* of a vertex v in a graph G, denoted by $\epsilon(v)$, is the distance of a vertex farthest from v in G. The *degree* of a vertex v in a graph G, denoted by $d(v)$, is the number of edges incident to it.

A topological index is a numerical graph invariants used for Quantitative Structure-Activity Relationship (QSAR) and Quantitative Structure-Property Relationship (QSPR) studies. The Wiener index, introduced in 1947 by Herold Wiener [9], is the first non-trivial topological index in Chemistry. Then after many topological indices have been defined such as Zagreb index, PI-index etc. and successfully used to study the chemical, pharmaceutical and other properties of molecules. More recently, a new adjacent-cum-distance based topological index, the *eccentric connectivity index* (or ECI for short) denoted by $\xi^c(G)$ has been introduced by Sharma *et al.* [7] which is defined as

$$\xi^c(G) = \sum_{v \in V(G)} \epsilon(v) \cdot d(v) \tag{1}$$

© Springer International Publishing AG 2018
B. S. Panda and P. P. Goswami (Eds.): CALDAM 2018, LNCS 10743, pp. 179–187, 2018.
https://doi.org/10.1007/978-3-319-74180-2_15

The eccentric connectivity index has been employed successfully for the development of numerous mathematical models for the prediction of biological activities of diverse nature. Recently, many results on the eccentric connectivity index have been obtained and some of them have been applied as means for modeling chemical, pharmaceutical and other properties of molecules, for details see [1,2,5–7,10].

The common trend of research for the topological indices and its variants is to determine the extremal graphs for the given topological index or its variant. Also the trend is to determine the extremal trees for the given topological index or its variant. For most of introduced topological indices or its variants, the extremal graphs or trees are determined except Wiener index; the origin of all topological indices and its variants. The same approach is considered by Morgan *et al.* for the eccentric connectivity index of graphs in [3,4].

In [3], Morgan *et al.* noted the eccentric connectivity index of some basic graph families and determined the eccentric connectivity index of other three classes of graphs namely broom graph $B_{n,d}$ (a graph which consists a path P_d, together with $(n - d)$ end vertices all adjacent to the same end vertex of P_d), lollipop graph $L_{n,d}$ (a graph obtained from a complete graph K_{n-d} and a path P_d, by joining one of the end vertices of P_d to all the vertices of K_{n-d}) and volcano graph $V_{n,d}$ (a graph obtained from a path P_{d+1} and a set S of $n - d - 1$ vertices, by joining each vertex in S to a central vertex/vertices of P_{d+1}). Note that for a fixed value of n, when d is even, the volcano graph $V_{n,d}$ is unique; whereas when d is odd, there may be several non-isomorphic volcano graphs $V_{n,d}$. The readers are advised to see [3,4] for figures and details on these graph families. The eccentric connectivity index of path P_{d+1} and volcano graph $V_{n,d}$ is given as follows.

Proposition 1. *Let $n \geq 2$ be an integer. Then*

$$\xi^c(P_n) = \begin{cases} \frac{1}{2}(3n^2 - 6n + 4), & \text{for } n \text{ even,} \\ \frac{3}{2}(n - 1)^2, & \text{for } n \text{ odd.} \end{cases} \tag{2}$$

Proposition 2. *Let n, d be non-negative integers. Then*

$$\xi^c(V_{n,d}) = \begin{cases} nd + n + \frac{d^2}{2} - 2d - 1, & \text{for } d \text{ even,} \\ nd + 2n + \frac{d^2}{2} - 3d - \frac{3}{2}, & \text{for } d \text{ odd.} \end{cases} \tag{3}$$

In [3], Morgan *et al.* gave a lower bound for the eccentric connectivity index of trees and proved that $\xi^c(T) \geq \xi^c(V_{n,d})$ for any tree T of order $n \geq 3$ and diameter d. Later, in [4], Morgan *et al.* extended this lower bound for the eccentric connectivity index to arbitrary connected graph of order n and diameter $d \geq 3$ but we emphasize that the proof is lengthy and complicated. In this paper, we give a short proof of this result by considering the connectivity of vertices in graph which is a very simple approach. Moreover, this approach can also be extend to prove similar types of results for other topological indices.

2   Preliminaries

We follow [8] for graph theoretic definition and notation. A tree T is a connected graph that contains no cycle. A caterpillar is a tree in which all the vertices are within distance one from central path. The diameter of a graph G, denoted by $\mathrm{diam}(G)$ or simply d, is $\max\{d(u,v) : u,v \in V(G)\}$. The *degree* of a vertex in a graph G, denoted by $\mathrm{d}(v)$, is defined as the number of edges incident to it. Let $H \subseteq V(G)$ then for any $v \in V(G)$, define $\mathrm{d}(v|H)$ is the degree of a vertex v in a subgraph induced by H of G. Note that if $H = V(G)$ then $\mathrm{d}(v|H) = \mathrm{d}(v)$; otherwise $\mathrm{d}(v|H) \leq \mathrm{d}(v)$. The center of a graph G, denoted by $C(G)$, is the set of vertices with minimum eccentricity. Note that for any $v \in V(G), \epsilon(v) \geq \lceil d/2 \rceil$ and $\epsilon(v) = \lceil d/2 \rceil$ if and only if $v \in V(C(G))$. We notice the following well known results about center of graphs.

Proposition 3. *The center $C(G)$ of a graph G is contained in a block of G.*

Proposition 4. *The center $C(T)$ of a tree T consists of a single vertex or two adjacent vertices.*

The following lemma characterize the vertices with minimum eccentricity in a graph G which is useful for our main result.

Lemma 1. *Let G be any graph and $P_{d+1} = v_0 - v_1 - \ldots - v_d$ be a fixed diametral path joining v_0 and v_d.*

(a) *If $\epsilon(v) = d/2$ for $v \in V(G \setminus P_{d+1})$ then v is on other diametral path joining v_0 and v_d, where $C(P_{d+1}) = \{w\}$.*
(b) *If $\epsilon(v) = (d+1)/2$ for $v \in V(G \setminus P_{d+1})$ then either v is on other diametral path joining v_0 and v_d or v is adjacent to both w and w', where $C(P_{d+1}) = \{w, w'\}$.*

Proof. (a) Let G be any graph and $P_{d+1} = v_0 - v_1 - \ldots - v_d$ be a diametral path joining v_0 and v_d such that $C(P_{d+1}) = \{w\}$. Let $v \in V(G \setminus P_{d+1})$ such that $\epsilon(v) = d/2$. If possible then assume that v is not on any diametral path joining v_0 and v_d. Then it is clear that either $d(v, v_0) > d/2$ or $d(v, v_d) > d/2$; otherwise the shortest path $P' = v_0 - \ldots - v - \ldots - v_d$ is a diametral path as $\epsilon(v) = d/2, d(v_0, v_d) = d$ and $\epsilon(u) \geq d/2$ for any $u \in V(G)$. But note that one of $d(v, v_0) > d/2$ or $d(v, v_d) > d/2$ gives $\epsilon(v) > d/2$, a contradiction. Hence v is on diametral path joining v_0 and v_d.

(b) Let G be any graph and $P_{d+1} = v_0 - v_1 - ... - v_d$ be a diametral path joining v_0 and v_d such that $C(P_{d+1}) = \{w, w'\}$. It is clear that if v is adjacent to both w and w' then $\epsilon(v) = (d+1)/2$. So assume that $\epsilon(v) = (d+1)/2$ for some $v \in V(G \setminus P_{d+1})$ and v is not adjacent to both w and w' then as in case (a) one can prove that v is on other diametral path joining v_0 and v_d.

Lemma 2. *Let G be any graph and $v \in V(G)$ such that $\epsilon(v) = \lceil d/2 \rceil$ then* $\mathrm{d}(v) \geq 2$.

Proof. By Lemma 1 if $\epsilon(v) = \lceil d/2 \rceil$ then either $v \in C(P_{d+1})$ or v is on other diametral path joining v_0 and v_d where $P_{d+1} = v_0 - v_1 - ... - v_d$ is a diametral path joining v_0 and v_d or v is adjacent to both w and w' in the case when $C(P_{d+1}) = \{w, w'\}$. Hence in any case $\mathrm{d}(v) \geq 2$.

3   Main Result

In this section, we continue to use the terminology and notation defined in previous section. First we prove the following Theorem.

Theorem 1. *Let $G = (V, E)$ be a connected graph of order n, and diameter $d \geq 3$ such that every spanning tree T of G is a caterpillar of diameter d. Then*

$$\xi^c(G) \geq \xi^c(V_{n,d}). \tag{4}$$

Proof. Let G be a connected graph of order n and diameter $d \geq 3$ such that every spanning tree T of G is a caterpillar of diameter d. Let $P_{d+1} = v_0 - v_1 - ... - v_d$ be a fixed diametral path of a graph G. It is clear that a caterpillar T contains P_{d+1} and it is the central path of T. Then define
$P = \{v \in V(G) : v \in V(P_{d+1})\}$;
$P' = \{v \in V(G) : v \notin V(P_{d+1})\}$;
$P_c = \{v \in P : \epsilon(v) = \lceil d/2 \rceil\}$;
$P_{c'} = \{v \in P : \epsilon(v) > \lceil d/2 \rceil\}$;
$P'_c = \{v \in P' : \epsilon(v) = \lceil d/2 \rceil\}$;
$P'_{c'} = \{v \in P' : \epsilon(v) > \lceil d/2 \rceil\}$;
$P'_{cc} = \{v \in P'_c : v$ is adjacent to $C(P_{d+1})\}$;
$P'_{cc'} = \{v \in P'_c : v$ is not adjacent to $C(P_{d+1})\}$;
$P'_{c'c} = \{v \in P'_{c'} : v$ is adjacent to $C(P_{d+1})\}$;
$P'_{c'c'} = \{v \in P'_{c'} : v$ is not adjacent to $C(P_{d+1})\}$.
     Let $|P'_c| = n_1$ and $|P'_{c'}| = n_2$, where $0 \leq n_1, n_2 \leq n - d - 1$; $|P'_{cc}| = n_{11}$ and $|P'_{cc'}| = n_{12}$, where $0 \leq n_{11}, n_{12} \leq n_1$; $|P'_{c'c}| = n_{21}$ and $|P'_{c'c'}| = n_{22}$, where $0 \leq n_{21}, n_{22} \leq n_2$. Note that $n_1 + n_2 = n - d - 1$, $n_{11} + n_{12} = n_1$ and $n_{21} + n_{22} = n_2$. Moreover, $V(G) = P \cup P' = P_c \cup P_{c'} \cup P'_c \cup P'_{c'} = P_c \cup P_{c'} \cup P'_{cc} \cup P'_{cc'} \cup P'_{c'c} \cup P'_{c'c'}$.

We consider the following two cases.

Case 1: d is even.

$$\xi^c(G) = \sum_{v \in v(G)} \epsilon(v)\mathrm{d}(v)$$

$$= \sum_{v \in P} \epsilon(v)\mathrm{d}(v) + \sum_{v \in P'} \epsilon(v)\mathrm{d}(v)$$

$$= \sum_{v \in P} \epsilon(v)(\mathrm{d}(v|P) + \mathrm{d}(v|P')) + \sum_{v \in P'_c} \epsilon(v)\mathrm{d}(v) + \sum_{v \in P'_{c'}} \epsilon(v)\mathrm{d}(v)$$

$$= \sum_{v \in P} \epsilon(v)\mathrm{d}(v|P) + \sum_{v \in P} \epsilon(v)\mathrm{d}(v|P') + \sum_{v \in P'_c} \epsilon(v)\mathrm{d}(v) + \sum_{v \in P'_{c'c}} \epsilon(v)\mathrm{d}(v)$$

$$+ \sum_{v \in P'_{c'c'}} \epsilon(v)\mathrm{d}(v)$$

$$= \sum_{v \in P} \epsilon(v)\mathrm{d}(v|P) + \sum_{v \in P} \epsilon(v)\mathrm{d}(v|P') + \sum_{v \in P_{c'}} \epsilon(v)\mathrm{d}(v|P') + \sum_{v \in P'_c} \epsilon(v)\mathrm{d}(v)$$

$$+ \sum_{v \in P'_{c'c}} \epsilon(v)\mathrm{d}(v) + \sum_{v \in P'_{c'c'}} \epsilon(v)\mathrm{d}(v)$$

$$\geq \frac{3}{2}d^2 + n_{21}\left(\frac{d}{2}\right)(1) + (2n_1 + n_{22})\left(\frac{d}{2} + 1\right)(1) + n_1\left(\frac{d}{2}\right)(2)$$

$$+ n_{21}\left(\frac{d}{2} + 1\right)(1) + n_{22}\left(\frac{d}{2} + 1\right)(1)$$

$$\geq \frac{3}{2}d^2 + n_{21}\left(\frac{d}{2}\right) + n_1\left(\frac{d}{2} + 1\right) + n_{22}\left(\frac{d}{2} + 1\right) + n_1\left(\frac{d}{2}\right)$$

$$+ n_{21}\left(\frac{d}{2} + 1\right) + n_{22}\left(\frac{d}{2} + 1\right)$$

$$\geq \frac{3}{2}d^2 + n_{21}\left(\frac{d}{2}\right) + n_1\left(\frac{d}{2} + 1\right) + n_{22}\left(\frac{d}{2}\right) + n_1\left(\frac{d}{2}\right) + n_{21}\left(\frac{d}{2} + 1\right)$$

$$+ n_{22}\left(\frac{d}{2} + 1\right)$$

$$= \frac{3}{2}d^2 + n_1\left(\frac{d}{2}\right) + n_1\left(\frac{d}{2} + 1\right) + n_2\left(\frac{d}{2}\right) + n_2\left(\frac{d}{2} + 1\right)$$

$$= \frac{3}{2}d^2 + (n_1 + n_2)\left(\frac{d}{2}\right) + (n_1 + n_2)\left(\frac{d}{2} + 1\right)$$

$$= \frac{3}{2}d^2 + (n - d - 1)\left(\frac{d}{2}\right) + (n - d - 1)\left(\frac{d}{2} + 1\right)$$

$$= nd + n + \frac{d^2}{2} - 2d - 1$$

$$= \xi^c(V_{n,d}).$$

Case 2: d is odd.

$$\xi^c(G) = \sum_{v \in v(G)} \epsilon(v) \mathrm{d}(v)$$

$$= \sum_{v \in P} \epsilon(v) \mathrm{d}(v) + \sum_{v \in P'} \epsilon(v) \mathrm{d}(v)$$

$$= \sum_{v \in P} \epsilon(v)(\mathrm{d}(v|P) + \mathrm{d}(v|P')) + \sum_{v \in P'_c} \epsilon(v) \mathrm{d}(v) + \sum_{v \in P'_{c'}} \epsilon(v) \mathrm{d}(v)$$

$$= \sum_{v \in P} \epsilon(v) \mathrm{d}(v|P) + \sum_{v \in P} \epsilon(v) \mathrm{d}(v|P') + \sum_{v \in P'_c} \epsilon(v) \mathrm{d}(v) + \sum_{v \in P'_{c'c}} \epsilon(v) \mathrm{d}(v)$$

$$+ \sum_{v \in P'_{c'c'}} \epsilon(v) \mathrm{d}(v)$$

$$= \sum_{v \in P} \epsilon(v) \mathrm{d}(v|P) + \sum_{v \in P_c} \epsilon(v) \mathrm{d}(v|P') + \sum_{v \in P_{c'}} \epsilon(v) \mathrm{d}(v|P') + \sum_{v \in P'_c} \epsilon(v) \mathrm{d}(v)$$

$$+ \sum_{v \in P'_{c'c}} \epsilon(v) \mathrm{d}(v) + \sum_{v \in P'_{c'c'}} \epsilon(v) \mathrm{d}(v)$$

$$\geq \frac{3}{2} d^2 + \frac{1}{2} + (2n_{11} + n_{21}) \left(\frac{d+1}{2} \right) (1) + (2n_{12} + n_{22}) \left(\frac{d+3}{2} \right) (1)$$

$$+ n_1 \left(\frac{d}{2} \right) (2) + n_{21} \left(\frac{d+3}{2} \right) (1) + n_{22} \left(\frac{d+3}{2} \right) (1)$$

$$> \frac{3}{2} d^2 + \frac{1}{2} + n_{11} \left(\frac{d+3}{2} \right) + n_{12} \left(\frac{d+3}{2} \right) + n_{21} \left(\frac{d+1}{2} \right)$$

$$+ n_{22} \left(\frac{d+3}{2} \right) + n_1 \left(\frac{d+1}{2} \right) + (n_{21} + n_{22}) \left(\frac{d+3}{2} \right)$$

$$= \frac{3}{2} d^2 + \frac{1}{2} + (n_{11} + n_{12}) \left(\frac{d+3}{2} \right) + (n_{21} + n_{22}) \left(\frac{d+1}{2} \right)$$

$$+ n_1 \left(\frac{d+1}{2} \right) + n_2 \left(\frac{d+3}{2} \right)$$

$$= \frac{3}{2} d^2 + \frac{1}{2} + n_1 \left(\frac{d+3}{2} \right) + n_2 \left(\frac{d+1}{2} \right) + n_1 \left(\frac{d+1}{2} \right) + n_2 \left(\frac{d+3}{2} \right)$$

$$= \frac{3}{2} d^2 + \frac{1}{2} + (n_1 + n_2) \left(\frac{d+3}{2} \right) + (n_1 + n_2) \left(\frac{d+1}{2} \right)$$

$$= \frac{3}{2} d^2 + \frac{1}{2} + (n - d - 1) \left(\frac{d+3}{2} \right) + (n - d - 1) \left(\frac{d+1}{2} \right)$$

$$= \frac{3}{2} d^2 + \frac{1}{2} + (n - d - 1)(d + 1) + (n - d - 1)$$

$$= nd + 2n - \frac{d^2}{2} - 3d - \frac{3}{2}$$

$$= \xi^c(V_{n,d}).$$

Theorem 2. *Let G be a connected graph of order n and diameter $d \geq 2$. Then*

$$\xi^c(G) \geq \xi^c(V_{n,d}). \tag{5}$$

Proof. Let $G_0 \subseteq G_1 \subseteq \ldots \subseteq G_k = G$ be a sequence of subgraphs such that G_0 is a connected subgraph of G which contain a diametral path P_{d+1} and every spanning tree of G_0 is a caterpillar of diameter d, and G_{i+1} $(0 \leq i \leq k-1)$ is a induced subgraph of G with vertex set $V(G_{i+1}) = V(G_i) \cup \{v\}, v \in V(G \setminus G_i)$. Let the order of G_i is n_i then $|G_k| = n_k = n$. Then by Theorem 1, we obtain $\xi^c(G_0) \geq V_{n_0,d}$. Now consider the graph $G_1 = G_0 \cup \{v\}$ where $v \in V(G \setminus G_0)$. Note that for a newly added vertex v in G_0, either $\epsilon(v) > \lceil d/2 \rceil$ or $\epsilon(v) = \lceil d/2 \rceil$. If $\epsilon(v) > \lceil d/2 \rceil$ then it contribute one degree for some vertex of G_0 and hence v contribute at least $(\lceil d/2 \rceil)(1) + (\lceil d/2 \rceil + 1)(1)$ for $\xi^c(G_1)$ as G_1 is connected and $\epsilon(u) \geq d/2$ for every $u \in V(G_0)$. Hence we obtain, $\xi^c(G_1) \geq \xi^c(G_0) + (\lceil d/2 \rceil)(1) + (\lceil d/2 \rceil + 1)(1) = \xi^c(V_{n_0,d}) + (\lceil d/2 \rceil)(1) + (\lceil d/2 \rceil + 1)(1) = \xi^c(V_{n_1,d})$. If $\epsilon(v) = \lceil d/2 \rceil$ then by Lemma 2, $d(v) \geq 2$ and it adjacent to at least two vertices of G_0. Hence v contribute at least $4(\lceil d/2 \rceil) > (\lceil d/2 \rceil) + (\lceil d/2 \rceil + 1)$ for $\xi^c(G_1)$. Hence we obtain $\xi^c(G_1) \geq \xi^c(G_0) + 4(\lceil d/2 \rceil) \geq \xi^c(V_{n_0,d}) + (\lceil d/2 \rceil) + (\lceil d/2 \rceil + 1) = \xi^c(V_{n_1,d})$.

Continuing in this way, finally we obtain $\xi^c(G_k) \geq \xi^c(G_{k-1}) + (\lceil d/2 \rceil + 1)(1) + (\lceil d/2 \rceil)(1) = \xi^c(V_{n_{k-1},d}) + (\lceil d/2 \rceil + 1)(1) + (\lceil d/2 \rceil)(1) = \xi^c(V_{n_k,d}) = \xi^c(V_{n,d})$.

Note that equality holds if at each step of above procedure equality holds and hence we obtain that volcano graph $V_{n,d}$ attain a lower bound which completes the proof.

Corollary 1. *Let T be a tree of order n and diameter $d \geq 3$. Then*

$$\xi^c(T) \geq \xi^c(V_{n,d}).$$

Example 1. The readers are advised to refer the following example for the procedure used in Theorems 1 and 2 to give a lower bound for the eccentric connectivity index of graphs.

In Fig. 1, the graph G of order 19 and diameter 7 is shown in which $P_8 = v_0 - v_1 - \ldots - v_7$ is a fixed diametral path and the vertices with circle are vertices with minimum eccentricity.

In Fig. 2, a sequence of subgraphs $G_0 \subset G_1 \subset G_2 \subset G_3 = G$ of G with ordered pair whose first coordinate denote vertex degree and second coordinate denote eccentricity of that vertex in $G_i, 0 \leq i \leq 3$ is shown. It is clear that G_0 is a graph whose each spanning tree is a caterpillar of diameter 7 and $G_{i+1} = G_i \cup \{v\}(0 \leq i \leq 2)$ for some $v \in G \setminus G_i$. Moreover, $|G_0| = 16, |G_1| = 17, |G_2| = 18, |G_3| = |G| = 19$ and $\text{diam}(G_i) = 7$ for $0 \leq i \leq 3$. Note that $\xi^c(G_0) = 182$, $\xi^c(G_1) = 195, \xi^c(G_2) = 204$ and $\xi^c(G_3) = \xi^c(G) = 213$ (The readers can calculate it using the ordered pair at each vertex in G_i). Using (3), it is easy to calculate that $\xi^c(V_{16,7}) = 146, \xi^c(V_{17,7}) = 155, \xi^c(V_{18,7}) = 164$ and $\xi^c(V_{19,7}) = 173$. It is clear from above that $\xi^c(G_0) \geq \xi^c(V_{16,7}), \xi^c(G_1) \geq \xi^c(V_{17,7}), \xi^c(G_2) \geq \xi^c(V_{18,7})$ and $\xi^c(G_3) = \xi^c(G) \geq \xi^c(V_{19,7})$.

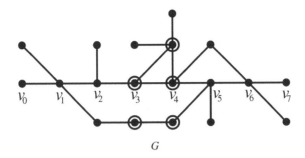

Fig. 1. Graph G of order 19 and diameter 7.

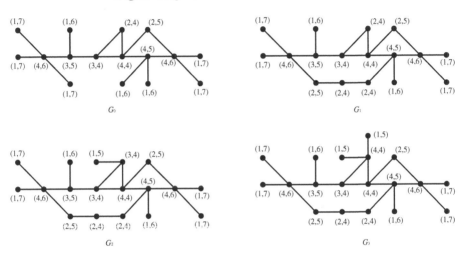

Fig. 2. Graphs $G_0 \subset G_1 \subset G_2 \subset G_3 = G$.

4 Concluding Remarks

The determination of extremal graphs for topological indices is remain interest of many researchers due to its various application in chemical, pharmaceutical and other properties of molecules. Various approaches are followed by researchers to determine the extremal graphs for the topological indices and its variants. In this work, we considered the adjacency relation of vertices to determine a lower bound for the eccentric connectivity index of graphs. This approach can also be useful to determine extremal graphs for other topological indices and its variants.

Acknowledgements. I want to express my deep gratitude to anonymous referees for kind comments and constructive suggestions.

References

1. Gupta, S., Singh, M.: Application of graph theory: relationship of eccentric connectivity index and Wiener's index with anti-inflammatory. J. Math. Anal. Appl. **266**, 259–268 (2002)
2. Ilić, A., Gutman, I.: Eccentric connectivity index of chemical trees. MATCH Commun. Math. Comput. Chem. **65**, 731–744 (2011)
3. Morgan, M.J., Mukwembi, S., Swart, H.C.: On the eccentric connectivity index of a graph. Discrete Math. **311**, 1229–1234 (2011)
4. Morgan, M.J., Mukwembi, S., Swart, H.C.: A lower bound on the eccentric connectivity index of a graph. Discrete Appl. Math. **160**, 248–258 (2012)
5. Sardana, S., Madan, A.K.: Application of graph theory: relationship of antimycobacterial activity of quinolone derivatives with eccentric connectivity index and Zagreb group parameters. MATCH Commun. Math. Comput. Chem. **45**, 35–53 (2002)
6. Sardana, S., Madan, A.K.: Application of graph theory: Relationship of molecular connectivity index, Wiener's index and Eccentric connectivity index with diuretic activity. MATCH Commun. Math. Comput. Chem. **43**, 85–98 (2000)
7. Sharma, V., Goswami, R., Madan, A.K.: Eccentric connectivity index: a novel highly discriminating topological descriptor for structure-activity studies. J. Chem. Inf. Comput. Sci. **37**, 273–282 (1997)
8. West, D.B.: Introduction to Graph Theory. Prentice-Hall of India (2001)
9. Wiener, H.: Structural determination of parafin boiling points. J. Am. Chem. Soc. **69**, 17–20 (1947)
10. Zhou, B., Du, Z.: On eccentric connectivity index. MATCH Commun. Math. Comput. Chem. **63**, 181–198 (2010)

On the Tractability of (k, i)-Coloring

Saurabh Joshi, Subrahmanyam Kalyanasundaram$^{(\boxtimes)}$, Anjeneya Swami Kare, and Sriram Bhyravarapu

Department of Computer Science and Engineering,
IIT Hyderabad, Sangareddy, India
{sbjoshi,subruk,cs14resch01002,cs16resch11001}@iith.ac.in

Abstract. In an undirected graph, a proper (k, i)-coloring is an assignment of a set of k colors to each vertex such that any two adjacent vertices have at most i common colors. The (k, i)-coloring problem is to compute the minimum number of colors required for a proper (k, i)-coloring. This is a generalization of the classic graph coloring problem. Majumdar et al. [CALDAM 2017] studied this problem and showed that the decision version of the (k, i)-coloring problem is fixed parameter tractable (FPT) with tree-width as the parameter. They asked if there exists an FPT algorithm with the size of the feedback vertex set (FVS) as the parameter without using tree-width machinery. We answer this in positive by giving a parameterized algorithm with the size of the FVS as the parameter. We also give a faster and simpler exact algorithm for $(k, k-1)$-coloring, and make progress on the NP-completeness of specific cases of (k, i)-coloring.

1 Introduction

In an undirected graph $G = (V, E)$, $|V| = n$, a *proper vertex coloring* is to color the vertices of the graph such that adjacent vertices get different colors. The classic graph coloring problem asks to compute the minimum number of colors required to properly color the graph. The minimum number of colors required is called the *chromatic number* of the graph, denoted by $\chi(G)$. This is a well known NP-hard problem and has been studied in multiple directions.

Many variants and generalizations of the graph coloring problem have been studied in the past. In this paper we address a generalization of the graph coloring problem called (k, i)-coloring problem. For a proper (k, i)-coloring, we need to assign a set of k colors to each vertex such that the adjacent vertices share at most i colors. The (k, i)-coloring problem asks to compute the minimum number of colors required to properly (k, i)-color the graph. The minimum number of colors required is called the (k, i)-*chromatic number*, denoted by $\chi_k^i(G)$. Note that $(1, 0)$-coloring is the same as the classic graph coloring problem.

(k, i)-COLORING PROBLEM
Instance: An undirected graph $G = (V, E)$.
Output: The (k, i)-chromatic number of G, $\chi_k^i(G)$.

© Springer International Publishing AG 2018
B. S. Panda and P. P. Goswami (Eds.): CALDAM 2018, LNCS 10743, pp. 188–198, 2018.
https://doi.org/10.1007/978-3-319-74180-2_16

We also define below the (q, k, i)-coloring problem, the decision version of the (k, i)-coloring problem.

(q, k, i)-COLORING PROBLEM
Instance: An undirected graph $G = (V, E)$.
Question: Does G have a proper (k, i)-coloring using at most q colors?

The (k, i)-coloring problem was first studied by Méndez-Díaz and Zabala in [1]. For arbitrary k and i, the (k, i)-coloring problem is NP-hard because $(1, 0)$-coloring is NP-hard. Apart from studying the basic properties, they also gave an integer linear programming formulation of the problem. Stahl [2] and independently Bollobás and Thomason [3] introduced the $(k, 0)$-coloring problem under the names of k-tuple coloring and k-set coloring respectively. The k-tuple coloring problem has been studied in detail [4,5], and Irving [6] showed that this problem is NP-hard as well. Some of the applications for the $(k, 0)$-coloring problem include construction of pseudorandom number generators, randomness extractors, secure password management schemes, aircraft scheduling, biprocessor tasks and frequency assignment to radio stations [7,8]. Brigham and Dutton [9] studied another variant of the problem, where k colors have to be assigned to each vertex such that the adjacent vertices share exactly i colors.

Bonomo et al. [10] studied the connection between the (k, i)-coloring problem on cliques and the theory of error correcting codes. In coding theory, a (j, d, k)-constant weight code represents a set of codewords of length j with exactly k ones in each codeword, with Hamming distance at least d. Bonomo et al. observed a direct connection between $A(j, d, k)$, the largest possible size of a (j, d, k)-constant weight code, and the (k, i)-colorability of cliques and used the existing results from coding theory (such as the Johnson bound [11]) to infer results on the (k, i)-colorability of cliques. Finding bounds on $A(j, d, k)$ is a well-studied problem in coding theory, and lots of questions on $A(j, d, k)$ are still open. This indicates the difficulty of the (k, i)-coloring problem even on graphs as simple as cliques.

Since the (k, i)-coloring problem is NP-hard in general, it is natural to study the tractability for special classes of graphs. Polynomial time algorithms are only known for a few of such classes namely bipartite graphs, cycles, cacti and graphs with bounded vertex cover or tree-width [10,12]. From the NP-hardness perspective, it is interesting to ask if the (k, i)-coloring problem is NP-hard for specific values of i. Except for the cases $i = k$, where the problem is trivial, and $i = 0$, where the problem is NP-hard [6], the NP-hardness remains open for all other values of i.

Recently, Majumdar et al. [12] studied the (k, i)-coloring problem and gave exact and parameterized algorithms for the problem. They showed that the problem is fixed parameter tractable (FPT) when parameterized by tree-width. As the tree-width is at most $(|S| + 1)$, where S is a feedback vertex set (FVS) of the graph, their algorithm also implies that (k, i)-coloring is FPT when parameterized by the size of FVS. As an open question, they asked to devise an FPT algorithm parameterized by the size of FVS, without going through tree-width. In this paper we answer this question.

Our results are:

– An $O((\binom{q}{k})^{|S|+2} n^{O(1)})$ time algorithm for the (q, k, i)-coloring problem that does not use tree-width machinery. Here S is an FVS of the graph.
– We make progress on the NP-hardness of the (k, i)-coloring problem. We show that $(k, 1)$-coloring and $(k, k-1)$ coloring are NP-complete, in addition to other NP-completeness results. This partially answers questions posed in [1,12].
– We give a $2^n n^{O(1)}$ time exact algorithm for the $(k, k-1)$-coloring problem. This is a direct improvement to the algorithm given in [12] for the same problem.

2   Preliminaries

A *parameterized problem* is a language $B \subseteq \Sigma^* \times \mathbb{N}$ where Σ is a fixed, finite alphabet. For example $(x, \ell) \in \Sigma^* \times \mathbb{N}$, here ℓ is called the parameter. A parameterized problem $B \subseteq \Sigma^* \times \mathbb{N}$ is called *fixed-parameter tractable* (FPT) if there is an algorithm $\mathcal{A}$, a computable function $f : \mathbb{N} \to \mathbb{N}$, and a constant c such that, given $(x, \ell) \in \Sigma^* \times \mathbb{N}$, the algorithm $\mathcal{A}$ correctly decides whether $(x, \ell) \in B$ in time bounded by $f(\ell)|x|^c$.

We assume that the graph is simple and undirected. We use n to denote $|V|$, the number of vertices of the graph. We say that the vertices u and v are *adjacent* (*neighbors*) if $\{u, v\} \in E$. For $v \in V$, we let $N(v)$ denote the set of neighbors of v. For $S \subseteq V$, the sub graph induced by S is denoted by $G[S]$. We use $O^*(f(n))$ to denote $O(f(n)n^{O(1)})$. We use the set of natural numbers for coloring the graph. We use the standard notations $[q] = \{1, 2, \ldots, q\}$ and $\binom{[q]}{k}$ to denote the set of all k-sized subsets of $[q]$. In the rest of the paper, we use the term *coloring* of a set $X \subseteq V$ to denote a mapping $h : X \to \binom{[q]}{k}$. We say that h is a *proper* (q, k, i)-coloring (or proper (k, i)-coloring) of X if any pair of adjacent vertices in X have no more than i colors in common.

3 (q, k, i)-Coloring Parameterized by Size of FVS

In this section, we assume that q, k, i are fixed values and focus on the decision problem of (q, k, i)-coloring. Let $G = (V, E)$ be an undirected graph. Let $G[V']$ denote a subgraph of G induced by $V' \subseteq V$. A *Feedback Vertex Set (FVS)* is a set of vertices $S \subseteq V$, removal of which from the graph G makes the remaining graph $(G[V \setminus S])$ acyclic. Many NP-hard problems have been shown to be tractable for graphs with bounded FVS [13].

In [12], Majumdar et al., gave an algorithm for the (q, k, i)-coloring problem in $O((\binom{q}{k})^{tw+1} n^{O(1)})$ time[1], where tw denotes the tree-width of the graph. Let S be a smallest FVS of G. It is known that $tw \leq |S| + 1$, see for instance [14].

---

[1] Even though [12] claims a running time of $O((\binom{q}{k})^{tw} n^{O(1)})$ for their algorithm, there is an additional factor of $\binom{q}{k}$ that is omitted, presumably because $\binom{q}{k}$ is treated as a constant.

In this section, we present an algorithm for (q, k, i)-coloring that runs in $O((\binom{q}{k})^{|S|+2} n^{O(1)})$ time, where $|S|$ is the size of the FVS of the graph. Our algorithm does not use the tree-width machinery. Note that, FVS has a 2-approximation algorithm [15], but there is no known polynomial time algorithm that approximates tree-width within a constant factor [16]. Computing the size of the smallest FVS is also known to be FPT parameterized by $|S|$, the size of the smallest FVS. There has been a series of results improving the running time, the fastest known algorithm [17] runs in $O(3.619^{|S|} n^{O(1)})$ time.

A brief description of our algorithm follows. Let S be an FVS of G. We start with a coloring of the vertices of S. Recall that $G[V \backslash S]$ is a forest. Each of the connected components of $G[V \backslash S]$ is a tree. For each of these components, we traverse the tree bottom-up and use a dynamic programming technique to compute the list of k-colorings that each vertex $w \in V \backslash S$ can take. For each $C \in \binom{[q]}{k}$, we include C in w's list if there is a coloring for the subtree rooted at w, consistent with the coloring of S, such that w receives color set C. We repeat this for all proper colorings of S.

Let $\Psi = \binom{[q]}{k}$ denote the family of all k-sized subsets of $[q]$. For any pair of sets $C, C' \in \Psi$, we say that (C, C') is *legal* if $|C \cap C'| \leq i$, and *illegal* if $|C \cap C'| > i$. Given two sets $C, C' \in \Psi$, it is easy to check if (C, C') is a legal pair. Formally, we have:

Proposition 1. *Given $C, C' \in \Psi$, it takes $O(k \log k)$ time to check if (C, C') is a legal pair.*

Definition 2. *Consider a partial coloring $h : S \to \Psi$ where only the vertices of the FVS S are colored. For a vertex $w \in V \backslash S$ and a set $C \in \Psi$, we say that (w, C) is h-compatible if for all $x \in S \cap N(w)$, the pair $(C, h(x))$ is legal.*

The set $\{C \in \Psi \mid (w, C) \text{ is } h\text{-compatible}\}$ is defined to be the set of h-compatible colorings of w.

Proposition 3. *Let $h : S \to \Psi$ be a coloring of the vertices in S. Let $w \in V \backslash S$ and $d_S(w) = |N(w) \cap S|$. Then the set of h-compatible colorings of w can be computed in time $O\left((\binom{q}{k}) d_S(w) k \log k\right)$.*

Proof. For each $C \in \Psi$, we check if (w, C) is h-compatible. For this, we need to check for all neighbors x of w in S, whether $(C, h(x))$ is legal. The total running time is $\binom{q}{k} \cdot d_S(w) \cdot O(k \log k)$. $\qquad\square$

Definition 4. *Given a graph $G = (V, E)$ and a coloring $h : X \to \Psi$ for some $X \subseteq V$, we say that the coloring $h' : V \to \Psi$ is an extension of h, or extends h if for all $v \in X$, we have $h(v) = h'(v)$.*

Lemma 5. *Given a proper (q, k, i)-coloring h of the vertices in a feedback vertex set S of the graph $G = (V, E)$, we can determine if h can be extended to a proper (q, k, i)-coloring of V in $O((\binom{q}{k})^2 n^{O(1)})$ time.*

Proof. The graph $G[V \setminus S]$ is a forest because S is a feedback vertex set. Therefore each connected component of $G[V \setminus S]$ is a tree. Below, we describe an algorithm that we can apply to each of these trees to yield a proper (q, k, i)-coloring extending h for the trees. Combining the colorings, we get a proper (q, k, i)-coloring of V, that is an extension of h.

Let T denote one of the trees in the forest. We will designate any one of the vertices (say r) of T as root. Let T_w denote the subtree rooted at a node $w \in T$.

Our plan is to maintain a table at each vertex w, indexed with the elements of Ψ. The entry at each color set C is denoted by $M_w(C)$. The entry $M_w(C)$ indicates whether there is a proper (q, k, i)-coloring of T_w, with w assigned the set C, consistent with the coloring h of S.

We will process T in a post order fashion as follows:

1. **When w is a leaf in T:** In this case, we set $M_w(C) = 1$ if (w, C) is h-compatible. Otherwise, we set $M_w(C) = 0$.
   For any leaf w, the values $M_w(C)$ corresponding to all $C \in \Psi$ can be computed in time $O(\binom{q}{k} d_S(w) k \log k)$ by Proposition 3. Here $d_S(w)$ denotes the number of neighbors of w in S.
2. **When w is an internal node in T:** Let $u_1, u_2, \ldots$ be the children of w in T. Recall that we process T in post order fashion. Before we process w, the M_{u_j} values for all the children of w would already have been computed. The value $M_w(C)$ is computed as follows:

   - If (w, C) is not h-compatible, we set $M_w(C) = 0$.
   - If (w, C) is h-compatible, we do the following:
     - If for each child u_j of w, there exists at least one coloring $C' \in \Psi$ such that $M_{u_j}(C') = 1$ and (C, C') is a legal pair, then set $M_w(C) = 1$.
     - Otherwise set $M_w(C) = 0$.

For each w and C, the h-compatibility check takes $O(d_S(w) k \log k)$ time. If (w, C) is h-compatible, we need to check all the children u_j, and the table entries $M_{u_j}(C')$ for all $C' \in \Psi$. Together with the check for (C, C') being a legal pair, the computation takes $d_T(w) \cdot \binom{q}{k} \cdot O(k \log k)$ time, where $d_T(w)$ is the number of children of w in the tree T.

Adding all up, the computation of the table entries for w takes time

$$O\left(\binom{q}{k} \cdot k \log k \cdot \left[d_S(w) + d_T(w) \binom{q}{k} \right] \right). \tag{1}$$

If for some $C \in \Psi$, $M_r(C) = 1$, then we know that there exists a proper (q, k, i)-coloring of T that is consistent with the coloring h of S.

The time complexity is obtained by adding the expression in (1) over all the vertices $w \in V \setminus S$. By using the bounds $d_S(w) \le n$ and $\sum_{\text{Trees } T} \sum_{w \in V(T)} d_T(w) \le \sum_{\text{Trees } T} |V(T)| \le n$, we get that the time complexity is upper bounded by

$$O\left(\binom{q}{k} \cdot k \log k \cdot \left[n^2 + n \binom{q}{k} \right] \right),$$

which is at most $O\left(\binom{q}{k}^2 \cdot n^2 \right)$, by noting that k is a constant. $\square$

The correctness of the procedure explained in the above lemma can be proved using an induction on the vertices of T according to its post order traversal. The inductive claim says that $M_w(C) = 1$ if and only if there is a proper (q,k,i)-coloring of T_w, with w assigned the set C, consistent with the given coloring of S.

Lemma 6. *Given a proper (q,k,i)-coloring h of the vertices in a feedback vertex set S of the graph $G = (V, E)$, we can determine if h can be extended to a proper (q,k,i)-coloring of V with space complexity $O(\binom{q}{k}n)$.*

Proof. Recall the algorithm explained in Lemma 5. At each vertex w in $G[V \setminus S]$, we need $O(\binom{q}{k})$ space to store values $M_w(C)$ for all $C \in \Psi$.   □

Theorem 7. *The (q,k,i)-coloring problem can be solved in time $O(\binom{q}{k}^{|S|+2}n^{O(1)})$ and $O(\binom{q}{k}n)$ space, where S is a feedback vertex set of G.*

Proof. For each coloring assignment h of S, we first determine if h is a proper (q,k,i)-coloring. This can be done in $O(|S|^2 k \log k)$ time. Then we determine whether there exists a proper (q,k,i)-coloring that extends h in $O(\binom{q}{k}^2 . n^{O(1)})$ time by Lemma 5. Since there are at most $\binom{q}{k}^{|S|}$ many colorings of S, we can determine whether there exists a proper (q,k,i)-coloring of G in $O(\binom{q}{k}^{|S|+2}n^{O(1)})$ time.

We need $O(|S|k \log q)$ space to store the coloring h of S. And by Lemma 6, we need $O(\binom{q}{k}n)$ space to determine if h can be extended to a proper coloring of G. The latter is the dominating term and determines the total space requirement of the algorithm.   □

On generating a proper (q,k,i)-coloring. We observe that we can modify Theorem 7 to obtain an algorithm that generates a proper (q,k,i)-coloring of G, if one exists. After executing the steps of the algorithm corresponding to Theorem 7, we traverse the tree in top-down fashion from the root, and find colorings for each vertex $w \in T$, consistent with its parent, subtree T_w and coloring of S. The latter two are already encoded in $M_w(C)$ value. The asymptotic time and space complexity are the same as that in Theorem 7.

We would like to observe a difference in the space usage of our FPT algorithm to the FPT algorithm for (q,k,i)-coloring parameterized by tree-width in [12]. We note that the algorithm in [12] can also be modified similarly to obtain an algorithm that generates a proper coloring. However, such an algorithm would require to store all feasible colorings at each bag of the tree-decomposition, resulting in a $O(\binom{q}{k}^{tw+1})$ space usage at each bag. Since there are $O(n)$ bags, total space required by the algorithm is $O(\binom{q}{k}^{tw+1}n)$, which is significantly larger than the $O(\binom{q}{k}n)$ space required by our algorithm.

Decision vs. search problem. We note that we could run the algorithm for (q,k,i)-coloring for $q = 1, 2, 3, \ldots$ till we reach $\chi_k^i(G)$, the smallest q for which the graph has a proper (q,k,i)-coloring. The running time of this procedure would be at most $O\left(\chi_k^i (\chi_k^i)^{|S|+2} n^{O(1)} \right)$. Thus an FPT algorithm parameterized by the size of the FVS for the (q, k, i)-coloring problem implies an FPT algorithm parameterized by combined parameters—the size of the FVS and the (k, i)-chromatic number.

3.1 Counting All Proper (q, k, i)-Colorings

Here we show that we can modify the algorithm described in Lemma 5 to count the number of proper (q, k, i)-colorings of G. Let a proper (q, k, i)-coloring h of FVS S be given. Instead of maintaining $M_w(C)$ for a vertex w in a rooted tree T, we maintain another value $M_w^\#(C)$.

$$M_w^\#(C) = \begin{cases} 0 & \text{if } (w, C) \text{ is not } h\text{-compatible.} \\ 1 & \begin{cases} \text{if } w \text{ is a leaf,} \\ \text{and } (w, C) \text{ is } h\text{-compatible.} \end{cases} \\ \displaystyle\prod_{\forall u_j \in \text{child}(w)} \sum_{\text{legal}(C,C')} M_{u_j}^\#(C') & \begin{cases} \text{if } w \text{ is a non-leaf vertex,} \\ \text{and } (w, C) \text{ is } h\text{-compatible.} \end{cases} \end{cases}$$

At each vertex w, $M_w^\#(C)$ maintains a count of the proper (q, k, i)-colorings of T_w, consistent with the coloring h of S, where w gets assigned the set C. The correctness can be verified by a straightforward induction on the tree vertices in post order traversal. If r is the root of T, $M_r^\#(C)$ gives the count of proper (q, k, i)-colorings of T, where r is colored C, consistent with the coloring h of S.

The total number of proper (q, k, i)-colorings of G is therefore computed by taking into account (i) all proper (q, k, i)-colorings h of S, (ii) all the trees T_j in $G[V \backslash S]$, and (iii) all color sets $C \in \Psi$ at the root of T_j. The full expression is as follows:

$$\text{No. of proper } (q,k,i)\text{-colorings} = \sum_{\substack{\text{proper } (q,k,i)\text{-} \\ \text{colorings of } S}} \left(\prod_{T_j \text{ in } G[V \backslash S]} \left(\sum_{C \in \Psi} M_{\text{root}(T_j)}^\#(C) \right) \right).$$

The above expression implies the following theorem. The asymptotic time complexity remains the same as Theorem 7, whereas the space complexity incurs a blowup of $nk \log q$, because of the maximum value $M_w^\#(C)$ can take.

Theorem 8. *There is an algorithm that computes the number of proper (q, k, i)-colorings of G, in $O((\binom{q}{k})^{|S|+2} n^{O(1)})$ time and $O((\binom{q}{k}) n^2 \log q)$ space, where S is a feedback vertex set of G.*

4   Faster Exact Algorithm for $(k, k-1)$-Coloring

In [12], Majumdar et al. gave an $O^*(4^n)$ time exact algorithm for the $(k, k-1)$-coloring problem. Their algorithm was based on running an exact algorithm for a set cover instance where the universe is the set of all the vertices V and the family of sets $\mathcal{F}$ is the set of all independent sets of vertices of G. To show correctness and running time, they used a claim (unnumbered) that relates $\chi_k^{k-1}(G)$ to the size of solution of the set cover instance, an $O(2^n \cdot n \cdot |\mathcal{F}|)$ time exact algorithm for the set cover problem [18] and an upper bound of 2^n on the size of the family of sets $\mathcal{F}$. Hence, the time complexity of their algorithm is $O(2^n \cdot n \cdot 2^n) = O(4^n \cdot n)$.

We first note that their algorithm also works when $\mathcal{F}$ is replaced by $\mathcal{F}'$, the set of all maximal independent sets of G. This is because any independent set $A \in \mathcal{F}$ is contained in a maximal independent set $A' \in \mathcal{F}'$. In any set covering of V using elements of $\mathcal{F}$, each set A can be replaced by an $A' \in \mathcal{F}'$, thus obtaining a set cover of V using elements of only $\mathcal{F}'$. By using the $3^{n/3}$ upper bound of Moon and Moser [19] on the number of maximal independent sets, the time complexity improves to $O(2^n \cdot n \cdot 3^{n/3}) = O(2.88^n \cdot n)$.

We now present a simpler and faster $O^*(2^n)$ algorithm to determine $\chi_k^{k-1}(G)$.

Lemma 9. *For any graph G, $\chi_k^{k-1}(G) = q$ where q is the smallest integer such that $\binom{q}{k} \geq \chi_1^0(G)$. Thus there is a polynomial time reduction from the $(k, k-1)$-coloring problem to the $(1, 0)$-coloring problem.*

Proof. The $(k, k-1)$-coloring problem asks to assign sets of k colors to each vertex, with the requirement that neighboring vertices must have distinct sets assigned to them. We may view each of the k-sized subsets as a color, and the $(1, 0)$-chromatic number $\chi_1^0(G)$ is the number of distinct k-sized subsets required.

Thus $\chi_k^{k-1}(G)$ is the smallest q that will provide $\chi_1^0(G)$ number of k-sized subsets. The polynomial time reduction is immediate. $\square$

Combining the above lemma with the $O^*(2^n)$ time algorithm of Koivisto [20] to compute $\chi_1^0(G)$, we get the following theorem.

Theorem 10. *There is an algorithm with $O^*(2^n)$ time complexity that computes the $(k, k-1)$ chromatic number of a given graph.*

Further, we can infer from Lemma 9 that for those graphs G where we can compute $\chi_1^0(G)$ in polynomial time, $\chi_k^{k-1}(G)$ can also be found in polynomial time. For instance, $\chi_k^{k-1}(K_n)$ can be computed in polynomial time as $\chi_1^0(K_n) = n$.

5   NP-Completeness Results

Since the (k, i)-coloring problem is a generalization of the $(1, 0)$-coloring problem, it follows that (k, i)-coloring is NP-hard. Méndez-Díaz and Zabala [1] conjectured that the (k, i)-coloring problem remains NP-hard even for specific values of i. In this section, we show NP-completeness results for some specific cases of (k, i)-coloring. We will only be proving the NP-hardness aspect of NP-completeness. Given a coloring, we can easily verify that it is a proper (k, i)-coloring in polynomial time.

Trivially, we have $\chi_k^k(G) = k$ for all graphs G. For the $(k, 0)$-coloring problem, we have the following result by Irving.

Theorem 11 ((k,0)-coloring is NP-complete [6]). *The $(2k+1, k, 0)$-coloring problem is NP-complete for all $k \geq 1$.*

The NP-completeness of the $(k, k-1)$-coloring problem is claimed by [1]. However, we are unable to follow and verify the proof. We provide an alternate NP-hardness proof as a consequence of the correspondence in Lemma 9.

Theorem 12. *The $(k, k-1)$-coloring problem is NP-complete for all $k \geq 1$.*

Proof. We use reductions from the $(1,0)$-coloring problem, for each value of $k \geq 2$. We show that the $(q, k, k-1)$-coloring problem is NP-complete for all values of $q > k \geq 2$. From the correspondence in Lemma 9, it follows that for any given $k \geq 1$, a graph G is $(q, k, k-1)$-colorable if and only if G is $\left(\binom{q}{k}, 1, 0\right)$-colorable. Since the $(r, 1, 0)$-coloring problems are NP-complete for all $r \geq 3$, it follows that $\left(\binom{q}{k}, 1, 0\right)$-coloring problems are NP-complete for all $q > k \geq 2$, and hence we get that the $(q, k, k-1)$-coloring problems are NP-complete for all $q > k \geq 2$. □

The following lemmas will help us in proving further NP-completeness results.

Lemma 13 (Complement trick). *For integers $k, i \geq 1$, any graph G is $(2k+i, k+i, i)$-colorable if and only if it is $(2k+i, k, 0)$-colorable.*

Proof. Let $f : V \to \binom{[2k+i]}{k}$ be a $(2k+i, k, 0)$-coloring of G. Consider the coloring f' where each vertex v is assigned the complement set $[2k+i] \setminus f(v)$. Notice that, every vertex is assigned $(k+i)$ colors, and any pair of adjacent vertices will share exactly i colors in the coloring f'. Thus we have a $(2k+i, k+i, i)$-coloring of G.

Similarly, if we start from a $(2k+i, k+i, i)$-coloring of G, we can get to a $(2k+i, k, 0)$-coloring by taking the complement coloring. □

Theorem 11 and the above lemma together imply the NP-completeness of $(k, 1)$-coloring for all $k \geq 2$.

Theorem 14 ((k,1)-coloring is NP-complete). *The $(2k+1, k+1, 1)$-coloring problem is NP-complete for all $k \geq 1$.*

Now we introduce another simple gadget, the universal vertex. Given a graph G, we can construct a graph G' by adding a new vertex v that is adjacent to all the vertices of G. It is straightforward to see that G has a $(q, k, 0)$-coloring if and only if the new graph G' has a $(q+k, k, 0)$-coloring. Thus we have the following:

Lemma 15 (Universal vertex). *There is a polynomial time reduction from the $(q, k, 0)$-coloring problem to the $(q+k, k, 0)$-coloring problem. Therefore, the $(q+k, k, 0)$-coloring problem is NP-complete when the $(q, k, 0)$-coloring problem is NP-complete.*

Combining Lemmas 13 and 15 yields a collection of NP-completeness results.

Theorem 16. *For any integers $p \geq 2$ and $\ell \geq 1$, the $(p\ell + 1, (p-1)\ell + 1, (p-2)\ell + 1)$-coloring problem is NP-complete.*

Proof. From Theorem 11, we have that the $(2\ell + 1, \ell, 0)$-coloring problem is NP-complete. By applying the universal vertex gadget of Lemma 15 a total of $(p-2)$ times, we get that the $(p\ell + 1, \ell, 0)$-coloring problem is NP-complete for all $p \geq 2$. Now we can use the complement trick of Lemma 13 to infer the NP-completeness of the $(p\ell + 1, (p-1)\ell + 1, (p-2)\ell + 1)$-coloring problem.   □

The above theorem gives us a collection of NP-completeness results. If we set $\ell = 2$, we get that problems $(5, 3, 1)$-coloring, $(7, 5, 3)$-coloring, $(9, 7, 5)$-coloring etc. are NP-complete. For other values of ℓ, we get similar sequence of NP-completeness results. But we cannot infer the NP-completeness of $(k, 3)$-coloring, or $(k, 5)$-coloring from this because all values of k are not covered. To show that $(k, 3)$-coloring is NP-hard, we need to exhibit for all relevant k, a value q such that $(q, k, 3)$-coloring is NP-hard. As conjectured in [1], we believe that the (k, i)-coloring problem is NP-complete for all values of i. As of now, the NP-completeness of (k, i)-coloring is still open for $2 \leq i \leq k - 2$.

Acknowledgment. The authors would like to thank the anonymous reviewer for helpful comments, and pointing out a flaw in the proof of Theorem 12 in an earlier version of the paper.

References

1. Méndez-Díaz, I., Zabala, P.: A generalization of the graph coloring problem. Investig. Oper. **8**, 167–184 (1999)
2. Stahl, S.: n-tuple colorings and associated graphs. J. Comb. Theor. Ser. B **20**(2), 185–203 (1976)
3. Bollobás, B., Thomason, A.: Set colourings of graphs. Discrete Math. **25**(1), 21–26 (1979)
4. Klostermeyer, W., Zhang, C.Q.: n-tuple coloring of planar graphs with large odd girth. Graphs Combinatorics **18**(1), 119–132 (2002)
5. Šparl, P., Žerovnik, J.: A note on n-tuple colourings and circular colourings of planar graphs with large odd girth. Int. J. Comput. Math. **84**(12), 1743–1746 (2007)
6. Irving, R.W.: NP-completeness of a family of graph-colouring problems. Discrete Appl. Math. **5**(1), 111–117 (1983)
7. Marx, D.: Graph colouring problems and their applications in scheduling. Period. Polytech. Electr. Eng. **48**(1–2), 11–16 (2004)
8. Beideman, C., Blocki, J.: Set families with low pairwise intersection. arXiv preprint arXiv:1404.4622 (2014)
9. Brigham, R.C., Dutton, R.D.: Generalized k-tuple colorings of cycles and other graphs. J. Comb. Theor. Ser. B **32**(1), 90–94 (1982)
10. Bonomo, F., Durán, G., Koch, I., Valencia-Pabon, M.: On the (k, i)-coloring of cacti and complete graphs. In: Ars Combinatoria (2014)
11. Johnson, S.: A new upper bound for error-correcting codes. IRE Trans. Inf. Theor. **8**(3), 203–207 (1962)

12. Majumdar, D., Neogi, R., Raman, V., Tale, P.: Exact and parameterized algorithms for (k, i)-coloring. In: Third International Conference on Algorithms and Discrete Applied Mathematics, CALDAM 2017, India, pp. 281–293 (2017)
13. Kratsch, S., Schweitzer, P.: Isomorphism for graphs of bounded feedback vertex set number. In: Kaplan, H. (ed.) SWAT 2010. LNCS, vol. 6139, pp. 81–92. Springer, Heidelberg (2010). https://doi.org/10.1007/978-3-642-13731-0_9
14. Jansen, B.M., Raman, V., Vatshelle, M.: Parameter ecology for feedback vertex set. Tsinghua Sci. Technol. **19**(4), 387–409 (2014)
15. Bafna, V., Berman, P., Fujito, T.: A 2-approximation algorithm for the undirected feedback vertex set problem. SIAM J. Discrete Math. **12**(3), 289–297 (1999)
16. Wu, Y.L., Austrin, P., Pitassi, T., Liu, D.: Inapproximability of treewidth, one-shot pebbling, and related layout problems. J. Artif. Intell. Res. **49**(1), 569–600 (2014)
17. Kociumaka, T., Pilipczuk, M.: Faster deterministic feedback vertex set. Inf. Process. Lett. **114**(10), 556–560 (2014)
18. Fomin, F.V., Kratsch, D.: Exact Exponential Algorithms. Texts in Theoretical Computer Science. An EATCS Series. Springer, Heidelberg (2010). https://doi.org/10.1007/978-3-642-16533-7
19. Moon, J.W., Moser, L.: On cliques in graphs. Isr. J. Math. **3**(1), 23–28 (1965)
20. Koivisto, M.: An $O^*(2^n)$ algorithm for graph coloring and other partitioning problems via inclusion-exclusion. In: Proceedings of the 47th Annual IEEE Symposium on Foundations of Computer Science. FOCS 2006, Washington, D.C., pp. 583–590. IEEE Computer Society (2006)

Window Queries for Problems on Intersecting Objects and Maximal Points*

Farah Chanchary$^{(\boxtimes)}$, Anil Maheshwari, and Michiel Smid

School of Computer Science, Carleton University,
Ottawa, ON K1S 5B6, Canada
farah.chanchary@carleton.ca,
{anil,michiel}@scs.carleton.ca

Abstract. We present data structures that can answer *window queries* for a sequence of geometric objects, such as points, line segments, triangles and convex c-gons. We first present data structures to solve windowed intersection decision problems using line segments, triangles and convex c-gons. We also present data structures to count points on maximal layer, to decide whether a given point belongs to a maximal layer, and to count k-dominant points for a fixed integer k for a sequence of points in $\mathbb{R}^d$, $d \geq 2$. All data structures presented in this paper answer queries in polylogarithmic time and use subquadratic space.

Keywords: Intersection decision problem · Maximal point
Maximal layer · Window query

1 Introduction

We construct data structures for various *geometric objects* (e.g., points, line segments, triangles and convex c-gons) to efficiently answer *window queries*. In a *window query* we are given two positive integers i and j, with $i < j$, such that the interval $[i, j]$ represents a query window of width $j-i+1$. Let $S = (s_1, s_2, \ldots, s_n)$ be a sequence of n geometric objects. For $1 \leq i < j \leq n$, let $S_{i,j}$ denote the subsequence $(s_i, s_{i+1}, \ldots, s_j)$. We want to preprocess S into some data structures such that given a query interval $q = [i, j]$ and a predicate $\mathcal{P}$, we can answer window queries using the objects in $S_{i,j}$ that match $\mathcal{P}$.

Recently the same model of data structure has been considered in various studies (see [1–3,5–7]), where the authors mapped a sequence of geometric objects (or graph edges) to a sequence of *timestamped events*, where for each k, with $1 \leq k \leq n$, an object s_k has a unique timestamp k.

In this paper, we present new results for windowed intersection decision problems and a variety of windowed reporting problems using points on maximal layers. We define the *windowed intersection decision problem* as 'Given a pair

This research work was supported by NSERC Research Grants and Ontario Graduate Scholarship.

B. S. Panda and P. P. Goswami (Eds.): CALDAM 2018, LNCS 10743, pp. 199–213, 2018.
https://doi.org/10.1007/978-3-319-74180-2_17

of indices (i, j), where $1 \leq i < j \leq n$, report whether there is any intersection between objects in $(s_i, \ldots, s_j)$'. In [5], Chan and Pratt presented orthogonal segment intersection decision problems, whereas our algorithms can preprocess sequences of objects, i.e., line segments, triangles, and convex c-gons, with arbitrary orientations. For our second set of problems, we consider a sequence of n points $P = (p_1, p_2, \ldots, p_n)$ in $\mathbb{R}^d$, where $d \geq 2$, and we answer queries related to maximal layers and dominance. More specifically we solve three types of windowed queries of the following forms: Given a query interval $[i, j]$ (i) count the number of maximal points in $P_{i,j} = (p_i, p_{i+1}, \ldots, p_j)$, (ii) given an integer k, with $i \leq k \leq j$, decide if point p_k is on the maximal layer L_γ of $P_{i,j}$, where $\gamma = 1$, or 2, or ≥ 3, and (iii) for a fixed integer $k \geq 1$, report all maximal points in $P_{i,j}$ such that each point dominates at least k points of $P_{i,j}$.

1.1   Previous Work

Bannister et al. [2] were the first to consider this window model for preprocessing timestamped graph edges into data structures that can answer windowed queries. Subsequently more results on windowed graph problems were presented in [6,7]. Similar time window model for geometric objects was first studied in [1], where the authors presented results for reporting the convex hull of points in the plane, and skyline and proximity relations of point sets in $\mathbb{R}^d$. They used a hierarchical decomposition in time to construct binary decomposition trees on given temporal points to answer windowed queries related to convex hull and proximity relations in polylogarithmic time. However, their skyline queries use a different preprocessing technique based on the rectangle stabbing data structures. Mouratidis et al. [12] considered problems of monitoring top-k maximal layers (mentioned as k-skyband in [12]) using fixed width sliding query windows. However, our results for points on maximal layers are different from those of [1,12].

Later more results have been presented by Bokal et al. [3], and Chan and Pratt [5]. In [3,5], the authors mainly focused on answering decision problems on hereditary properties, such as the convex hull area decision problem (in 2D), the diameter decision problem (in 2D and 3D), the width decision problem (in 2D) and the orthogonal segment intersection detection problem. In [3], Bokal et al. showed a sketch based general methodology for finding all maximal subsequences for a set of n points in plane, i.e., for all i, with $1 \leq i \leq n$, they find the largest index of the maximal interval starting at i that holds some hereditary property $\mathcal{P}$. They solved problems for finding all maximal subsequences with unit diameter, all maximal subsequences whose convex hull area is at most 1 and all maximal subsequences that define monotone paths in some (subpath-dependent) direction. Later, Chan and Pratt [5] improved some of their preprocessing times including diameter decision problems and convex hull area decision problems. In [5] the authors presented techniques to reduce the windowed problems into range successor problems such that a query can be performed by standard range searching techniques. The diameter decision problem in 2D and 3D, and the orthogonal line segment intersection detection problems follow this strategy.

As a second approach, they used dynamic data structures and a first-in-first-out sequence of processing geometric objects to find all maximal subsequences of intervals that satisfies some property $\mathcal{P}$. Authors named this process as *FIFO updates* and used this technique to solve the 2D convex hull area decision problem and the 2D width decision problem.

1.2   New Results

The main contributions of this paper are listed below, and are also summarized in Table 1.

1. *Intersection decision problems*: Given a sequence S of n geometric objects we can preprocess S for the windowed intersection decision problem in $O(n^{4/3} \cdot polylog(n))$ time using $O(n^{4/3} \cdot polylog(n))$ space so that queries can be answered in $O(\log n)$ time.
2. *Problems on points on maximal layers*: Given a sequence of n points $P = (p_1, p_2, \ldots, p_n) \subseteq \mathbb{R}^d$ we can preprocess P into data structures to report the following.

– Given a query interval $[i, j]$ and a point p_k with $i \le k \le j$, we can report whether p_k is on the maximal layer of the sequence of points $P_{i,j} = (p_i, \ldots, p_j)$ in $O(1)$ time. Preprocessing takes $O(n \log^{d-1} n)$ time using $O(n \log^{d-2} n)$ space .

Table 1. Summary of results. n is the number of input objects, $\gamma \ge 2$ is the number of maximal layer, w is the output size, and k is a fixed parameter.

Problems	Preprocessing time	Space	Query
Intersection decision			
Segment, Bichromatic segment, Triangle, c-gon (c is constant)	$O(n^{4/3} polylog(n))$	$O(n^{4/3} polylog(n))$	$O(\log n)$
Points on maximal layers			
p_k on maximal layer L_1	$O(n \log^{d-1} n)$	$O(n \log^{d-2} n)$	$O(1)$
p_k on max layer $L_\gamma, \gamma = 2, \ge 3$	$O(n \log^{d+1} n)$	$O(n \log^{d+1} n)$	$O(\log^{d+1} n)$
Count maximal points: $d = 2, 3$	$O(n \log^2 n)$	$O(n \log^2 n)$	$O(\log^2 n)$
Count maximal points: $d \ge 4$	$O(n \log^{d-1} n)$	$O(n \log^{d-2} n)$	$O(\log^2 n)$
k-dom points: $2 \le d \le 5$	$O(kn \log^4 n)$	$O(kn \log^3 n)$	$O(\log^4 n + kw)$
k-dom points: $d \ge 6$	$O(kn \log^{d-1} n)$	$O(kn \log^{d-2} n)$	$O(\log^4 n + kw)$

- Given a query interval $[i, j]$ and a point p_k with $i \leq k \leq j$, we can report whether p_k is on layer 2 or ≥ 3 of the sequence $P_{i,j}$ in $O(\log^{d+1} n)$ time. Preprocessing takes $O(n \log^{d+1} n)$ time using $O(n \log^{d+1} n)$ space.
- We can count the total number of maximal points of $P_{i,j}$ in $O(\log^2 n)$ time. Preprocessing takes $O(n \log^{d-1} n)$ time using $O(n \log^{d-2} n)$ space when $d \geq 4$. However, when $d = 2$ and 3, preprocessing takes $O(n \log^2 n)$ time and $O(n \log^2 n)$ space.
- Given a fixed integer k, we can report all maximal points in $P_{i,j}$ each dominating at least k points (we call this the 'k-dominant points' problem) in $O(\log^4 n + kw)$ time, where w is the size of the output. Preprocessing takes $O(kn \log^4 n)$ time using $O(kn \log^3 n)$ space when $2 \leq d \leq 5$. For $d \geq 6$, preprocessing takes $O(kn \log^{d-1} n)$ time and $O(kn \log^{d-2} n)$ space.

1.3 Organization

This paper is organized as follows. In Sect. 2, we present algorithms for windowed intersection decision problems using geometric objects. Section 3 presents more results for windowed queries using points on maximal layers. Section 4 concludes this paper.

2   Geometric Object Intersections

In this section, we discuss *windowed intersection decision problems* on a given sequence of geometric objects (e.g., line segments, triangles, and constant size polygons) within a query interval $[i, j]$. Input consists of a sequence of n geometric objects $S = (s_1, s_2, \ldots, s_n)$. We will represent the sequence S in an array A, where $A[i] = s_i$, for $i = 1, 2, \ldots, n$. The *windowed intersection decision problem* is 'Given a pair of indices (i, j), where $1 \leq i < j \leq n$, report whether there is any intersection between objects in $(s_i, \ldots, s_j)$'.

2.1 Overview of Our Data Structure

Before we discuss our data structure, we define a *valid pair* of indices (α, β) with $1 \leq \alpha < \beta \leq n$ as follows: For each $1 \leq \alpha \leq n$, let β be the smallest index larger than α such that the object $A[\beta]$ intersects $A[\alpha]$. If there is no $A[\beta]$ that intersects $A[\alpha]$ then $\beta = \infty$. Suppose, there exists a data structure that can find all valid pairs (α, β). Then we can reduce the windowed intersection decision problem into a range query problem as follows. For each valid pair (α, β), we store a point $(\alpha, \beta) \in \mathbb{R}^2$ using a priority search tree (PST) data structure [10]. A PST takes linear space to store $O(n)$ points in the plane and it can be built in $O(n \log n)$ time. For a given query interval $[i, j]$, we perform a range search in PST with the query rectangle $R_q = [i, \infty) \times (-\infty, j]$. Note that, there will be an intersecting pair of objects $(A[\alpha], A[\beta])$ in query interval $[i, j]$ if and only if there is a point $(\alpha, \beta) \in R_q$. Hence, if the range searching query returns a positive count of points in R_q, then we report that some objects intersect in the interval $[i, j]$. This query can be answered in $O(\log n)$ time. Thus we obtain the following lemma.

Lemma 1. *Suppose a sequence of n geometric objects is stored in an array $A[1..n]$, where i is the index of the object stored in $A[i]$. Given all valid pairs (α, β) for every $1 \leq \alpha \leq n$ in A, we can build a data structure of size $O(n)$ that can answer windowed intersection decision queries in $O(\log n)$ time.*

Next we show how to find all the valid pairs. Suppose X is a set of n geometric objects and we assume that we have a data structure $DS(X)$ that can find whether a query object q intersects any member of X. Furthermore, $DS(X)$ takes $M(n)$ space, $P(n)$ preprocessing time and $Q(n)$ query time, where $M(n)/n$, $P(n)/n$ and $Q(n)$ are all non-decreasing functions. To find all the valid pairs, we maintain a tree T defined as follows. The leaves of T store objects $A[1], A[2], \ldots, A[n]$ in order from left to right. For each internal node v of T, let $P[v]$ be the set of objects at the leaves of the subtree rooted at v. Each node v of T stores all the objects in its subtree in a secondary data structure $DS[P[v]]$. Each level i of T has 2^i nodes. So each node at level i requires $M(n/2^i)$ space. The total space requirement is $O(n) + \sum_{i=0}^{\log n} 2^i \cdot M(n/2^i) = O(n) + \sum_{i=0}^{\log n} n \cdot \frac{M(n/2^i)}{n/2^i} \leq O(n) + \sum_{i=0}^{\log n} n \cdot \frac{M(n)}{n} \leq O(n) + \sum_{i=0}^{\log n} M(n) = O(M(n) \cdot \log(n))$.

The total preprocessing time can analogously be computed as $O(P(n) \log n)$. Now for any $1 \leq \alpha \leq n$, we can find β in time $O(Q(n) \log n)$ as follows. To identify a valid pair (α, β), we first search T to find the leaf v' containing α. It requires $O(\log n)$ time using a standard binary search. Then we move up from v' towards the root node and at every step perform the following search. Each time we move from a child node v' towards its parent node $p(v')$, we query the secondary structure stored at the right child of $p(v')$ to decide whether it contains an object β that intersects with α. If the search is unsuccessful we move upwards one more level in T, and repeat the process. Otherwise, we find the node that contains the intersecting object and we continue descending from $p(v')$ to locate the leaf node containing β (see Fig. 1). In this way, the total time required to find all valid pairs is $O(n \cdot Q(n) \log n)$. From Lemma 1 we obtain the following result.

Theorem 1. *Given a sequence S of n geometric objects, we can preprocess S into a data structure of size $O(M(n) \log n)$ in time $O(P(n) \log n + n \cdot Q(n) \log n)$ so that it can answer windowed intersection decision queries in $O(\log n)$ time.*

Next, we discuss the construction of the secondary data structure $DS(X)$ for different problems.

Segment Intersections: Given a sequence of n line segments $S = (s_1, s_2, \ldots, s_n)$ in the plane, we want to preprocess S to answer windowed queries for segment intersections. As we have described previously, our primary data structure T stores n input segments at the leaf nodes sorted in order from left to right. At each node v of T we build a multi-level partition tree that answers the queries of the form '*Given a query segment s_q, does s_q intersect any segment of*

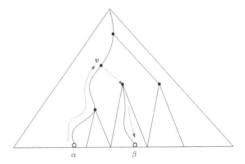

Fig. 1. Query path from α to β in our multilevel data structure.

$\{s_a, s_{a+1}, \ldots, s_b\}$, *where* $1 \le a < b \le n$?'. For a sequence of n line segments in the plane, we can obtain a data structure of $O(n \log^3 n)$ preprocessing time and $O(n \log^2 n)$ space such that we can report whether a query line segment s_q intersects any input segment in $O(\sqrt{n} \log^2 n)$ expected time by applying Corollary 7.3(i) in [4] three times, where $d = 2$. Finally, by repeated applications of Corollary 7.3(i) and Corollary 7.8 in [4] with $d = 2$, we can build a data structure to answer above mentioned queries that requires preprocessing time $P(n) = O(m \cdot polylog(n))$ and query time $Q(n) = O(n/\sqrt{m} \cdot polylog(n))$, where $m = n^{4/3}$. So by Theorem 1, the total time required for segment intersection preprocessing is $O(n^{4/3} \cdot polylog(n) + n \cdot n^{1/3} \cdot polylog(n)) = O(n^{4/3} \cdot polylog(n))$.

Corollary 1. *Given a sequence S of n segments, we can preprocess S for the windowed segment intersection decision problem in $O(n^{4/3} \cdot polylog(n))$ time using $O(n^{4/3} \cdot polylog(n))$ space.*

Bichromatic Segment Intersection: Let $S = (B \cup R)$ be a sequence of bichromatic line segments, where B is a sequence of b pairwise disjoint *blue* segments, R is a sequence of r pairwise disjoint *red* segments, and $N = b + r$. Our data structure for segment intersection problem can be extended for reporting windowed bichromatic segment intersection problem (intersections of red segments with blue segments) using the same preprocessing time and space bound. We assume that every segment in S has a unique timestamp. We build two sets of the same data structure we presented in this section. Let T_B be one structure where we store b blue segments, make queries with r red segments, and find *valid pairs* (s_r, s_b), where a red segment s_r intersects with a blue segment s_b. T_R is the analogous structure that gives us all *valid pairs* (s_b, s_r). The only minor change occurs during searching the primary data structures with a query segment. For example, when we query T_B with any red segment s_r, first we have to find the leaf node containing a blue segment with the smallest timestamp such that $t(s_b) > t(s_r)$. The rest of the search technique remains unchanged.

Corollary 2. *Given a sequence $S = (B \cup R)$ of N bichromatic line segments, where B is a sequence of b pairwise disjoint blue segments, R is a sequence* *of r pairwise disjoint red segments, and $N = b + r$. We can preprocess S for the windowed bichromatic segment intersection decision problem in $O(N^{4/3} \cdot polylog(N))$ time using space $O(N^{4/3} \cdot polylog(N))$.*

Triangle Intersections: The input for this problem is a sequence of n triangles $T = (t_1, t_2, \ldots, t_n)$ and we want to preprocess them to answer queries for windowed triangle intersections. First, we categorize all possible orientations of triangle intersections. Figure 2 illustrates three orientations of a query triangle t_q that intersects with a triangle t_i. We describe inputs and query types for each of the cases.

Case (a): A sequence of triangles is stored and we ask the query: Given a point p, is p contained in some triangle?
Case (b): A sequence of points $(p_1, p_2, \ldots, p_n)$ (one vertex of each triangle in $(t_1, t_2, \ldots, t_n)$) is stored and we ask the query: Given a triangle t_q, does t_q contain some point p_i?
Case (c): A sequence of triangles $(t_1, t_2, \ldots, t_n)$ is stored and we ask the query: Given a triangle t_q, does t_q overlap some triangle t_i?

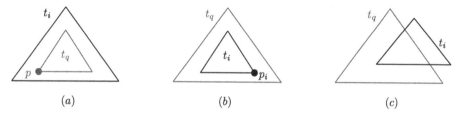

(a)   (b)   (c)

Fig. 2. Possible orientations of intersections of a query triangle t_q (blue) with some triangle t_i (black). (Color figure online)

For cases (a) and (c), by repeated applications of Corollary 7.3(i) and Corollary 7.5 in [4] we can build a data structure that can answer such queries with preprocessing time $O(m \cdot polylog(n))$ and query time $O(n/\sqrt{m} \cdot polylog(n))$, where $m = n^{4/3}$. For case (b), we build a data structure by applying Corollary 7.5 and Corollary 7.7(i) in [4], which also requires the same preprocessing and query time as mentioned for the previous two cases. Finally, we put together all cases in a single data structure $DS(T)$ that we use as the secondary data structure stored at each node of our main search tree.

Corollary 3. *Given a sequence T of n triangles, we can preprocess T for the windowed triangle intersection decision problem in $O(n^{4/3} \cdot polylog(n))$ time using $O(n^{4/3} \cdot polylog(n))$ space.*

We observe that our data structure for the windowed triangle intersection problem can be extended to any convex polygon with c sides, where c is a constant. Solving the *windowed c-gon intersection decision problem* will add some extra levels to our structure, and thus increase the *polylog* factors in the preprocessing time. Hence we obtain the following result.

Corollary 4. *For some constant c, given a sequence S of n convex c-gons (poly-gons with c sides), we can preprocess S for the windowed c-gon intersection decision problem in $O(n^{4/3} \cdot polylog(n))$ time using $O(n^{4/3} \cdot polylog(n))$ space.*

3   Points on Maximal Layers

In this section, we present some results for windowed queries on points on maximal layers of a given sequence of points in $\mathbb{R}^d$, $d \geq 2$. Let P be a set of n points in $\mathbb{R}^2$. The first maximal layer L_1 of P is defined to be the maximal points under the dominance relation where a point p is said to be dominated by a point p' if $p[x] \leq p'[x]$ and $p[y] \leq p'[y]$, and $p \neq p'$. For $\gamma > 1$, the γ'th maximal layer L_γ is the set of maximal points in $P - \cup_{l=1}^{\gamma-1} L_l$ [8]. For a point $q = (q_1, q_2) \in \mathbb{R}^2$, we define $NE(q)$ to be the set of points in $\mathbb{R}^2$ that lie in the North-East quadrant of q, i.e., $NE(q) = \{(a, b) \in \mathbb{R}^2 : a > q_1, \text{ and } b > q_2\}$, and $SW(q)$ to be the set of points in $\mathbb{R}^2$ that lie in the South-West quadrant of q, i.e., $SW(q) = \{(c, d) \in \mathbb{R}^2 : c < q_1, \text{ and } d < q_2\}$ (see Fig. 3).

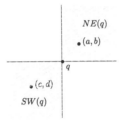

Fig. 3. A point $(a, b) \in NE(q)$ and a point $(c, d) \in SW(q)$.

We present results on windowed queries to count points on the maximal layer, and to report whether a given point is on the layer $\gamma = 1$ or 2 or ≥ 3. We also solve the problem of reporting all maximal points within a query window that dominate at least k points, where k is a fixed integer. We call it the *k-dominant points* problem. In this paper we present solutions to all problems in $\mathbb{R}^2$. All generalized solutions to $\mathbb{R}^d$, where $d \geq 2$ (see Table 1), can be found in the full version of the paper.

3.1 p_k on the Maximal Layer L_1

Suppose $P = (p_1, p_2, \ldots, p_n)$ is a sequence of n points in $\mathbb{R}^2$. We want to preprocess P into a data structure such that given a query interval $[i, j]$ and an integer k with $i \leq k \leq j$ we can report whether the point p_k is on the maximal layer L_1 of $P_{i,j} = (p_i, p_{i+1}, \ldots, p_j)$. We assume that no two points have the same x-coordinate or the same y-coordinate. Let $p_0 = (\infty, \infty)$ and $p_{n+1} = (\infty, \infty)$ be two new points added to P. Let $A[0 .. n + 1]$ be an array, where $A[i] = p_i$ for all $0 \leq i \leq n + 1$. For any k with $1 \leq k \leq n$, we define the following two functions: $\alpha(k) = \min\{i : i > k$ and $p_i \in NE(p_k)\}$ and $\beta(k) = \max\{i : i < k$ and $p_i \in NE(p_k)\}$

A point p_k is on the maximal layer L_1 of a sequence of points $P_{i,j} = (p_i, p_{i+1}, \ldots, p_j)$ if none of these points dominates p_k in $[i, j]$. We have the following lemma.

Lemma 2. *Suppose $1 \leq i \leq k \leq j \leq n$. The point p_k is on the maximal layer L_1 of $P_{i,j} = (p_i, p_{i+1}, \ldots, p_j)$ if and only if $\alpha(k) > j$ and $\beta(k) < i$ (Fig. 4).*

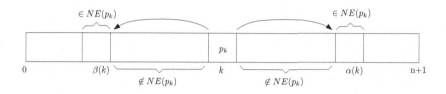

Fig. 4. $A[\alpha(k)]$ and $A[\beta(k)]$ of a point p_k in array $A[0 \ldots n + 1]$.

Suppose, we have a data structure that computes $\alpha(k)$ and $\beta(k)$ for each $p_k \in P$. Now we augment array A such that every element $A[k]$ stores two pointers pointing to $A[\alpha(k)]$ and $A[\beta(k)]$, respectively. This data structure requires $O(n)$ space. Now according to Lemma 2, we can answer the query whether a given point p_k is on the maximal layer of $P_{i,j}$ with $i \leq k \leq j$ in $O(1)$ time by checking $A[\alpha(k)]$ and $A[\beta(k)]$.

Lemma 3. *Suppose a sequence of n points $P = (p_1, p_2, \ldots, p_n)$ in $\mathbb{R}^2$ is given and there exists a data structure that computes $\alpha(k)$ and $\beta(k)$ for each $p_k \in P$ using $S(n)$ space and $T(n)$ time. P can be preprocessed into a data structure of size $O(n)$ in $O(T(n))$ time such that given a query interval $[i, j]$ and a point p_k with $i \leq k \leq j$, we can report whether p_k is on the maximal layer L_1 of $P_{i,j}$ in $O(1)$ time.*

Data structure for computing $\alpha(.)$ and $\beta(.)$ for points in $\mathbb{R}^2$: Given a sequence of n points $P = (p_1, p_2, \ldots, p_n)$ in plane, we want to build a data structure to compute $\alpha(k)$ and $\beta(k)$ for each p_k, $1 \leq k \leq n$. We present the technique for computing $\alpha(k)$ here (see Algorithm 1). We initialize an empty priority search tree (PST) T. For $1 \leq i \leq n$, we query T with $q = (-\infty, p_{i,x}] \times (-\infty, p_{i,y}]$, and find points that appear before p_i in the sequence and are dominated by p_i. Let this set of points be S. According to the definition of $\alpha(.)$, i becomes the $\alpha(.)$ value for all these points. For each $p_k \in S$ we set $\alpha(k) = i$ and delete p_k from T. Now we insert p_i into T. More specifically we maintain the following invariant.

- For each k with $1 \leq k \leq i$: if p_k is in T, then $\alpha(k) \geq i$. If p_k is not in T, then $\alpha(k) < i$ and $\alpha(k)$ has been determined.

Algorithm 1. SetAlpha(P)

Input   : A sequence of n points $P = (p_1, p_2, \ldots, p_n) \in \mathbb{R}^2$.
1 Initialize an empty PST T.
2 **for** $i = 1$ *to* n **do**
3 $\quad$ Query T with $q = (-\infty, p_{i,x}] \times (-\infty, p_{i,y}]$.
4 $\quad$ Let S be the output of this query.
5 $\quad$ **for** *each* $p_k \in S$ **do**
6 $\quad\quad$ Set $\alpha(k) = i$.
7 $\quad\quad$ Delete p_k from T.
8 $\quad$ Insert p_i into T.

This data structure requires $O(n)$ space and $O(n \log n)$ time to set $\alpha(k)$ for all $p_k \in P$, where $1 \leq k \leq n$. Similarly we can compute $\beta(k)$ for all p_k by reversing their order of insertion to T.

Lemma 4. *Given a sequence of n points $P = (p_1, p_2, \ldots, p_n)$ in $\mathbb{R}^2$, we can compute the values of $\alpha(k)$ and $\beta(k)$ for all $p_k \in P$, in $O(n \log n)$ total time using $O(n)$ space.*

Remark: Bannister et al. [1] used a dynamic data structure for dominance queries by Mortensen [11] to compute $\alpha(.)$ and $\beta(.)$ values for all points in $\mathbb{R}^d$. Our data structure for computing all $\alpha(.)$ and $\beta(.)$ improves the amount of time by a factor of $O(\log n)$ and the amount of space by a factor of $O(\log^2 n)$.

Finally from Lemmas 3 and 4 we obtain the following theorem for points in $\mathbb{R}^2$.

Theorem 2. *A sequence of n points $P = (p_1, p_2, \ldots, p_n)$ in $\mathbb{R}^2$ can be preprocessed into a data structure of size $O(n)$ in $O(n \log n)$ time such that given a query interval $[i, j]$ and a point p_k with $i \leq k \leq j$, we can report whether p_k is on the maximal layer L_1 of points $P_{i,j}$ in $O(1)$ time.*

3.2   Count Points on Maximal Layer L_1

Given a sequence of n points $P = (p_1, p_2, \ldots, p_n)$ in $\mathbb{R}^2$, and a query interval $[i, j]$ with $1 \leq i \leq j \leq n$, we want to count the total number of points on the maximal layer L_1 of $P_{i,j} = (p_i, p_{i+1}, \ldots, p_j)$.

Following Lemma 2, we transform each point $p_k = (p_{k,x}, p_{k,y}) \in P$ into a point $p'_k = (k, \alpha(k), \beta(k)) \in \mathbb{R}^3$ for $1 \leq k \leq n$. Now we have a set of n points in $\mathbb{R}^3$. We can compute all $\alpha(k)$ and $\beta(k)$ in $O(n \log n)$ time by Lemma 4. We build a standard 3-dimensional range tree [9], where the first level of the tree is based on the time of the points. At the second level of the tree, for each canonical node we build a range tree using the second ($\alpha(k)$) and the third coordinates ($\beta(k)$) of each point p'_k. The total space requirement is $O(n \log^2 n)$ and this data structure can be built in $O(n \log^2 n)$ time [9].

We transform a given query interval $[i, j]$ into a query box $[i, j] \times [j+1, +\infty) \times (-\infty, i-1]$. The first level of the range tree is queried using interval $[i, j]$. This requires $O(\log n)$ query time. For each canonical subset in the second level, we query using $[j+i, +\infty) \times (-\infty, i-1]$. This step requires $O(\log n)$ time for each canonical node on the search path. Thus, given any query interval $q = [i, j]$ we can report the total number of maximal points in $P_{i,j}$ in $O(\log^2 n)$ time.

Theorem 3. *A sequence of n points $P = (p_1, p_2, \ldots, p_n)$ in $\mathbb{R}^2$ can be preprocessed into a data structure of size $O(n \log^2 n)$ in $O(n \log^2 n)$ time such that given a query interval $[i, j]$ we can report the total number of maximal points of $P_{i,j}$ in $O(\log^2 n)$ time.*

3.3 p_k on Maximal Layer L_γ, Where $\gamma = 2$ or $\gamma \geq 3$

Given a sequence of n points $P = (p_1, p_2, \ldots, p_n)$ in $\mathbb{R}^2$, a query interval $[i, j]$ and an integer k with $1 \leq i \leq k \leq j \leq n$, we want to report if point p_k is on maximal layer L_γ, where $\gamma = 2$ or ≥ 3 of $P_{i,j} = (p_i, p_{i+1}, \ldots, p_j)$. First we solve the problem for $\gamma \geq 3$ and then show that the result for $\gamma = 2$ follows.

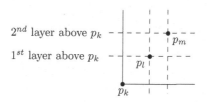

Fig. 5. Point p_k on some maximal layer ≥ 3 iff $p_l \in NE(p_k)$ and $p_m \in NE(p_l)$.

Lemma 5. *Recall the definitions of $\alpha(.)$ and $\beta(.)$ from Sect. 3.1. Suppose $1 \leq i \leq k \leq j \leq n$. The point p_k is on layer L_γ, where $\gamma \geq 3$ of $P_{i,j} = (p_i, p_{i+1}, \ldots, p_j)$ if and only if at least one of the following is true.*

1. *There exists some l such that $l \geq i$, $p_l \in NE(p_k)$, and $\alpha(l) \leq j$*
2. *There exists some l such that $l \leq j$, $p_l \in NE(p_k)$, and $\beta(l) \geq i$.*

Proof (Lemma 5). The 'if' part is obvious from the definitions of $\alpha(l)$ and $\beta(l)$, and by Lemma 2. To prove the converse, we assume p_k to be a point on some maximal layer L_γ, where $\gamma \geq 3$, of points in $P_{i,j}$. Then there must exist some l and m such that $i \leq l \leq j$, $i \leq m \leq j$, $p_l \in NE(p_k)$ and $p_m \in NE(p_l)$ (see Fig. 5). Now, point p_m can come at one of two positions with respect to p_l. *Case 1*: Suppose p_m comes after p_l, i.e., $m > l$. Since $\alpha(l) \leq m$ by the definition of $\alpha(l)$, we obtain $\alpha(l) \leq m \leq j$. *Case 2*: Suppose p_m comes before p_l, i.e., $m < l$. By the similar argument since $\beta(l) \geq m$ by the definition of $\beta(l)$, we obtain $\beta(l) \geq m \geq j$. $\square$

We map each point $p_l = (p_{l,x}, p_{l,y}) \in P$, where $1 \leq l \leq n$, to a point in $\mathbb{R}^4$ as follows. For part (1) of Lemma 5, we define a function $f(l) = (p_{l,x}, p_{l,y}, l, \alpha(l)) \in \mathbb{R}^4$. We set $S = \{f(l) : 1 \leq l \leq n\}$. Similarly, for part (2) of Lemma 5, we define a function $g(l) = (p_{l,x}, p_{l,y}, l, \beta(l)) \in \mathbb{R}^4$. We set $T = \{g(l) : 1 \leq l \leq n\}$. For each $1 \leq l \leq n$, $\alpha(l)$ and $\beta(l)$ can be computed in $O(n \log n)$ time. Now we have two sets S and T each having n points in $\mathbb{R}^4$. We store S and T using two 4-dimensional range trees. A standard 2-dimensional range tree requires $O(n \log n)$ space and can be built in $O(n \log n)$ time. For each additional level the required time and space increase by a logarithmic factor. Therefore our 4-dimensional range tree can be built using $O(n \log^3 n)$ space in $O(n \log^3 n)$ time. Now to answer the query whether some point p_k is on layer ≥ 3 in $P_{i,j}$, we define two functions to map our original query (i, j, k) to equivalent queries in $\mathbb{R}^4$ as follows. For $1 \leq i \leq k \leq j \leq n$, let $F(i, j, k)$ be a function such that $F(i, j, k) = [p_{k,x}, \infty) \times [p_{k,y}, \infty) \times [i, \infty) \times (-\infty, j]$. Similarly, let $G(i, j, k)$ be a function such that $G(i, j, k) = [p_{k,x}, \infty) \times [p_{k,y}, \infty) \times (-\infty, j] \times [i, \infty)$. It gives us the following lemma.

Lemma 6

1. *There exists some l such that $l \geq i$, $p_l \in NE(p_k)$, and $\alpha(l) \leq j$ if and only if $F(i, j, k) \cap S \neq \emptyset$.*
2. *There exists some l such that $l \leq j$, $p_l \in NE(p_k)$, and $\beta(l) \geq i$ if and only if $G(i, j, k) \cap T \neq \emptyset$.*

Combining Lemmas 5 and 6, the query of the form 'Given $1 \leq i \leq k \leq j \leq n$, decide if p_k is on maximal layer $\gamma \geq 3$ of $P_{i,j}$' becomes an equivalent query of the form 'Given a set of points in $\mathbb{R}^4$, count the number of points in the range of 4-dimensional quadrants'. This new query can be answered by querying the data structures on the point sets S and T using 4-dimensional quadrants defined by $F(i, j, k)$ and $G(i, j, k)$, respectively. If at least one of these queries returns some point (i.e., the range count is non-zero) then p_k is on layer $\gamma \geq 3$. Each query takes $O(\log^3 n)$ time. The following theorem summarizes the results.

Theorem 4. *Suppose $P = (p_1, p_2, \ldots, p_n)$ is a sequence of n points in $\mathbb{R}^2$. P can be preprocessed into a data structure of size $O(n \log^3 n)$ in $O(n \log^3 n)$ time such that given a query interval $[i, j]$ and k, where $i \leq k \leq j$, we can answer whether p_k is on some maximal layer L_γ, where $\gamma \geq 3$, of $P_{i,j} = (p_i, p_{i+1}, \ldots, p_j)$ in $O(\log^3 n)$ time.*

Corollary 5 follows from the results of Theorems 2 and 4.

Corollary 5. *Suppose $P = (p_1, p_2, \ldots, p_n)$ is a sequence of n points in $\mathbb{R}^2$. P can be preprocessed into a data structure of size $O(n \log^3 n)$ in $O(n \log^3 n)$ time such that given a query interval $[i, j]$ and k, where $i \leq k \leq j$, we can answer whether p_k is on the second maximal layer L_2 of $P_{i,j} = (p_i, p_{i+1}, \ldots, p_j)$ in $O(\log^3 n)$ time.*

3.4   Report k-Dominant Points

Given a sequence of n points $P = (p_1, p_2, \ldots, p_n)$ in $\mathbb{R}^2$, and a fixed constant k, we want to report all k-*dominant points* in $P_{i,j}$ in the query interval $[i, j]$ (i.e., we want to report all maximal points of $P_{i,j} = (p_i, p_{i+1}, \ldots, p_j)$ such that each point dominates at least k other points of $P_{i,j}$). For any l with $1 \leq l \leq n$, we define the following. Let $p_{l_1}, p_{l_2}, \ldots, p_{l_k}$ be the first k points that are dominated by p_l and come after l according to this sequence. Similarly, let $p_{l'_k}, \ldots, p_{l'_2}, p_{l'_1}$ be the last k points that are dominated by p_l and come before l according to the given sequence (see Fig. 6). Then each interval (l'_{k-a}, l_a) with $0 \leq a \leq k$ represents k points that are dominated by point p_l. Here $l'_0 = l_0 = l$. There exist at most $k + 1$ such intervals for each point in P. We obtain the following lemma.

Lemma 7. *Suppose $1 \leq i < j \leq n$, and k is a fixed integer. A point $p_l \in P_{i,j}$ dominates at least k points in $P_{i,j}$ if and only if there exists some a such that $i \leq l \leq j$, $l_a \leq j$ and $l'_{k-a} \geq i$, where $0 \leq a \leq k$.*

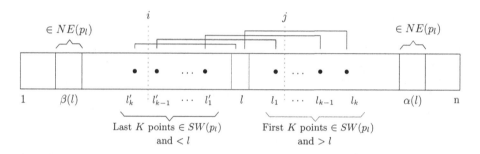

Fig. 6. Proof of Lemma 8. For an example, the highlighted interval (l'_{k-1}, l_1) satisfies the query interval $[i, j]$.

Now for each $1 \leq l \leq n$, we map each point p_l to at most $k + 1$ points $(l, \alpha(l), \beta(l), l_a, l'_{k-a}) \in \mathbb{R}^5$, where $0 \leq a \leq k$. This gives us a set of at most $(k + 1)n$ points. From Lemmas 2 and 7 we obtain the following.

Lemma 8. *Suppose $1 \leq i \leq l \leq j \leq n$ and k is a fixed integer. The point p_l is a maximal point in $P_{i,j}$ that dominates at least k points in $P_{i,j}$, if and only if $i \leq l \leq j$, there are at least k points in $SW(p_l)$ and $\alpha(l) > j$, $\beta(l) < i$.*

We use our data structure presented in Algorithm 1 (see Sect. 3.1) to find the points $p_{l_1}, p_{l_2}, \ldots, p_{l_k}$ and $p_{l'_k}, \ldots, p_{l'_2}, p_{l'_1}$ for each p_l. This time we only insert points to the structure and no points are deleted from it. Starting with $l = 1$ we insert point p_l to T and store the last k points that are dominated by p_l. Next we repeat inserting points to T in the reverse order starting from p_n to p_1 and store the first k points that are dominated by p_l. This process takes $O(n \log n)$ time and $O(n)$ space (by Lemma 4).

We build a 3-dimensional range tree for the first three coordinates of $(k+1)n$ points. At the last level, for each canonical node we add a PST that is built on the last two coordinates of the points stored in that node. It takes $O(kn\log^4 n)$ time and $O(kn\log^3 n)$ space in total. We next map our query interval $q = [i,j]$ to $q' = [i,j] \times [j+1,\infty) \times (-\infty, i-1] \times (-\infty, j] \times [i,\infty))$. If we query our data structure with q', each point p_l will be reported at most $k+1$ times. To report each point exactly once we build an array $R[1..n]$ initially storing 0 in all $R[l]$, $1 \leq l \leq n$. Each time a point p_l is reported during query, we first check the value stored in $R[l]$. If $R[l]$ contains 0 then p_l is seen for the first time; we report p_l and update $R[l] \leftarrow 1$. If $R[l]$ contains 1 then p_l is already reported before and we do not report it this time. The entire query takes $O(\log^4 n + kw)$ time, where w is the output size.

Theorem 5. *Suppose $P = (p_1, p_2, \ldots, p_n)$ is a sequence of n points in $\mathbb{R}^2$ and k is a fixed integer. P can be preprocessed into a data structure of size $O(kn\log^3 n)$ in $O(kn\log^4 n)$ time such that given a query interval $[i,j]$ we can report all k-dominant points in $P_{i,j}$ in $O(\log^4 n + kw)$ time, where w is the output size.*

4   Conclusion

In this paper, we present data structures to solve a number of *windowed* problems using geometric objects. Our window data structures can answer windowed intersection decision queries for line segments, triangles and convex c-gons in plane. We also present solutions for windowed queries for counting maximal points, and reporting whether a given point is on the maximal layers L_γ ($\gamma = 1, 2, \geq 3$) and all k-dominant points in $\mathbb{R}^d$, where $d \geq 2$.

References

1. Bannister, M.J., Devanny, W.E., Goodrich, M.T., Simons, J.A., Trott, L.: Windows into geometric events: data structures for time-windowed querying of temporal point sets. In: Proceedings of the 26th Canadian Conference on Computational Geometry (CCCG) (2014)
2. Bannister, M.J., DuBois, C., Eppstein, D., Smyth, P.: Windows into relational events: data structures for contiguous subsequences of edges. In: Proceedings of the 24th ACM-SIAM SODA, pp. 856–864. SIAM (2013)
3. Bokal, D., Cabello, S., Eppstein, D.: Finding all maximal subsequences with hereditary properties. In: 31st International Symposium on Computational Geometry, SoCG 2015, pp. 240–254 (2015)
4. Chan, T.M.: Optimal partition trees. Discrete Comput. Geom. **47**(4), 661–690 (2012)
5. Chan, T.M., Pratt, S.: Two approaches to building time-windowed geometric data structures. In: 32nd International Symposium on Computational Geometry, SoCG 2016, pp. 28:1–28:15 (2016)
6. Chanchary, F., Maheshwari, A.: Counting subgraphs in relational event graphs. In: Kaykobad, M., Petreschi, R. (eds.) WALCOM 2016. LNCS, vol. 9627, pp. 194–206. Springer, Cham (2016). https://doi.org/10.1007/978-3-319-30139-6_16

7. Chanchary, F., Maheshwari, A., Smid, M.: Querying relational event graphs using colored range searching data structures. In: Gaur, D., Narayanaswamy, N.S. (eds.) CALDAM 2017. LNCS, vol. 10156, pp. 83–95. Springer, Cham (2017). https://doi.org/10.1007/978-3-319-53007-9_8

8. Cormen, T.H.: Introduction to Algorithms. MIT press, Cambridge (2009)

9. De Berg, M., Van Kreveld, M., Overmars, M., Schwarzkopf, O.C.: Computational Geometry. Springer, Heidelberg (2000)

10. McCreight, E.M.: Priority search trees. SIAM J. Comput. **14**(2), 257–276 (1985)

11. Mortensen, C.W.: Fully dynamic orthogonal range reporting on RAM. SIAM J. Comput. **35**(6), 1494–1525 (2006)

12. Mouratidis, K., Bakiras, S., Papadias, D.: Continuous monitoring of top-k queries over sliding windows. In: Proceedings of the 2006 ACM SIGMOD International Conference on Management of Data, pp. 635–646. ACM (2006)

Bounded Stub Resolution for Some Maximal 1-Planar Graphs

Michael Kaufmann[1], Jan Kratochvíl[2], Fabian Lipp[3(✉)],
Fabrizio Montecchiani[4], Chrysanthi Raftopoulou[5], and Pavel Valtr[2]

[1] Universität Tübingen, Tübingen, Germany
mk@informatik.uni-tuebingen.de
[2] Charles University, Prague, Czech Republic
honza@kam.mff.cuni.cz
[3] Universität Würzburg, Würzburg, Germany
fabian.lipp@uni-wuerzburg.de
[4] Università degli Studi di Perugia, Perugia, Italy
fabrizio.montecchiani@unipg.it
[5] National Technical University of Athens, Athens, Greece
crisraft@mail.ntua.gr

Abstract. The resolution of a drawing plays a crucial role when defining criteria for its quality and readability. In the past, grid resolution, edge-length resolution, angular resolution and crossing resolution have been investigated. We continue the study of the recently introduced *stub resolution* as an additional aesthetic criterion for nonplanar drawings of graphs. A crossed edge is divided into parts, called stubs, which should not be too short for the sake of readability. Thus, the stub resolution of a drawing is defined as the minimum ratio between the length of a stub and the length of the entire edge containing that stub, over all the edges of the drawing. As a meaningful graph class, where crossings are naturally involved, we consider 1-planar graphs (i.e., graphs that allow planar drawings in which every edge is crossed at most once). In an attempt to prove the conjecture that the stub resolution of 1-planar graphs is bounded, we closely investigate a class of maximal 1-planar graphs arising from double-wheels. We show that each such graph allows a straight-line 1-planar drawing with stub resolution $\frac{1}{5}$.

1 Introduction

A *straight-line drawing* of an undirected graph G is a mapping of its vertices to distinct points in the Euclidean plane, with edges being represented by

This research was initiated at the Bertinoro Workshop on Graph Drawing 2017. Research by J. Kratochvíl and P. Valtr was supported by project CE-ITI no. P202/12/G061 of the Czech Science Foundation (GAČR). F. Lipp was partially supported by Cusanuswerk. Research of Fabrizio Montecchiani supported in part by the project: "Algoritmi e sistemi di analisi visuale di reti complesse e di grandi dimensioni"- Ricerca di Base 2017, Dipartimento di Ingegneria dell'Universita degli Studi di Perugia".

© Springer International Publishing AG 2018
B. S. Panda and P. P. Goswami (Eds.): CALDAM 2018, LNCS 10743, pp. 214–220, 2018.
https://doi.org/10.1007/978-3-319-74180-2_18

straight-line segments connecting the points representing their end-vertices. We will only consider drawings in which no edge passes through a vertex (other than its end-points) and no three edges cross in the same point of the plane.

An edge e of a drawing Γ that is crossed k times is divided into $k+1$ parts called *stubs*. Let l_e and s_e be the length of e and of its shortest stub, respectively. The *stub resolution* of e is $\mathrm{sr}_e = \frac{s_e}{l_e}$. The *stub resolution* of Γ is the minimum stub resolution over all edges of Γ, i.e., $\mathrm{sr}_\Gamma = \min_{e \in \Gamma} \mathrm{sr}_e$.

A drawing is aesthetically pleasing if all stubs of an edge are as equal in length as possible. That is, ideally, $\mathrm{sr}_e = \frac{1}{k+1}$ if e is crossed by k other edges. In this paper we consider *straight-line 1-planar graphs*, i.e., graphs that allow straight-line drawings in which every edge is crossed by at most one other edge. For such graphs the optimal stub resolution might be $\frac{1}{2}$, but this cannot be always attained. In a companion paper [2] we show that each straight-line drawing of the complete graph on five vertices K_5 has stub-resolution strictly smaller than $\frac{1}{2}$. A natural question raised in [2] asks whether the stub-resolution of straight-line 1-planar graphs is bounded from below, i.e., if there is a constant $\delta > 0$ such that every straight-line 1-planar graph allows a straight-line 1-planar drawing such that $\mathrm{sr}_e > \delta$ for every edge e of G.

Straight-line 1-planar graphs have been studied by Didimo in [1] who showed that every straight-line 1-planar graph on n vertices has at most $4n - 9$ edges and that this bound is achieved by infinitely many graphs. A large class of these *optimal straight-line 1-planar graphs* is obtained by deleting a single edge from the (extended) double wheel, which is the graph obtained by taking the second power of an even cycle C_k and adding a pair of vertices connected to all vertices of C_k (see also the more formal definition at the beginning of the next section). The main result of this paper is to show that these graphs have straight-line 1-planar drawings with stub resolution bounded (from below) by $\frac{1}{5}$. We find this result quite surprising. It seems crucial for our construction that some edges in it are much longer than others — the ratios between the lengths of some pairs of edges are exponentially large.

Straight-line 1-planar graphs form an important proper subclass of the class of *1-planar graphs* which are graphs that allow (not necessarily straight-line) drawings in which every edge is crossed by at most one other edge (see the recent survey [3] on 1-planar graphs). In the above mentioned companion paper [2] we show that 1-planar graphs have 1-planar drawings with stub resolution bounded from below by a positive constant when we allow a bounded number of bends on the edges.

2   Constructions with Large Stub Resolution

For $k \geq 3$, we define the *double-wheel* DW_k (frequently called the *extended double-wheel* in the literature) as the graph obtained by taking the second power $(C_k)^2$ of a cycle C_k of an even length k and adding two *special* vertices connected to all vertices of C_k. Thus, DW_k has $n := k + 2$ vertices and $4n - 8 = 4k$ edges.

2.1   Semi-Double-Wheel

We first establish constant lower bounds on the stub resolution of certain straight-line 1-planar drawings having almost as many edges as possible.

Theorem 1. *For every $n \geq 4$, there is a 1-plane straight-line drawing of a graph on n vertices with $4n - 10$ edges and with stub resolution $1/5$.*

Proof. Consider a double-wheel DW_k with vertices along the k-cycle denoted by $a_1, \ldots, a_k$ and with the two special vertices denoted s and t.

The *semi-double-wheel* SDW_k is obtained in the same way as DW_k with the exception that we use the path $P_k = a_1 a_2 \ldots a_k$ instead of C_k during the construction. Equivalently, SDW_k is obtained from DW_k by removing the three edges $a_1 a_k$, $a_1 a_{k-1}$ and $a_2 a_k$. The graph SDW_k has $n := k + 2$ vertices and $4k - 3 = 4n - 11$ edges.

To prove Theorem 1 we show that SDW_k has a straight-line 1-plane drawing with stub resolution $1/5$, even after we add the edge st to it. We determine the drawing by setting the coordinates of the vertices of SDW_k (see Fig. 1). For $i := 1, \ldots, k$, we put $a_i := (2^i, (-1)^{i+1})$. Further, we put $s := (0, -9)$ and $t := (0, 9)$. For each odd i, $3 \leq i \leq k - 1$, the edges $a_i s$ and $a_{i-1} a_{i+1}$ cross in the point $((4/5)2^i, -1)$ and the stub resolution of each of them is exactly $1/5$. Similarly, for each even i, $i \leq k - 1$, the edges $a_i t$ and $a_{i-1} a_{i+1}$ cross in the point $((4/5)2^i, 1)$ and the stub resolution of each of them is also $1/5$. All the other pairs of edges do not intersect. This is true also for the additional edge st. This finishes the proof of Theorem 1.   □

2.2   Double-Wheel Without an Edge

Here we show that there are arbitrarily large straight-line 1-planar drawings with the maximal number of edges and with the stub resolution bounded from below by a positive constant.

There are three types of edges in the double-wheel DW_k for an even $k \geq 6$. The three types are:

   C-*edges*: the k edges of the cycle C_k,
   B-*edges*: the k edges of $(C_k)^2$ not lying on the cycle C_k, and
   A-*edges*: the $2k$ edges connecting the two special vertices with the k vertices of C_k.

Theorem 2. *If we remove from DW_k an A-edge or a B-edge, the obtained graph is a 1-planar graph with n vertices and $4n - 9$ edges which has a 1-planar straight-line drawing with stub resolution $1/5$ for infinitely many values of k.*

Proof. **Case 1:** Removing an A-edge from DW_k:

We assume that k is equal to 2 (mod 4) and that it is at least 10. We use vertices with the following coordinates (see Fig. 2):

   $a_i := (2^i, (-1)^{i+1} + 17 \cdot 2^{i+1-k/2})$, for $i = 1, 2, ..., k/2 - 1$,
   $a_0 := (0, -1)$,

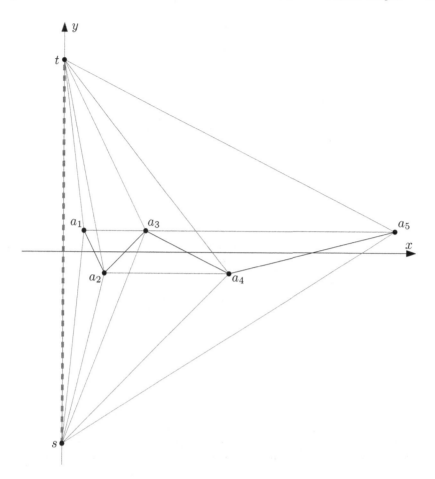

Fig. 1. Drawing of SDW_5 with stub resolution $1/5$. The edges of P_5 are drawn black, the edges of $(P_5)^2$ not in P_5 are drawn blue, and the edges from a path vertex to a special vertex are drawn red. Adding the dashed green edge st does not change the stub resolution of the drawing. (Color figure online)

$a_i := (-2^{-i}, (-1)^{-i+1} + 17 \cdot 2^{-i+1-k/2})$, for $i = 1 - k/2, 2 - k/2, ..., -2, -1$, [thus, a_i and a_{-i} are always reflections of each other along the y-axis]
$z := (0, 15)$,
$s := (0, -9)$,
$t := (0, 9)$.
The vertices of the double wheel DW_k are embedded in the plane such that the k-cycle C_k is the cycle $C_k = a_{1-k/2}, a_{2-k/2}, ..., a_{-1}, a_0, a_1, ..., a_{k/2-1}, z$, and the vertices s and t are the two special vertices of DW_k.

Now, a straight-line graph DW_k^{-A} with $n = k + 2$ vertices and $4n - 9 = 4k - 1$ edges is obtained by drawing all the edges of DW_k, except the A-edge sz, as the straight-line segments between the corresponding pairs of vertices.

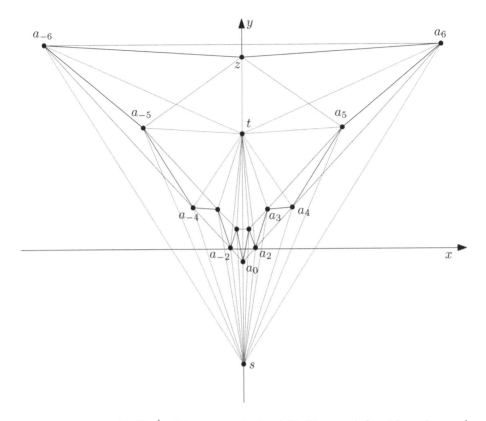

Fig. 2. Drawing of DW_{14}^{-A} with stub resolution 1/5. The graph has 16 vertices and $55 = 4 \cdot 16 - 9$ edges. The fourteen C-edges are drawn black, the fourteen B-edges are drawn blue, and the twenty-seven A-edges are drawn red. (Color figure online)

The stub resolution of DW_k^{-A} is 1/5, which can be seen as follows. For most of the pairs of crossing edges, we can use the same argument as in the construction of the semi-double-wheel because the drawing is symmetric about the y-axis, and on the right of the y-axis, the point/vertex set (including s and t, without a_0 and z) is just an affine transformation of the construction of the semi-double-wheel with $(k/2 - 1) + 2$ vertices. [The affine transformation keeps the x-coordinates fixed and for each point it adds a (small) constant multiple of the x-coordinate to the y-coordinate. More precisely, it is the affine transformation which maps (x, y) to $(x, y + 17 \cdot 2^{1-k/2} \cdot x)$.] Observe that affine transforms do not change stub resolutions. Thus, we only have to look at two types of pairs of crossing edges:

(i) those pairs where at least one of the edges properly crosses the y-axis,
(ii) those pairs which involve at least one of the vertices a_0 and z.

There is one pair of both types (which is the only pair of type (i)):

$$ta_0 \ vs. \ a_{-1}a_1.$$

Additionally, there are the following four pairs of type (ii):

$$ta_{k/2-1} \text{ vs. } za_{k/2-2},$$

$$ta_{1-k/2} \text{ vs. } za_{2-k/2},$$

$$a_0a_2 \text{ vs. } sa_1,$$

$$a_0a_{-2} \text{ vs. } sa_{-1}.$$

In the first two pairs of type (ii), the stub resolution of each edge is $1/3$ (due to the choice of the y-coordinate of z). In the last two pairs of type (ii), the stub resolution of the edge a_0a_2 (a_0a_{-2}, resp.) is bigger than $1/5$ and the stub resolution of the edge sa_1 (sa_{-1}, resp.) is exactly $1/5$. In the pair of both types, the edge $a_{-1}a_1$ has stub resolution $1/2$ and the stub resolution of the other edge is approaching $1/5$ from above (for k large enough) as k goes to infinity. This finishes the argument in Case 1 (double-wheel minus an A-edge).

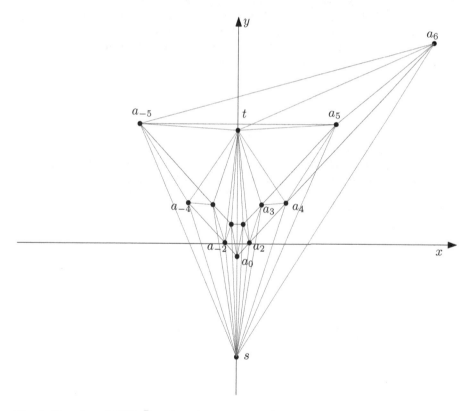

Fig. 3. Drawing of DW_{12}^{-B} with stub resolution $1/5$. The graph has 14 vertices and 47 edges. The twelve C-edges are drawn black, the eleven B-edges are drawn blue, and the twenty-four A-edges are drawn red. (Color figure online)

Case 2: Removing a B-edge from DW_k:

We use almost the same construction as above with the constant 17 replaced by 20. We remove the two points $a_{1-k/2}$ and z from the cycle C_k and keep the circular order of the remaining $k-2$ points unchanged (see Fig. 3). This gives us a $(k-2)$-cycle. The points s and t have the same coordinates as above. Then we consider the corresponding double-wheel DW_{k-2} on $(k-2)+2 = k$ vertices and remove the B-edge $a_{3-k/2}a_{k/2-1}$. The constant 20 was chosen in such a way that the stub resolution of the edges in the crossing pair $ta_{k/2-1}$ $vs.$ $a_{2-k/2}a_{k/2-2}$ is exactly 1/5 resp. 3/10. For k large enough, the stub resolution of other edges is at least 1/5 by the reasoning as above. This finishes the argument in Case 2 (double-wheel minus a B-edge). This finishes the proof of the theorem.   □

3   Conclusion

In this paper we continued the study of the stub resolution as an aesthetic criterion for drawings of nonplanar graphs, in particular of 1-planar graphs. We showed that the stub resolution is bounded for a certain large class of optimal straight-line 1-planar graphs arising from the double-wheel. The result may seem somewhat isolated, but we feel that it is crucial for understanding the stub resolution of 1-planar graphs and that it will shed light onto the general conjecture.

References

1. Didimo, W.: Density of straight-line 1-planar graph drawings. Inf. Process. Lett. **113**(7), 236–240 (2013)
2. Kaufmann, M., Kratochvíl, J., Lipp, F., Montecchiani, F., Raftopoulou, C., Valtr, P.: The stub resolution of 1-planar graphs (manuscript)
3. Kobourov, S.G., Liotta, G., Montecchiani, F.: An annotated bibliography on 1-planarity. CoRR, abs/1703.02261 (2017)

On Structural Parameterizations of Firefighting

Bireswar Das, Murali Krishna Enduri, Neeldhara Misra,
and I. Vinod Reddy[✉]

IIT Gandhinagar, Gandhinagar, India
{bireswar,endurimuralikrishna,neeldhara.m,reddy_vinod}@iitgn.ac.in

Abstract. The FIREFIGHTING problem is defined as follows. At time
$t = 0$, a fire breaks out at a vertex of a graph. At each time step $t \geqslant 0$, a
firefighter permanently defends (protects) an unburned vertex, and the
fire then spread to all undefended neighbors from the vertices on fire.
This process stops when the fire cannot spread anymore. The goal is to
find a sequence of vertices for the firefighter that maximizes the number
of saved (non burned) vertices.

The FIREFIGHTING problem turns out to be NP-hard even when
restricted to bipartite graphs or trees of maximum degree three. We study
the parameterized complexity of the FIREFIGHTING problem for various
structural parameterizations. All our parameters measure the distance to
a graph class (in terms of vertex deletion) on which the FIREFIGHTING
problem admits a polynomial time algorithm. Specifically, for a graph
class $\mathcal{F}$ and a graph G, a vertex subset S is called a *modulator* to $\mathcal{F}$ if
$G \setminus S$ belongs to $\mathcal{F}$. The parameters we consider are the sizes of modulators to graph classes such as threshold graphs, bounded diameter graphs,
disjoint unions of stars, and split graphs.

To begin with, we show that the problem is W[1]-hard when parameterized by the size of a modulator to diameter at most two graphs and
split graphs. In contrast to the above intractability results, we show that
FIREFIGHTING is fixed parameter tractable (FPT) when parameterized
by the size of a modulator to threshold graphs and disjoint unions of
stars, which are subclasses of diameter at most two graphs. We further
investigate the kernelization complexity of these problems to find that
FIREFIGHTING admits a polynomial kernel when parameterized by the
size of a modulator to a clique, while it is unlikely to admit a polynomial
kernel when parameterized by the size of a modulator to a disjoint union
of stars.

1 Introduction

The FIREFIGHTING problem was introduced by Hartnell [15] to model the spread
of diseases and computer viruses. It is a turn-based game between two players
(the "fire" and the "firefighter"), which is played on a graph G as follows. Initially,
at time $t = 0$, a fire starts at a vertex s, at each following time step the following

M. K. Enduri—Supported by Tata Consultancy Services (TCS) research fellowship.
N. Misra—Supported by a DST-INSPIRE Fellowship.

© Springer International Publishing AG 2018
B. S. Panda and P. P. Goswami (Eds.): CALDAM 2018, LNCS 10743, pp. 221–234, 2018.
https://doi.org/10.1007/978-3-319-74180-2_19

happens. A firefighter defends one vertex which is not on fire, and the fire then spreads from each burning vertex to all its undefended neighbors. Once a vertex is defended it remains so for all time intervals. The process stops when the fire can no longer spread. The natural algorithmic question associated with this game is to find a strategy that optimizes some desirable criteria, for instance, maximizing the number of saved vertices [3], minimizing the number of rounds, the number of firefighters per round [4], or the number of burned vertices [3,9], and so on. These questions are well-studied in the literature, and while most variants are NP-hard, approximation and parameterized algorithms have been proposed for various scenarios. In this work, we will focus on the goal of finding a sequence of defending vertices that maximizes the number of saved (not burned) vertices and we refer to this as the FIREFIGHTING problem. We also use SAVING k-VERTICES to refer to the decision version of this problem, where we are given a demand k and the goal is to save at least k vertices.

We study the parameterized complexity of FIREFIGHTING with respect to various structural parameters. In particular, our focus is on distance-to-triviality parameterizations, wherein we identify classes of graphs on which the FIRE-FIGHTING problem is solvable in polynomial time, and understand the parameterized complexity of the problem parameterized by the distance of a graph to these graph classes. In this paper, our notion of distance to a graph class in the vertex deletion distance. More precisely, for a class $\mathcal{F}$ of graphs, we say that X is an $\mathcal{F}$-modulator of a graph G if $G \setminus X \in \mathcal{F}$. If the size of a smallest modulator to $\mathcal{F}$ is k, we also say that the distance of G to the class $\mathcal{F}$ is k. Throughout this paper, we will assume that a modulator is given to us as a part of the input. This assumption is without loss of generality since such modulators can be computed in FPT time. We are now ready to describe our results.

Our Contributions. The FIREFIGHTING problem is FPT when parameterized by the vertex cover and distance to a clique parameterizations. On the other hand, it is para-NP-hard when parameterized by feedback vertex set, tree-width and clique-width [10]. However, the parameter vertex cover is very restrictive and significantly large for dense graphs. This motivates us to consider parameters that are intermediate between vertex cover and clique-width. In this spirit, Ganian [14] studied the parameterized complexity of FIREFIGHTING problem for parameter twin-cover which is a generalization of vertex cover and showed that FIREFIGHTING is FPT with respect to twin-cover. Recently Chlebíková et al. [5] showed that the problem is FPT parameterized by distance to cluster graphs which is a generalization of twin-cover.

We study the parameterized complexity of the FIREFIGHTING problem with respect to the distance from following graph classes: threshold graphs, disjoint union of stars, disjoint union of graphs of diameter at most two, and split graphs. Studying the parameterized complexity of FIREFIGHTING with respect to these parameters improves the understanding of the boundary between tractable and intractable parameterizations. For instance, the parameterization by distance to cluster graphs (as studied by [5]) directly generalizes both vertex cover and distance to clique Fig. 1. Observe that cluster graphs are precisely the graphs whose

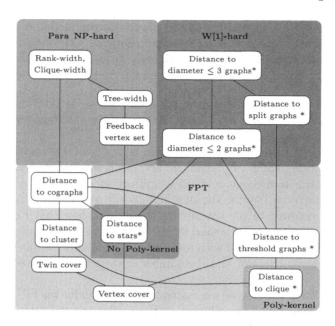

Fig. 1. A schematic showing the parameterized complexity of FIREFIGHTING with respect to various structural graph parameters. There is a line between two parameters if the parameter below is larger than the parameter above. Results shown in this paper are marked by an asterisk (*).

connected components have diameter one, and a natural generalization to consider is the class of graphs whose connected components have diameter two. Here, we show that there is a transition in complexity: the problem becomes W[1]-hard. As a natural intermediate problem, we consider the subclass of graphs where every connected component is a star. Here, using ideas similar to the ones that lead to the FPT algorithm for the distance to cluster parameter, we obtain a FPT algorithm. The case analysis here is more delicate because we have to distinguish between the central vertex and the leaves.

On the other hand, the distance to threshold graphs parameter directly generalizes the distance to clique parameter, while the distance to stars parameter is a generalization of vertex cover. For both of these parameters, we establish that FIREFIGHTING is FPT. These being smaller parameters, our results improve several known algorithms. Note that the next "natural" parameter to consider after distance to stars hierarchy is the feedback vertex number, or the distance to forests; however here the problem is already NP-hard on trees, leading to para-NP-hardness. Similarly, a natural next step from distance to threshold graphs is the distance to split graphs, but here also we demonstrate W[1]-hardness. Finally, a promising generalization from the distance to cluster graphs is the distance to cographs, and here we leave the parameterized complexity of the problem open.

We also consider the kernelization complexity of the problem and make the following advances: for the distance to clique parameterization, we demonstrate a quadratic kernel, while for the distance to stars parameterization, we show that a polynomial kernel is unlikely under standard complexity-theoretic assumptions. The kernelization complexity of the problem relative to vertex cover, however, remains an interesting open problem. We summarize our results below.

- We show that the problem is fixed parameter tractable (FPT) when parameterized by the size of a modulator to threshold graphs, cluster graphs and disjoint unions of stars.
- We further investigate the kernelization complexity of these problems to find that Firefighting admits a polynomial kernel when parameterized by the size of a modulator to a clique, while it is unlikely to admit a polynomial kernel when parameterized by the size of a modulator to a disjoint union of stars.
- Finally, in contrast to the tractability results, we show that FIREFIGHTING is W[1]-hard when parameterized by the distance to split graphs. In fact, the problem remains W[1]-hard with respect to the combined parameter involving the size of the modulator and the number of vertices to be saved.

Methodology. By and large, we use a standard approach for the FPT algorithms: we guess the behavior of the solution on the modulator and attempt to find a solution consistent with the guessed behavior. The second part relies on exploiting the structural properties of $G \setminus X$, which is the part of the graph outside the modulator. Usually one is able to group the vertices of $G \setminus X$ based on the structure of their neighborhoods in the modulator, and argue that all vertices of the same "type" have a similar behavior, which leads to a controlled search space. In the case of threshold graphs, we are able to prove that simple greedy techniques work within a particular type. On the other hand, for disjoint unions of stars, we have to account for several scenarios, and the classification of $G \setminus X$ is more intricate, and accordingly, we have to account for more cases in the analysis. The hardness results follow from reductions using standard techniques, while the kernelization algorithm uses the fact that when $G \setminus X$ is a large clique, several vertices behave in a similar fashion, and this observation allows us to replace the large clique with a much smaller one—a careful argument is required, however, to demonstrate that the instance we constructed in this fashion is indeed equivalent to the original.

Related Work. The FIREFIGHTING problem is known to be NP-hard even for special classes of graphs, including bipartite graphs [18], trees of maximum degree three [10] and cubic graphs [16]. The firefighter problem can be solved in linear time on split graphs and co-graphs [13]. From the parameterized complexity point of view, the firefighting problem is W[1]-hard when parameterized by the number of saved vertices for bipartite graphs [1]. Cai et al. [3] propose several FPT algorithms and polynomial kernels, and they consider the following parameters: the number of saved vertices, the number of saved leaves, and the number of protected vertices. Leung [17] use the random separation method to give FPT algorithms on general graphs parameterized by the number of burnt vertices and on degree bounded graphs and unicyclic graphs parameterized by the number of protected vertices. We refer the reader to the survey [11], as well as the references within, for more details.

2   Preliminaries

In this section, we introduce the notation and the terminology that we will need to describe our algorithms. Most of our notation is standard. We use $[k]$ to denote the set $\{1, 2, \ldots, k\}$. All graphs we consider in this paper are undirected, connected, finite and simple. For a graph $G = (V, E)$, let $V(G)$ and $E(G)$ denote the vertex set and edge set of G respectively. An edge in E between vertices x and y is denoted as xy for simplicity. For a subset $X \subseteq V(G)$, the graph $G[X]$ denotes the subgraph of G induced by vertices of X. Also, we abuse notation and use $G \setminus X$ to refer to the graph obtained from G after removing the vertex set X. For a vertex $v \in V(G)$, $N(v)$ denotes the set of vertices adjacent to v and $N[v] = N(v) \cup \{v\}$ is the closed neighborhood of v.

Graph Classes. We now define the graph classes that we will encounter frequently.

- A graph is a *split graph* if its vertices can be partitioned into a clique and an independent set. Split graphs are P_5-free [12].
- The class of P_4-free graphs are called *co-graphs*.
- A graph is a *threshold graph* if it can be constructed from the one-vertex graph by repeatedly adding either an isolated vertex or a universal vertex.
- A cluster graph is a disjoint union of complete graphs. Cluster graphs are also P_3-free graphs.

It is easy to see that a graph that is both split and co-graph is a threshold graph. We denote threshold graph (or a split graph) with $G = (C, I)$ where C and I denote a partition of G into a clique and an independent set. For any two vertices x, y in a threshold graph G we have either $N(x) \subseteq N[y]$ or $N(y) \subseteq N[x]$. For a class of graphs $\mathcal{F}$, the distance to $\mathcal{F}$ of a graph G is the minimum number of vertices to be deleted from G to get a graph in $\mathcal{F}$.

Parameterized Complexity. A parameterized problem is a pair $Q \subseteq \Sigma^* \times \mathbb{N}$, where Σ is fixed finite alphabet. For an instance $(x, k) \in \Sigma^* \times \mathbb{N}$, k is called the parameter. We say that a parameterized problem Q is *fixed parameter tractable* (FPT) if there exists an algorithm and a computable function $f : \mathbb{N} \to \mathbb{N}$ such that given $(x, k) \in \Sigma^* \times \mathbb{N}$ the algorithm correctly decides whether $(x, k) \in Q$ in $f(k)|(x, k)|^{O(1)}$ time. The function f is usually superpolynomial and only depends on parameter k. The class XP contains the problems which are solvable in time $|(x, k)|^{f(k)}$, the exponent of the running time depends on the parameter k. A problem is *para-NP-hard* if it is NP-hard for some fixed values of the parameter. The complexity class of parameterized intractability is called $W[1]$ (see [8] for the definition). A *kernelization* algorithm is a polynomial time algorithm that takes an instance (x, k) of a parameterized problem Q as input and outputs an equivalent instance (x', k') of P such that $|x'| \leqslant h(k)$ for some computable function h and $k' \leqslant k$. If h is polynomial then we say that (x', k') is a polynomial kernel. Let P and Q be two parameterized problems. A parameterized reduction from P to Q is a mapping $g : \Sigma^* \to \Sigma^*$ such that (i) for all $x \in \Sigma^*$ we have x is a yes instance of $P \Leftrightarrow g(x)$ is a yes instance of Q. (ii) g can be computed in $f(k)|x|^{O(1)}$ time, where f is computable function and k is the parameter of x. (iii) $k' \leqslant h(k)$ for some computable function h, where k and k' are parameters of x and $g(x)$ respectively. For more details on parameterized complexity the reader is referred to [6,8].

We now briefly justify our assumption about the modulators being given as a part of the input to our problems. Consider the following result.

Lemma 1 *[2].* *Let $\mathcal{F}$ be a graph class characterized by the finite forbidden induced subgraphs $H_1, \cdots, H_l$. Given a graph G and integer k, there is an* FPT *algorithm that finds a subset $X \subseteq V(G)$ of size at most k such that $G \setminus X \in \mathcal{F}$ in* $O(d^k n^d)$, *where d is the size of largest forbidden subgraph.*

The class of cluster graphs, threshold graphs and split graphs are P_3, P_4 are P_5 free respectively. By using above lemma, Given a graph G and integer k, the problem of deciding whether there exists a set X of vertices of size at most k whose deletion results in a cluster graph, threshold graph and split graph is fixed parameter tractable.

The Firefighting Problem. Finally, we formally define the problems that we consider in this paper, where $\mathcal{F}$ is a class of graphs or a graph property.

FIREFIGHTING $[\mathcal{F}]$
    Input: A graph G, a vertex s, a modulator $X \subseteq V(G)$ such
        that $G \setminus X \in \mathcal{F}$.
Parameter: The size $k := |X|$ of the modulator to $\mathcal{F}$.
  Question: Find a strategy that maximizes the number of saved
        vertices when a fire starts at s?

SAVING k-VERTICES$[\mathcal{F}]$
    Input: A graph G, a vertex s, a modulator $X \subseteq V(G)$ such
        that $G \setminus X \in \mathcal{F}$, and an integer k.
Parameter: The size $\ell := |X|$ of the modulator to $\mathcal{F}$.
  Question: Does there exists a strategy that saves at least k ver-
        tices when a fire starts at s?

Whenever $\mathcal{F}$ is clear from the context, we drop the explicit mention of it in the name of the problem. We also abuse notation and use k differently in the two definitions, to retain consistency with standard notation when we present reductions in the context of SAVING k-VERTICES.

3   The Parameterized Complexity of Firefighting

Let (G, s) be an instance of FIREFIGHTING problem. The vertices that have not been burned by the fire at the end of the process are called *saved* (including defended vertices). A vertex is *burned* if it is on fire. A strategy for FIREFIGHTING instance (G, s) is a sequence S of vertices $\{v_1, \cdots, v_l\}$ where v_i represents the position of the firefighter at time step i. We say that sequence S is a *valid strategy* for (G, s) if vertex v_i is not burning at the start of time step i and the process stops at time step l. A strategy S is called *minimal* if no subset of S yields a strategy that saves same number of vertices as S. A strategy is *optimal* if it is minimal and saves maximum number of vertices.

The following results from the literature will be useful for our algorithms. We give full proofs for the sake of completeness. The proofs of statements marked with a $(\star)$ are deferred to the full version [7], due to lack of space.

Lemma 2 *[13]. Let G be a graph and s be a vertex of G. Given an ordered set S of vertices of G, we can verify whether S is a valid strategy for the* FIREFIGHTING *problem on (G, s) and count the number of vertices saved by S in $O(n + m)$ time, where n and m denote the number of vertices and edges in G respectively.*

Proof. Let $S = \{v_1, \cdots, v_k\}$ is strategy in this order. To verify whether the sequence S is a valid strategy, we do BFS on the graph G starting from s. Find the distance $d(s, v)$ from the source s to each vertex v of S on graph $G[(V(G) \setminus S) \cup \{v\}]$. For $i = 1$ to k, If S is a valid strategy then distance from source s to the vertex $v_i, d(s, v_i) \geq i$, otherwise vertex v_i will be burned before time step i. Observe that the number of vertices burned by S in G equal to the number of vertices reachable from s in $G \setminus S$, which can be found by applying BFS on the graph $G \setminus S$ starting from s.   □

The proof of the following Lemma follows from the Lemma 2.

Lemma 3. *Let S be an optimal strategy for the* FIREFIGHTING *problem on (G, s) then the number of vertices burned by S in G is equal to the number of vertices reachable from s in $G \setminus S$.*

Lemma 4 *[13]. Let (G, s) be an instance of the* FIREFIGHTING *problem, and let l be the length of a longest induced path in G starting from s. Then any optimal strategy can defend at most l vertices.*

Proof. Let $S = \{v_1, \cdots, v_t\}$ be an optimal strategy defended in this order. Since S is a valid strategy there is an induced path P from s to v_t such that all the vertices on P are burned except v_t. Let P be a shortest such path, then P contains at least $t + 1$ vertices: otherwise v_t will be burned before time t. Since the length of longest induced path in G starting from s is l, we have $t + 1 \leq l + 1$ which implies $t \leq l$.   □

Lemma 5 *[13]. The* FIREFIGHTING *problem can be solved in $O(n^l)$ time on graphs with length of longest induced path is at most $l - 1$.*

Proof. Let (G, s) be an instance of firefighting problem. Since the length of the longest induced path in G is at most l. From Lemma 4 any optimal strategy can defend at most l vertices in G. We list out all possible subsets S of $V(G)$ of size at most l in $O(n^l)$ time. For each such subset S using Lemma 2 test whether S is valid strategy and count the number of vertices saved by S. The optimal strategy is the one which saves maximum number of vertices.   □

3.1   Parameterization by Distance to Threshold Graphs

In this section we give an FPT algorithm for the FIREFIGHTING problem parameterized by the distance to threshold graphs. Without loss of generality, we assume that threshold graph is connected, otherwise all the connected components that do not contain the source vertex are trivially saved.

Lemma 6. *Let* (G, s) *be an instance of* FIREFIGHTING *problem. Then any optimal strategy can defend at most* $2k+2$ *vertices, where* k *is the distance to threshold graphs.*

Proof. We know that fire always spreads along an induced path in G. As the length of the longest induced path in G is at most $2k + 2$, from Lemma 4 any optimal strategy can defend at most $2k + 2$ vertices in G.   □

Let G be a graph and $X \subseteq V(G)$ of size k such that $G \setminus X = (C, I)$ is a threshold graph. We partition the vertices of clique C and independent set I in $G \setminus X$ based on their neighborhoods in X. In particular, for every subset $Y \subseteq X$, let: $T_Y^C := \{x \in C \mid N(x) \cap X = Y\}$ and $T_Y^I := \{x \in I \mid N(x) \cap X = Y\}$.

Notice that in this way we can partition vertices of $G \setminus X$ into at most 2^{k+1} subsets (called types), two for each $Y \subseteq X$. Observe that all vertices in a type have same neighbors in X, where as they may have different neighbors inside the threshold graph. The following result shows that when we need to choose a defending vertex from a type the best strategy is to defend a highest degree vertex in that type. For a strategy S, let $sav(S)$ denote the number of vertices saved by the strategy S.

Lemma 7. *Let* v_1, v_2 *be two vertices in a type* T *such that* $N(v_2) \subseteq N[v_1]$. *Let* S *be a strategy containing* v_2, *which is defended at time step* i. *If* $v_1 \notin S$ *and not burning at the start of time step* i *then* $sav(S) \leqslant sav(S')$ *where* S' *is obtained from* S *by replacing* v_2 *with* v_1 *at time step* i.

Proof. Using Lemma 3, the vertices burned by S in G are the vertices which are reachable from s in $G \setminus S$. We show that every vertex u which is reachable from s in $G \setminus S'$ is also reachable from s in $G \setminus S$. Let P be a path between s and u in $G \setminus S'$. If $v_2 \notin P$ then P is a path in $G \setminus S$. If $v_2 \in P$, then using the fact that $N(v_2) \subseteq N[v_1]$, we can see that $P' = P \setminus \{v_2\} \cup \{v_1\}$ is a path between s and u in $G \setminus S$. Therefore the number of vertices burned by S is at least the number of vertices burned by S', which implies $sav(S) \leqslant sav(S')$.

Our FPT algorithm now follows by guessing, for each step in the defending sequence, if the vertex is from the modulator or the type of the vertex from $G \setminus X$. Then it simulates the sequence (by substituting for each guess of a type, a greedily chosen vertex from that type) to check if it is a valid solution.

Theorem 1. *The* FIREFIGHTING *problem can be solved in* $O((2^{k+1}+k)^{2k+2}(n+m))$ *time when parameterized by size of modulator to threshold graphs.*

Proof. Partition the vertices of $G \setminus X$ into at most 2^{k+1} sets. Each time when we want to defend a vertex we only choose from 2^{k+1} (types) $+k$ (size of X). From Lemma 7 we know which vertex has to be defended in a given type.

From Lemma 6 it is clear that we only need to defend at most $2k + 2$ times, therefore there are at most $(2^{k+1} + k)^{2k+2}$ possible firefighting strategies. For each such strategy S, using Lemma 2, test whether S is valid strategy and count the number of vertices saved by S. The strategy which saves maximum number of vertices is the optimal strategy. Therefore this procedure takes $O((2^{k+1} + k)^{2k+2}(n + m))$ time. □

3.2 Parameterization by Distance to Stars

In this section we design an FPT algorithm for the FIREFIGHTING problem parameterized by the distance to stars. Recall that this is the minimum number of vertices to be deleted from G to get a disjoint union of stars. Let X be a k-sized modulator to disjoint union of stars. Our first observation follows easily from the bound on the length of the longest induced path.

Lemma 8. *Let* (G, s) *be an instance of* FIREFIGHTING *problem. Any optimal strategy can defend at most* $4k + 2$ *vertices, where* k *is distance to disjoint union of stars.*

Proof. We know that fire always spreads along an induced path in G. As the length of maximum induced path in G is at most $4k + 2$, from Lemma 4 any optimal strategy can defend at most $4k + 2$ vertices in G. □

We now define a notion of equivalent stars, which will lead us to partitioning of the stars in $G \setminus X$ into types as before. For a star $\mathbb{S}$ with center c in $G \setminus X$, we use $B(\mathbb{S})$ to denote the set of vertices in $\mathbb{S}$ that have a neighbor in X, and call these the *border vertices*. Further, for a nonempty subset $Y \subseteq X$, we use $B_Y(\mathbb{S})$ to denote the set of vertices in $\mathbb{S}$ whose neighborhood in X is exactly Y.

Definition 1. *Let* $X \subseteq V(G)$ *such that* $G \setminus X$ *is a disjoint union of stars. We call two stars* $\mathbb{S}_i$ *and* $\mathbb{S}_j$ *are equivalent if, (a)* $N(c_i) = N(c_j)$, *where* c_i *and* c_j *are the centers of stars* $\mathbb{S}_i$ *and* $\mathbb{S}_j$ *respectively. (b)* $N(\mathbb{S}_i) = N(\mathbb{S}_j)$, *(c)* $|B(\mathbb{S}_i)| = |B(\mathbb{S}_j)|$ *and (d) For every non-empty subset* $Y \subseteq X$, *we have that* $|B_Y(\mathbb{S}_i)| = |B_Y(\mathbb{S}_j)|$.

For an equivalence class T, we use b_T to denote the size of the border for any star in T. Our next result bounds the number of equivalence classes. The bound follows roughly from the fact that one can associate a signature with an equivalence class based on condition (c) in Definition 1, which, in turn, can be put in one-one correspondence with strings of length 2^k over an alphabet of size ℓ, where ℓ is the maximum possible value of $|B(\mathbb{S})|$ in $G \setminus X$.

Lemma 9. *Let ℓ be defined as above and let $X \subseteq V(G)$ be a modulator to disjoint union of stars of size k. Then, the stars of $G \setminus X$ can be partitioned into at most $O(2^{2^k}\ell^{2^k})$ equivalence classes.*

Proof. First, partition the stars in $G \setminus X$ into at most 2^{2^k} sets such that all stars in each set satisfies conditions (a) and (b) of Definition 1. Now each set of the partition can be further divided based on the value of $|B(\mathbb{S}_i)|$ for each star $\mathbb{S}_i$ in that set. As $1 \leqslant |B(\mathbb{S}_i)| \leqslant \ell$ for all $\mathbb{S}_i \in G \setminus X$, each set can be partitioned into at most ℓ sets. In order to satisfy condition (d) in Definition 1 each set further partitioned into ℓ^{2^k-1} sets. Combining all, there are at most $O(2^{2^k}\ell^{2^k})$ equivalent star partitions of stars in $G \setminus X$. □

Our FPT algorithm begins by guessing the behavior of the solution on the modulator, and builds on the fact that there are a bounded number of equivalence classes. We defer the details of this algorithm to a full version because of space constraints, and note that it is similar—in spirit—to the approach used for the cluster vertex deletion parameter in [5].

Theorem 2. *[⋆]. The* FIREFIGHTING *problem is* FPT *when parameterized by the distance from the class of the disjoint union of stars.*

3.3 Parameterization by Distance to Diameter Two Graphs

In this section, we show that SAVING k-VERTICES is W[1]-hard when parameterized by k and distance to diameter two graphs by giving a reduction from the k-clique problem. The reduction is similar, in spirit, to the one used in [1].

Theorem 3 *[⋆].* SAVING k-VERTICES *is* W[1]-*hard parameterized by* $(k + l)$, *where l is the distance from the class of graphs with diameter two.*

We can obtain the hardness of SAVING k-VERTICES by a reduction from the k-clique problem as well, in fact by making minor changes to the reduction used to prove Theorem 3.

Corollary 1 *[⋆].* SAVING k-VERTICES *is* W[1]-*hard parameterized by* $(k + l)$, *where l is distance to split graphs.*

4   Kernelization Complexity

4.1 Parameterization by Distance to Clique

In this section, we give a polynomial kernel for SAVING k-VERTICES when parameterized by distance to clique, as summarized in the following theorem.

Theorem 4. SAVING k-VERTICES *admits a polynomial kernel of size at most* $O(l^2)$, *where* l *is the distance to clique.*

Let (G, s, k) be an instance of SAVING k-VERTICES and $X \subseteq V(G)$ of size l such that $G \setminus X = C$ is a clique. With out loss of generality we assume that $s \in X$, otherwise define $X' = X \cup \{s\}$ such that $G \setminus X'$ is a clique with size of modulator $l + 1$. Since the length of longest induced path in G is at most $l + 1$, using Lemma 4 we get the following Corollary.

Corollary 2. *Given an instance of* FIREFIGHTING, *then any optimal strategy can defend at most* $l + 1$ *vertices, where* l *is size of the clique modulator* X.

Let $X_L := \{x \in X : |N(x) \cap C| \leqslant l + 1\}$, $X_H = X \setminus X_L$. Let $J := \{y \in C \mid \exists x \in X_L, y \in N(x)\}$, $|J| \leqslant l(l+1)$. Our kernelization algorithm is based on the following observation. We show that replacing the clique $G[C \setminus J]$ with another clique of small size does not affect the solution.

Lemma 10. *Let* S *be a valid strategy containing a vertex* $v \in C \setminus J$, *then all the vertices of* $N(v) \setminus S$ *are burned.*

Proof. Since S is a valid strategy, there is an induced path P from s to v such that all vertices on P are burned, except v. Let $X_v := N(v) \cap X$ and for every $x \in X_v \setminus S$, we have $|N(x) \cap C| > l + 1$: suppose there is $x \in N(v) \cap X$ such that $|N(x) \cap C| \leqslant l + 1$ then $v \in J$, contradiction to $v \in C \setminus J$.

Let u be a burned neighbor of v on P. If $u \in C$ then $C \setminus S$ is burned and for every $x \in X_v \setminus S$, we have $|N(x) \cap C| > l + 1$. From Corollary 2 we can only defend at most $l + 1$ vertices, therefore all vertices of $X_v \setminus S$ are burned.

If $u \in X_v \setminus S$ then $|N(u) \cap C| > l + 1$, therefore all vertices $C \setminus S$ are burned. From the first case we can see that all vertices of $X_v \setminus S$ are also burned.   □

Reduction rules

1. Delete vertices of $C \setminus J$ from G. Add a clique K of size $l + 2$ and make each vertex of K adjacent to all vertices in $X_H \cup J$.
2. Add another clique L of size $\min\{l + 1, |C \setminus J|\}$ and for each vertex $u \in L$, add edges between u and $J \cup K$.

Let H be the graph obtained after applying above reduction rules (see Fig. 2). It is easy to see that $X \subseteq V(H)$ such that $H \setminus X$ is a clique. The size of the reduced instance H is at most $l^2 + 4l + 3$ and the reduction can be done in polynomial time.

Let $C = G \setminus X$ and $C' = H \setminus X$. We may assume that $|C \setminus J| > 2l + 3$; otherwise, trivially we get a kernel of size at most $l^2 + 4l + 3$.

Remark 1. Let G be a graph and S be a valid strategy. If S defends a subset S_1 of vertices in $C \setminus J$, then defending any subset S_2 of vertices in L instead of S_1, with $|S_1| = |S_2|$ is also a valid strategy S' for H.

Conversely let S' be a strategy on H. If S' defends a subset S_1 of vertices in $K \cup L$, then defending any subset S_2 of vertices in $C \setminus J$ instead of S_1, with $|S_1| = |S_2|$ is also a valid strategy S for G.

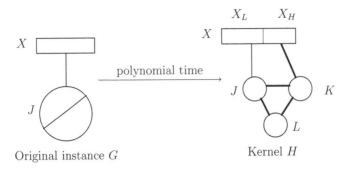

Fig. 2. A schematic view of kernelization algorithm of FIREFIGHTING parameterized by distance to a clique. A bold edge between two sets of vertices indicates the presence of all possible edges between the two sets.

Lemma 11. *Let* G *and* H *be graphs as defined above and* S, S' *be corresponding valid strategies as defined in the above remark. At least one vertex of* C *is burned in* G *by strategy* S *iff at least one vertex of* C' *is burned in* H *by strategy* S'.

Proof. Let $u \in C'$ is burned by strategy S' in H. If $u \in J$, then $u \in C$ will also burned by S in G. If $u \in K$ and no vertex of J burned by S' in H then, at least one vertex $x \in X_H$ gets burned and $|N(x) \cap C| > l+1$, therefore at least one vertex in $N(x) \cap C$ gets burned by S in G.

Let $u \in C$ is burned by strategy S in G. If $u \in J$ then it will be burned by S' in H. If $u \in C \setminus J$ and no vertex of J burned by S in G then there exists burned vertex $x \in X_H$. Since $|N(x) \cap C'| > l+1$, at least one vertex of K gets burned. □

Claim 1. *If a strategy* S *saves at least* $2l+1$ *vertices in* G *then* S *saves entire clique.*

Proof. Let S be a valid strategy which saves at least $2l+1$ vertices in G. Suppose assume that there exists a vertex $v \in C$ burned by strategy S, then no vertex in C is saved except the defended vertices. The strategy S can save at most $l-1$ vertices in X and $l+1$ vertices in C, total S saves at most $2l$ vertices which is a contradiction to the fact that S saves at least $2l+1$ vertices. □

Claim 2. *Saving one undefended vertex in* C *is equivalent to saving entire clique.*

Proof. Let v be an undefended saved vertex of C. If any vertex of C is burned by S then v gets burned contradicting the fact that v is a saved vertex. □

Lemma 12. *Let* G *be a graph and* H *be the graph obtained after applying reduction rules. A strategy* S *saves at least* k *vertices in* G *if and only if there exists a strategy* S' *that saves at least* k' *vertices in* H.

Proof. Let S be any valid strategy on G then we can find a corresponding strategy S' on H as follows. If S defends a vertex v in $C \setminus J$ which is deleted in reduction procedure, then by Remark 1 instead of v we can defend any other non-defended vertex in L. Since S can defend at most $l+1$ vertices in G and $|L| = l+1$, we can always replace vertices of $C \setminus J$ in S by applying Remark 1.

Conversely, let S' be any valid strategy on H then we can find its corresponding strategy S on G as follows. If S' defends a vertex v in $K \cup L$ which is added in reduction procedure, then by Remark 1 instead of v we can defend any other non-defended vertex in $C \setminus J$. Since S' can defend at most $l + 1$ vertices in H and $|C \setminus J| > 2l + 3$, we can always replace vertices of $K \cup J$ in $4S'$ by applying Remark 1.

The parameter k' is defined as follows.

1. If $k \in [|C|, |C| + l - 1]$, then set $k' = k - |C| + |J| + |K| + |L| \leqslant l^2 + 4l + 3$.
2. If $k \in [2l + 1, |C| - 1]$, then set $k' = 2l + 1$.
3. If $k \in [1, 2l]$, then set $k' = k$.

Using Lemma 11, Claims 1 and 2 we can see that the strategy S saves at least k vertices in G if and only if the strategy S' saves at least k' vertices in H.   □

4.2   Parameterization by Distance to Stars

Next, we show that SAVING k-VERTICES does not admit a polynomial kernel parameterized by distance to stars unless $\mathsf{NP} \subseteq \mathsf{coNP/poly}$. We obtain this result by a polynomial parameter transformation from the problem of finding a maximum clique parameterized by the size of the vertex cover.

Theorem 5 *[⋆]. SAVING k-VERTICES parameterized by the distance to disjoint union of stars does not admit a polynomial kernel, unless* $\mathsf{NP} \subseteq \mathsf{coNP/poly}$.

5   Conclusion

In this paper, we studied the parameterized complexity of FIREFIGHTING problem for various structural parameters. We considered the size of a modulator to threshold graphs, cluster graphs and disjoint unions of stars as parameters, and showed FPT algorithms in all cases. We also established that FIREFIGHTING admits a polynomial kernel when parameterized by the size of a modulator to a clique, while it is unlikely to admit a polynomial kernel when parameterized by the size of a modulator to a disjoint union of stars. In contrast to the tractability results, we found that FIREFIGHTING is W[1]-hard when parameterized by the distance to split graphs. The following problems remain open.

- Does SAVING k-VERTICES admit a polynomial kernel when parameterized by k and the size of a vertex cover?
- Is the FIREFIGHTING problem FPT when parameterized by distance to cographs?

References

1. Bazgan, C., Chopin, M., Cygan, M., Fellows, M.R., Fomin, F.V., van Leeuwen, E.J.: Parameterized complexity of firefighting. J. Comput. Syst. Sci. **80**(7), 1285–1297 (2014)
2. Cai, L.: Fixed-parameter tractability of graph modification problems for hereditary properties. Inf. Process. Lett. **58**(4), 171–176 (1996)
3. Cai, L., Verbin, E., Yang, L.: Firefighting on trees: $(1-1/e)$–approximation, fixed parameter tractability and a subexponential algorithm. In: Hong, S.-H., Nagamochi, H., Fukunaga, T. (eds.) ISAAC 2008. LNCS, vol. 5369, pp. 258–269. Springer, Heidelberg (2008). https://doi.org/10.1007/978-3-540-92182-0_25
4. Chalermsook, P., Chuzhoy, J.: Resource minimization for fire containment. In: Proceedings of the Twenty-First Annual ACM-SIAM Symposium on Discrete Algorithms, pp. 1334–1349. Society for Industrial and Applied Mathematics (2010)
5. Chlebíková, J., Chopin, M.: The firefighter problem: further steps in understanding its complexity. Theoret. Comput. Sci. **676**, 42–51 (2017)
6. Cygan, M., Fomin, F.V., Kowalik, L., Lokshtanov, D., Marx, D., Pilipczuk, M., Pilipczuk, M., Saurabh, S.: Parameterized Algorithms. Springer, Cham (2015). https://doi.org/10.1007/978-3-319-21275-3
7. Das, B., Enduri, M.K., Misra, N., Reddy, I.V.: On structural parameterizations of firefighting. arXiv preprint arXiv:1711.10227 (2017)
8. Downey, R.G., Fellows, M.R.: Parameterized Complexity, vol. 3. Springer, Heidelberg (1999). https://doi.org/10.1007/978-1-4612-0515-9
9. Finbow, S., Hartnell, B., Li, Q., Schmeisser, K.: On minimizing the effects of fire or a virus on a network. J. Comb. Math. Comb. Comput. **33**, 311–322 (2000)
10. Finbow, S., King, A., MacGillivray, G., Rizzi, R.: The firefighter problem for graphs of maximum degree three. Discrete Math. **307**(16), 2094–2105 (2007)
11. Finbow, S., MacGillivray, G.: The firefighter problem: a survey of results, directions and questions. Australas. J. Comb. **43**, 57–77 (2009)
12. Foldes, S., Hammer, P.L.: Split graphs. Universität Bonn, Institut für Ökonometrie und Operations Research (1976)
13. Fomin, F.V., Heggernes, P., van Leeuwen, E.J.: The firefighter problem on graph classes. Theoret. Comput. Sci. **613**, 38–50 (2016)
14. Ganian, R.: Improving vertex cover as a graph parameter. Discrete Math. Theor. Comput. Sci. **17**(2), 77–100 (2015)
15. Hartnell, B.: Firefighter! An application of domination. In: 25th Manitoba Conference on Combinatorial Mathematics and Computing (1995)
16. King, A., MacGillivray, G.: The firefighter problem for cubic graphs. Discrete Math. **310**(3), 614–621 (2010)
17. Leung, M.L.: Fixed parameter tractable algorithm for firefighting problem. arXiv preprint arXiv:1104.1044 (2011)
18. MacGillivray, G., Wang, P.: On the firefighter problem. J. Comb. Math. Comb. Comput. **47**, 83–96 (2003)

On the Simultaneous Minimum Spanning Trees Problem

Matěj Konečný, Stanislav Kučera, Jana Novotná, Jakub Pekárek,
Martin Smolík, Jakub Tětek, and Martin Töpfer[(✉)]

Department of Applied Mathematics, Faculty of Mathematics and Physics,
Charles University, Prague, Czech Republic
matejkon@gmail.com, stanislav.kucera@outlook.com, janca@kam.mff.cuni.cz,
edalegos@gmail.com, smolik.ma@gmail.com, j.tetek@gmail.com,
mtopfer@gmail.com

Abstract. Simultaneous Embedding with Fixed Edges (SEFE) [1] is a problem where given k planar graphs we ask whether they can be simultaneously embedded so that the embedding of each graph is planar and common edges are drawn the same. Problems of SEFE type have inspired questions of Simultaneous Geometrical Representations and further derivations. Based on this motivation we investigate the generalization of the simultaneous paradigm on the classical combinatorial problem of minimum spanning trees. Given k graphs with weighted edges, such that they have a common intersection, are there minimum spanning trees of the respective graphs such that they agree on the intersection? We show that the unweighted case is polynomial-time solvable while the weighted case is only polynomial-time solvable for $k = 2$ and it is NP-complete for $k \geq 3$.

1 Introduction

The problem of finding a minimum spanning tree is one of the most important and most well-studied problems in graph algorithms. We consider a variant of this problem inspired by the following motivation.

In a Sunflower land, there is a capital city and several smaller cities around it. In the past, there was a telecommunication company based in the capital city, but it is now bankrupt. The inhabitants of each of the small cities want to establish their own telecommunication company that would connect all of the houses in their city as well as all of the houses in the capital. The representatives of each city meet to coordinate their soon-to-be networks so that they all agree on the capital and can split the cost of covering the capital evenly. However, all of the companies are so afraid of bankruptcy that none of them would accept a solution that would cost them a single dollar more than necessary. Is it always possible to plan all of the networks so that all of the companies reach their goal simultaneously while each of the individual costs is minimized? How hard is it to find such a plan, if it exists, or recognize that it does not exist?

Supported by project CE-ITI P202/12/G061 of GA ČR and grant SVV-2017-260452.

B. S. Panda and P. P. Goswami (Eds.): CALDAM 2018, LNCS 10743, pp. 235–248, 2018.
https://doi.org/10.1007/978-3-319-74180-2_20

Problem 1 (Simultaneous Minimum Spanning Trees). Let k be a positive integer and let $G_1 = (V_1, E_1)$, $G_2 = (V_2, E_2), \ldots, G_k = (V_k, E_k)$ be graphs and w a non-negative weight function of all of their edges ($w : \bigcup_{i=1}^{k} E_i \to \mathbb{R}_0^+$) such that there is a graph $\bar{G}$ satisfying that $\bar{G} = G_i[\bar{V}]$ for any i from 1 to k, where $\bar{V} = V_i \cap V_j$ for any $i \neq j$ from 1 to k (i.e. the graphs together form a "sunflower" shape with no lateral edges). Find minimum spanning trees $T_i \subseteq G_i$, such that they all coincide on $\bar{G}$, or answer NO if there are no such spanning trees. We shall abbreviate this problem as $SMST$.

Note that the T_i's do not have to induce a spanning tree on $\bar{G}$, nor does the union of T_i's have to be acyclic on the union of all of the G_i's. Indeed both of these situations necessarily happen in solutions of some instances of the $SMST$ problem. Unlike the minimum spanning tree problem, the $SMST$ problem does not always have a solution.

As an example, let G_1 be a triangle xzy, let G_2 be a triangle xwy and let xy be the heavies edge. Although $G_1 \cap G_2$ induces a connected graph (edge xy), we have a unique solution $\{xz, zy, yw, wx\}$ which is not connected on $G_1 \cap G_2$ and is not acyclic on $G_1 \cup G_2$. Furthermore, if we remove any light edge, e.g. xz, then there is no solution.

We show that $SMST$ is an NP-complete problem already for a fairly small number of graphs (more than 2) and even when limited to simplified instances. We present a scheme that allows us to solve any $SMST$ for two graphs in polynomial time using a tandem of reductions and multiple runs of matroid intersection algorithm.

1.1 Preliminaries

The problem of finding a minimum spanning tree for a single graph has been studied thoroughly since Borůvka [2], Jarník-Prim [3,4] and Kruskal [5]. See [6] for more details. Currently, the optimal algorithm is known [7], but its asymptotics is still an open problem.

We do not distinguish instances where the input graph is connected from instances where it is disconnected. The inclusion of disconnected instances is natural as many constructions work just as well under such circumstances. Furthermore, usual incremental and iterative approaches typically work on subsets of the input graph and it is therefore not strictly clear whether they maintain a spanning tree or a spanning forest. For convenience, we define the usual term *spanning tree* as a maximal acyclic subgraph. In doing so we include the disconnected case, where the more proper term would be *spanning forest*.

We focus mainly on the Kruskal's algorithm and use its known properties. Kruskal's algorithm starts by sorting the edges in a non-decreasing order (by weight) or obtains the edges in a non-decreasing order on input. Then it processes all the edges one by one in sorted order while greedily maintaining maximum acyclic subgraph which we refer to as *partial spanning tree*.

Definition 1. *Consider the run of Kruskal's algorithm. A* stage *is a collection of steps in which the algorithm processes edges of the same weight.*

Fact 1. *Let $G = (V, E, w)$ be a graph with weighted edges. Then all of the following holds for Kruskal's algorithm applied to graph G and a non-decreasing order of edges π:*

- *Kruskal's algorithm is complete (finishes) and correct (answers correctly) for any non-decreasing π, although the created spanning trees might be different.*
- *Let T be a minimum spanning tree of G and let π_T be the non-decreasing order such that all edges from T are ordered before all edges of the same weight that are not from T. Then Kruskal's algorithm using π_T outputs exactly T.*
- *After every stage, components of the partial spanning tree span across the same vertices for all non-decreasing π.*
- *Edges added to the partial spanning tree in each stage depend only on their ordering, not on the edges chosen in the previous stages.*
- *Kruskal's algorithm accesses π in a read-once fashion, accepting or refusing each edge before accessing the next one.*

2   Simultaneous Kruskal's Algorithm

Consider a $SMST$ task for a given k and graphs $G_1, G_2, \ldots, G_k$. Let us denote the union of all G_is as G and their intersection as $\bar{G}$. Suppose we order all the edges of G in a non-decreasing order π which we call a *universal order* and denote $\pi[E(G_i)]$ the restrictions of π to edges in G_i for every i. For a set of edges F, we also say that a universal order is *F-preferring* if all the edges from F are ordered before any other edges of the same weight.

Consider the following construction. First, we fix an arbitrary non-decreasing universal order π. We simulate k independent instances of Kruskal's algorithm, $K_1, K_2, \ldots, K_k$ where the job of each K_i is to find a minimum spanning tree T_i of G_i using the order $\pi[E(G_i)]$, not considering the other instances. In parallel with the instances of the Kruskal's algorithm we try to incrementally build a simultaneous minimum spanning tree.

In the beginning, we start with an empty simultaneous spanning tree T and process all the edges one by one according to the universal order. We present each edge e to all instances K_i such that $e \in G_i$. If we assume a sunflower intersection, we can rephrase this in the following way: if $e \in \bar{G}$ then we present e to all instances and if $e \notin \bar{G}$ then $e \in G_j$ for some unique j and we present e only to one instance K_j. If every invoked K_i adds e to its local T_i, we also add e to T. If every invoked instance K_i refuses to add e to its local T_i, we also throw e away. If the invoked instances do not agree, we fail. If the algorithm processes all edges without failing, we output T as a solution.

We call this construction *Simultaneous Kruskal's algorithm* or SKA in short. There are two natural versions of the SKA. If SKA expects the universal order π on input, then it is a deterministic algorithm. Alternatively, SKA may be formulated as a non-deterministic algorithm which guesses a correct universal order which avoids failure (if any such order exists), then we speak of a non-deterministic simultaneous Kruskal's algorithm or $NSKA$ in short. We naturally extend the definition of a stage from Kruskal's algorithm to the *(N)SKA* as the collection of steps in which the algorithm processes edges of the same weight.

Lemma 1. *Let I be an instance of the SMST problem. Then all of the following holds for simultaneous Kruskal's algorithm:*

- *NSKA is complete (finishes) and correct (answers correctly).*
- *Let T be a solution of I and a let π_T be a T-preferring universal order. Then SKA using π_T outputs exactly T.*
- *After every successful stage of SKA and NSKA, components of the partial simultaneous spanning tree after restriction to any G_i span across the same vertices for all choices of universal order π.*
- *Edges added to the partial simultaneous spanning tree in each stage depend only on their ordering, not on the edges chosen in the previous stages.*
- *SKA accesses π in a read-once fashion, accepting or refusing each edge before accessing the next one.*

Proof. Let us first prove the second point. Suppose we run the *SKA* using the T-preferring universal order. Let us analyze the behavior of an arbitrary K_i. Let T_i denote the restriction of T to G_i. By definition T_i is a minimum spanning tree of G_i and $\pi[G_i]$ is a T_i-preferring order. From the properties of the Kruskal's algorithm (Fact 1) we know that K_i constructs exactly T_i. Since every K_i would construct exactly T_i should it run on its own, we observe that all the invoked instances K_j accept each edge if and only if it belongs to T, and the whole algorithm never fails. At the end of the computation the algorithm gives exactly T as a solution.

To prove correctness, let us first suppose that the *NSKA* terminates with success. Then the set T on output is a union of the local spanning trees from all K_i algorithms. Since each algorithm K_j processes all the edges from G_j in a non-decreasing order of weight and Kruskal's algorithm is sound, each of the local spanning trees is a minimum spanning tree. Thus, T is a solution of the *SMST* problem. If *NSKA* terminates with a failure, then from the second point it follows that there was no solution T, as otherwise *NSKA* guesses a T-preferring universal order and terminates successfully.

The last three points are simple observations extending the Fact 1 into simultaneous setting using the previous two points.   □

3   Cases and Variants

Lemma 2. *Let I be a feasible instance of the SMST problem. Then any solution T' of I restricted to edges of weight at most w can be extended to a solution T of the whole I by adding some edges of weight greater than w. Furthermore, this extension does not depend on T'.*

Proof. Let T be a solution of the *SMST* problem. We choose any w and split the edges into a set of light edges L of weight at most w and a set of heavy edges H with weight strictly greater than w.

Consider running *SKA* on the instance I restricted to edges from L using any T-preferring universal order. Since *SKA* does not look ahead, it cannot distinguish whether it runs on a restricted instance or the full instance and therefore it does not fail and outputs T restricted to L (denoted $T[L]$), which is a solution of the restricted instance. Since T' is also a solution of the restricted instance, both T' and $T[L]$ define the same components on all individual graphs and have the same weight. Let us define $\bar{T} = T' \cup T[H]$. Clearly $\bar{T}$ is acyclic on each graph and has the same weight as T. Therefore $\bar{T}$ is a solution of the full instance, extending (any) T'. $\qquad\square$

Observation 2. *Let us have an SMST instance I where $m(I)$ denotes the number of edges and $R(I)$ denotes the maximum number of repeats of any weight. If $R(I)! \in m(I)^{O(1)}$, in other words $R(I)$ is asymptotically very small, then I can be solved in a polynomial time.*

Proof. Suppose we implement the *NSKA* deterministically and use backtracking to guess the next edge in the universal order. The previous lemma shows that it is sufficient to consider only backtracks within the current stage. If we ever need to backtrack into the previous stage, then the solution of the previous stage cannot be extended and therefore no solution exists.

If all of the weights in our instance of the *SMST* are either distinct, or the number of repeats of each value is asymptotically very small, then we can try all possible orders within each stage in polynomial time. More precisely whenever $R(I)! \in m(I)^{O(1)}$ we have at most polynomially many orderings in each stage and the algorithm finishes in polynomial time. If $R(I) \in \mathcal{O}(\log \log n)$ then there are at most linearly many possible orderings and the algorithm's running time differs by only a factor of $\mathcal{O}(m)$ from the *NSKA*'s running time on a non-deterministic machine. $\qquad\square$

Definition 2. *A* Simultaneous $\{0,1\}$ Minimum Spanning Tree *problem, or* 01-*SMST* *in short, is an instance of SMST where we restrict all the edge weights to be either 0 or 1.*

We show an equivalence of the general *SMST* and 01-*SMST* up to a polynomial factor of complexity.

Lemma 3. *Any algorithm solving* 01-*SMST in polynomial time can be used to solve general SMST problem in polynomial time.*

Proof. First let us consider an instance of *SMST* using at most two distinct values for weights. Then we can replace these by 0 and 1. From the point of view of the individual graphs, each subset of edges is a minimum spanning tree after the modification if and only if the same holds before the modification; and so the same applies to the simultaneous minimum spanning trees.

We continue via induction. Let us have an algorithm based on any 01-*SMST* algorithm that solves any *SMST* instance with at most k distinct values of weight. We will extend this algorithm to $k+1$ values. Let us have an instance that uses $k+1$ values and let w denote the highest one. We restrict G to G' by restricting to edges lighter than w. We already know how to solve *SMST* for G', acquiring a partial solution T' or showing that no solution exists in which case the original $SMST$ has no solution. If we have the solution T', then according to Lemma 2 T' can be extended by some edges of weight w to a full solution.

We once again modify G into $\bar{G}$ as follows. We restrict G to edges from T' and edges of weight w. We set the weight of all edges from T' to 0 and the weight of the remaining edges to 1. We now have an instance of 01-$SMST$ such that any solution contains all the edges from T' as they form a partial simultaneous spanning tree and the SKA would accept all of the edges regardless of the universal order used. Let $\bar{T}$ be a solution of the 01-$SMST$ problem on $\bar{G}$, then $\bar{T}$ is also a solution of the original $SMST$ problem and the algorithm outputs $\bar{T}$, otherwise we answer "no".

To show completeness, suppose that there exists a solution T. Then we necessarily obtain T' in the first step and T' can be extended to a solution of the whole problem (not necessarily T) and thus the 01-$SMST$ on $\bar{G}$ has a solution $\bar{T}$.   □

Definition 3. *An* Intersection-Heavy Simultaneous $\{0,1\}$ Minimum Spanning Tree *problem, or* ∩-01-$SMST$ *in short, is an instance of $SMST$ where we restrict all the edge weights to be either 0 or 1. Furthermore all the edges of weight 1 are only in the intersection of all the individual graphs.*

The motivation behind this restriction comes from a simple observation.

Observation 3. *Let I be an instance of* 01-$SMST$ *(for any number of graphs) where no edges of weight 1 appear in the intersection. Then after solving the first stage, the* SKA *algorithm always finishes for any universal order π.*

Proof. This is easy to see as each edge of weight 1 will be presented by the SKA to a single instance of the Kruskal's algorithm and therefore in no step can the algorithm fail (get two opposite answers). Furthermore, one can see that the order of edges of weight 1 no longer matters, though different orders may give different solutions.   □

This observation formalizes an intuition that it is in some sense harder to deal with weight 1 edges in the intersection than in the exclusive parts.

It might therefore seem that to solve a 01-$SMST$ problem, one might first greedily find a subset of edges from the intersection and then extend it to the exclusive parts. This approach fails on a simple example. Let us have exactly four vertices a_1, a_2, b_1, b_2 in the intersection. Let G_1 contain four weight 0 edges $a_1c_1, a_2c_2, b_1d_1, b_2d_2$, and let G_2 contain two weight 0 paths P_i connecting a_i and b_i for both values of i. Finally let a_1a_2, b_1b_2 and c_1c_2 be weight 1 edges where the last one is exclusive for G_1. Clearly the only solution takes exactly the weight-1 edges b_1b_2 and c_1c_2. However if the graphs contains d_1d_2 rather than c_1c_2 then picking the edge b_1b_2 is not correct. Therefore an algorithm may not be oblivious to the exclusive parts.

It seems logical to also consider the opposite approach, that is to first solve the exclusive parts where the solution seems rather fixed and then exploit the information from exclusive parts to extend the partial solution to the intersection.

It is no surprise that this approach is flawed as well. As an example, let us have two graphs G_1 and G_2 where G_1 is only one edge xy and G_2 is a triangle xyz. If we were to first find a maximum acyclic set of each exclusive part, we would get the subset $\{xz, yz\}$. However now we cannot extend this subset into a solution as there are only two solutions $\{xy, yz\}$ and $\{xy, xz\}$.

Both of these greedy approaches to a 01-$SMST$ are flawed, even under the assumption that we are able to solve the first stage correctly in polynomial time. However according to the Observation 3 limiting all of the edges of weight 1 to the intersection gives instances that are in some sense easier, as the hardness of the problem is focused in the intersection which can be solved without considering exclusive weight 1 edges, as there are none. Later we show that ∩-01-$SMST$ is actually equivalent to 01-$SMST$, which will be a key step in solving the 01-$SMST$ problem.

Definition 4. *A* Simultaneous Spanning Tree *problem, SST in short, is an unweighted version of the SMST problem, in other words a SMST problem using only one weight.*

The SST is clearly at most as hard problem as all of the previous versions of the $SMST$ and is an interesting problem on its own. We use the SST as a simple base case in our construction later on.

Observation 4. $SST \subseteq$ ∩-01-$SMST \subseteq$ 01-$SMST \subseteq SMST$

4   Case $k \geq 3$ Is **NP-complete**

Problem 2 (3D matching). Let U, V, W be disjoint finite sets such that $|U| = |V| = |W| = k$ and let T be a subset of $U \times V \times W$. Is there a set $M \subset T$ with $|M| = k$, such that for any $x \in U \cup V \cup W$ there is exactly one hyperedge $e \in M$ such that $x \in e$.

Fact 5 ([8]). *3D matching is* **NP***-complete.*

Theorem 6. *The problem of 3D matching can be polynomially reduced to* ∩-01-$SMST$ *problem for 3 graphs.*

Proof. Without loss of generality we assume that every element of U, V and W is element of at least one hyperedge in T, otherwise the original 3D matching trivially has no solution.

We define graphs G_1, G_2, G_3 and H where $H = G_1 \cap G_2 \cap G_3$ forming a "sunflower" intersection, that is $H = G_i \cap G_j$ for each $i \neq j$. We associate G_1 with U, G_2 with V and G_3 with W.

First put a central vertex c in H. For each hyperedge $e \in T$, put a vertex $v_e \in H$ and connect it to c by an edge in H of weight 1. For each element $x \in U$, put a vertex v_x into the exclusive part of G_1 $(G_1 \backslash H)$ and for every $e \in T$ such that $x \in e$, connect v_e and v_x by an edge of weight 0. Do the same for V and

W with graphs G_2 and G_3 respectively. By construction these graphs form the required "sunflower" configuration.

The structure of the graph H can be alternatively described as follows. The intersection H contains exactly a star with center c and all edges of weight 1 where each ray represents a different element from T.

Let us focus on G_1 and U, for the other graphs and sets the arguments are symmetrical. The graph G_1 is composed of the central star and exclusive vertices representing elements of the associated set U. Every vertex representing an element x is connected via edges of weight 0 to all vertices representing the hyperedges that contain x. So for every element x, v_x is a center of a weight-0 star in G_1. All of these weight-0 stars are disjoint as in each hyperedge there is at most one element from U. Since all the edges of weight 0 form an acyclic subgraph of G_1, every solution of this $SMST$ instance must contain all of them. Let S be a solution of the $SMST$ problem. As for each $x \in U$, the v_x is in the same component as c in G_1, it must also be in the same component of $S[E(G_1)]$ and therefore at least one edge $cv_e \in S$ for some hyperedge e such that $x \in e$. If it happened that $cv_f \in S$ for some other hyperedge f with $x \in f$, then $cv_e, v_e v_x, v_x v_f, v_f c$ form a cycle in G_1 and we get a contradiction.

This means that the hyperedges represented by the edges (where e is represented by edge cv_e) of weight 1 in S are a solution of the 3D-matching. This is true as each $x \in U$ belongs to exactly one of the hyperedges from S and the same applies to every $y \in V$ and every $z \in W$.

On the other hand, let M be a solution of the 3D-matching. Then we can construct a solution of the $SMST$ by simply picking all the edges of weight 0 and all the edges of weight 1 that represent the hyperedges edges from M. As previously, we observe that everything in G_1 is connected into a single component. If we only consider the edges of weigh 0 on the other hand, then for each $x, y \in U$ the vertices v_x and v_y are in distinct components and can only be connected via the central star. Therefore any solution must connect G_1 into a single component using at least $|U|$ edges of weight 1. Since $|U| = |M|$, the solution of the $SMST$ constructed from M is clearly minimal.   □

Corollary 1. *The problem $SMST$ and its variants 01-$SMST$ and $\cap$-01-$SMST$ are* NP-*complete for 3 and more graphs.*

5   Case $k = 2$ Is in P

In this section we show that the general $SMST$ problem is polynomially solvable. We progress via a tandem of reductions. We already know that the general $SMST$ can be solved using an algorithm for 01-$SMST$ for a cost of some polynomial factor. We further reduce instances of 01-$SMST$ to tasks that are more orderly and symmetrical in some sense. We then use this to reduce the task to $\cap$-01-$SMST$. Finally, we show that solving $\cap$-01-$SMST$ can be reduced to a problem of intersection of two matroids, which is a polynomial problem for two graphs. As an intermediate step, we will also solve the SST problem by reduction to a matroid intersection problem.

Definition 5. *Let G_1 and G_2 be two graphs intersecting in a common induced subgraph and let F be a subset of edges of G_1 and G_2. We say that F is simultaneously acyclic if F restricted to each of the two graphs G_1 and G_2 forms an acyclic subgraph.*

5.1   Reduction of 2-Graph 01-$SMST$ to 2-Graph $\cap$-01-$SMST$

For technical reasons we first want to get rid of all edges of weight 1 that cross the boundary in between the intersection and one of the exclusive parts.

Observation 7. *Every 01-$SMST$ instance can be transformed to an instance where all of the edges of weight 1 have either both ends in the intersection or both ends in an exclusive part of one graph. This transformation at most doubles the number of edges and vertices.*

Proof. This can be achieved by a simple operation that shifts the edges into the exclusive parts. We take each edge xy of weight 1 such that x is in the intersection and y in the exclusive part of one of the two graphs. We subdivide xy into two edges xz and zy where the vertex z lies in the exclusive part of the relevant graph. We set the weight of xz to 0 and the weight of zy to 1. Since the vertex z has degree two, the 0-weight edge xz is an element of each solution of the new 01-$SMST$. It is now easy to see that we can construct the solution of the original 01-$SMST$ instance from any solution of the new instance by removing xz and substitution of zy with xy (if it is part of the solution).   □

Another issue is that each of the two graphs may require a different number of edges of weight 1, while each edge from the intersection would increase the size of both solutions.

Observation 8. *Every instance of 01-$SMST$ with two graphs G_1 and G_2 can be transformed into an instance where every minimum spanning tree of G_1 and every minimum spanning tree of G_2 contain the same number of edges of weight 1. This transformation at most doubles the number of edges and vertices.*

Proof. Let G denote the union of G_1 and G_2 and let $\bar{G}$ denote their intersection. From the properties of the SKA (Lemma 1) we know that we can determine beforehand the components of G_1 and G_2 after all the edges of weight 0 are processed and after all the edges of weight 1 are processed. We also know that in order to compute the restriction of the solution to the edges of weight 1 we do not need to know the exact choice of edges of weight 0, they are in fact independent. By considering the number of components of G_1 and G_2 just after processing all the edges of weight 0 and after processing all edges, we deduce how many edges of weight 1 must be added into the minimum spanning tree of each graph, which is equal to the difference of the two values.

Suppose that the solution of 01-$SMST$ must contain j_1 edges of weight 1 from the graph G_1 and j_2 edges of weight 1 from the graph G_2. If $j_1 = j_2$ then we do not need to modify the instance, otherwise without loss of generality $j_1 > j_2$.

We pick an arbitrary vertex v from the exclusive part of G_2 and extend G_2 by $j_1 - j_2$ leaves attached to v. All the leaves are new vertices and lie in the exclusive part of G_2; and all of the new edges have weight 1. Every spanning tree of G_2 must now contain all of these edges, while every solution of the original instance can be extended by exactly these edges. After this modification, $j_1 = j_2'$ where j_2' denotes the new number of weight-one edges in the graph G_2 after modification. Note that this construction also works for the case $j_2 = 0$, although this can be solved directly using SKA.   □

Lemma 4. *The* 01-$SMST$ *problem for* $k = 2$ *is polynomially reducible to* ∩-01-$SMST$ *problem for* $k = 2$ *of asymptotically at most quadratic size. Furthermore if the set of edges of weight 0 of the original* 01-$SMST$ *instance is simultaneously acyclic, then the same is true for the new* ∩-01-$SMST$ *instance.*

Proof. Let us have an instance of 01-$SMST$ and let G_1, G_2 denote the two graphs and $\bar{G}$ their intersection. Using the previous observation we can assume without loss of generality that all of the weight-1 edges have either both ends contained in $\bar{G}$ or both ends contained in the exclusive part of one of the two graphs; and that there exists a positive integer j such that every solution of the 01-$SMST$ constrained to both G_1 or G_2 has exactly j edges of weight 1. This increases the size of the problem by a small multiplicative constant.

We modify the problem so that all of the edges from the exclusive parts are removed and equivalently modeled by gadgets that have edges of weight 1 only in $\bar{G}$. To do this, we consider all pairs $e = (e_1, e_2), f = (f_1, f_2)$ of edges of weight 1 such that e is from the exclusive part of G_1 and f is from the exclusive part of G_2. We create two new vertices x_1^{ef}, x_2^{ef} in $\bar{G}$ and add edges $e_1 x_1^{ef}, e_2 x_2^{ef}, f_1 x_1^{ef}, f_2 x_2^{ef}$ of weight 0 and an edge x_1^{ef}, x_2^{ef} of weight 1. After processing all pairs, we delete all the edges of weight 1 from the exclusive parts.

Let M be a solution of the modified instance of 01-$SMST$ (which is in fact ∩-01-$SMST$). First we observe that whenever $x_1^{ef} x_2^{ef} \in M$ for some removed edges e and f then $x_1^{eg} x_2^{eg} \notin M$ for any $g \neq f$ as otherwise $e_1, x_1^{ef}, x_2^{ef}, e_2, x_2^{eg}, x_1^{eg}, e_1$ forms a cycle in $M[G_1]$. To get a solution of the original instance, we remove all the extra edges of weight 0 and replace each edge $x_1^{ef} x_2^{ef}$ by edges e and f. Let us denote the resulting set of edges M'. Consider the graph G_1 and the components defined by M' restricted to G_1. It is easy to see that the components are the same as in M with the exception of the new vertices which are now isolated. Also, the total weight of M' restricted to each graph (of the original instance) is the same as the total weight of M restricted to each graph (of the modified instance). We conclude that M' is a minimum simultaneous spanning tree (Fig. 1).

On the other hand, let $\bar{M}$ be a solution of the original instance. Since there is the same amount of edges of weight 1 in $\bar{M}$ restricted to G_1 and G_2, we can pair all of the edges from $\bar{M}$ of weight 1 that are in the exclusive parts of G_1 and G_2. We can now replace each pair of edges e and f by $x_1^{ef} x_2^{ef}$. After adding all the new edges of weight 0, we get a solution of the modified instance. Therefore the total cost of $\bar{M}$ is at most a total cost of the given solution.

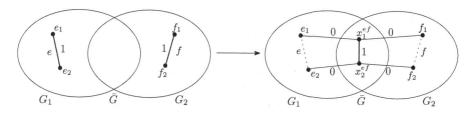

Fig. 1. Gadget replacing pairs of edges

Supposing that the original edges of weight 0 form a simultaneously acyclic set, we observe that the same is true after the reduction, as each new cycle contains an edge of weight 1. Furthermore we added at most a constant number of edges and vertices for all of the pairs of original edges, obtaining a problem of asymptotically at most quadratic size compared to the input problem.   □

5.2   Matroids

Definition 6. *A matroid M is a pair (E, I) where E is a set of elements and I is a family of independent sets (subsets of E) satisfying the following properties:*

1. $\emptyset \in I$
2. $\forall X, Y$ s.t. $X \in I$ and $Y \subset X : Y \in I$
3. $\forall X, Y \in I$ s.t. $|X| > |Y| : \exists x \in X \setminus Y$ s.t. $Y \cup \{x\} \in I$

Definition 7. *Let G be a graph with a set of edges E and I be a set of all acyclic subsets of E. Then (E, I) is a* graphic matroid *of G.*

Fact 9. *For any graph G (possibly multiphase with loops), the graphic matroid of G is a matroid and maximal independent sets of this matroid are exactly all possible spanning trees of G.*

Definition 8. *A* matroid intersection problem *of two matroids (E, I_1) and (E, I_2) on the same set of elements E is the problem of finding a maximum subset of E s.t. it is independent in both matroids.*

Fact 10 ([9]). *For a set E and two matroids (E, I_1) and (E, I_2) given as independence oracles, the matroid intersection problem is solvable in polynomial time and polynomially many oracle queries.*

Fact 11 ([10]). *There are specialized algorithms for graphic matroid intersection problem.*

Lemma 5. *Let G be a graph with edges divided into two disjoint subsets F and $\bar{E}$ where F is acyclic and $\bar{E} = E(G) \setminus F$. Let I be a set of all subsets X of $\bar{E}$ such that $F \cup X$ is an acyclic subgraph of G. Then $(\bar{E}, I)$ is a graphic matroid.*

Proof. Let H denote G with all edges from F contracted; we keep all the parallel edges and loops. We observe that the graphic matroid of H is exactly $(\bar{E}, I)$. □

5.3   Polynomiality

Theorem 12. $SST \in \mathsf{P}$ *for any number of graphs.*

Proof. To solve SST, it suffices to use Kruskal's algorithm (or any other MST algorithm) to first take a minimum spanning tree of the intersection, and then extend this partial solution to each individual graph using only exclusive edges. Clearly each exclusive edge may only create a cycle in its respective graph. On the other hand we are never forced to take an exclusive edge closing a cycle (in fact, Kruskal's algorithm refuses such edges by definition). □

Lemma 6. *Let I be an instance of $\cap$-01-$SMST$ for two graphs such that the edges of weight 0 form a simultaneously acyclic set. Then I can be solved in polynomial time using a matroid intersection algorithm.*

Proof. For each of the two graphs G_i for $i \in \{1, 2\}$ we define F_i as the set containing all edges of weight 0 and $\bar{E}$ the set of all edges of weight 1. Let I_i be a set of all subsets X of $\bar{E}$ such that $X \cup F_i$ is acyclic in G_i and let M_i denote the pair $(\bar{E}, I_i)$. According to Lemma 5 each M_i is a matroid. Furthermore, both of the matroids are defined on the same ground set $\bar{E}$.

Let $F = F_1 \cup F_2$ be all the edges of weight 0. By Lemma 2, F can be extended to a solution of the $\cap$-01-$SMST$ by a suitable subset of $\bar{E}$. We can now use a (graphic) matroid intersection algorithm to find a set X which is a maximum subset of $\bar{E}$ independent in both matroids M_1 and M_2. Therefore X is the maximum subset of $\bar{E}$ that extends F so that $X \cup F$ is simultaneously acyclic. If $X \cup F$ restricted to G_1 and G_2 spans all components, we output $X \cup F$, otherwise we answer "no". This is the same as to compare the size of $X \cup F$ to the size it should have.

Clearly if there exists a solution of the given $\cap$-01-$SMST$ instance, then according to Lemma 2 there exists a solution Y extending the set F. The set of edges $Y \backslash F$ is an independent set in both matroids M_1 and M_2 and therefore X exists and is of size $|Y \backslash F|$. This means that $X \cup F$ is a simultaneous spanning tree and the algorithm answers correctly. On the other hand, if no solution exists, then the set $X \cup F$ restricted to either G_1 or G_2 is acyclic but does not connect all the vertices connected in the original graph. We recognize this case and answer "no" correctly. □

Lemma 7. $\cap$-01-$SMST \in \mathsf{P}$ *for two graphs.*

Proof. Let I be an instance of the $\cap$-01-$SMST$ problem. We show that we can solve I using a (graphic) matroid intersection algorithm.

First suppose that the edges of weight 0 are not simultaneously acyclic. We simply restrict I to edges of weight 0, which gives us an instance of SST. We can solve this instance in polynomial time according to Theorem 12. If we obtain answer "no", then according to Lemma 2 there is no solution and we also answer "no".

Suppose we get a solution X. Then, by Lemma 2, we may delete all the edges of weight 0 except the edges from X and further assume that the edges of weight 0 are simultaneously acyclic. We use Lemma 6 to solve this reduced instance in polynomial time.   □

Theorem 13. *SMST* ∈ P *for two graphs.*

Proof. Let us have an instance of the *SMST* problem. According to Lemma 3, every instance of *SMST* can be solved by solving at most $O(m)$ 01-*SMST* problems, where m denotes the number of edges on input.

Any 01-*SMST* can be polynomially reduced to ∩-01-*SMST* as shown in Lemma 4; and according to Lemma 7, each ∩-01-*SMST* instance can be solved in polynomial time.   □

5.4   Complexity

Let us have an instance of *SMST* and let n denote the number of vertices, m the number of edges and w the number of weights in the given instance. We proceed according to Theorem 13.

The *SMST* problem is first decomposed into $(w-1)$ 01-*SMST* subproblems. We observe that each edge in these subproblems is either already fixed as a part of the solution of *SMST* or appears for the first time. The first kind of edges can be bound as at most $O(n)$ per 01-*SMST* subproblem, as they must form a simultaneously acyclic set. The second kind can be bound as at most $O(m)$ over all of the 01-*SMST* subproblems.

Each of the 01-*SMST* subproblems is reduced to a ∩-01-*SMST* problem of asymptotically at most quadratic size (by Lemma 4). Using the simultaneous acyclicity of edges of weight 0 we can use the approach of Lemma 6 in all but the first subproblem, and use the Lemma 7 to solve the first subproblem. Therefore we solve at most w (graphic) matroid intersection problems during the whole process and one instance of *SST* problem. The final complexity depends on the choice of algorithms used to solve the matroid intersection problems and the *SST*.

Furthermore, if w asymptotically approaches m, then some weight values have few representatives and more direct methods from Observations 2 and 3 using *SKA* may be applied to reduce the complexity.

Acknowledgements. This paper is the output of the 2017 Problem Seminar of Charles University. At this seminar undergraduate students attempt to solve open problems and learn to do research. We would like to thank Jan Kratochvíl and Pavel Valtr for their guidance, help and tea.

References

1. Bläsius, T., Kobourov, S.G., Rutter, I.: Simultaneous embedding of planar graphs. arXiv.org:1204.5853 (2015)
2. Borůvka, O.: O jistém problému minimílním (About a certain minimal problem) (Czech, German summary). Práce mor. přírodověd. spol. v Brně III **3**, 37–58 (1926)

3. Jarník, V.: O jistém problému minimálním. Práce Moravské Přírodovědecké Společnosti **6**, 57–63 (1930)
4. Prim, R.C.: Shortest connection networks and some generalizations. Bell Labs Tech. J. **36**(6), 1389–1401 (1957)
5. Kruskal, J.B.: On the shortest spanning subtree of a graph and the traveling salesman problem. Proc. Am. Math. Soc. **7**(1), 48–50 (1956)
6. Graham, R.L., Hell, P.: On the history of the minimum spanning tree problem. Ann. Hist. Comput. **7**(1), 43–57 (1985)
7. Pettie, S., Ramachandran, V.: An optimal minimum spanning tree algorithm. J. ACM (JACM) **49**(1), 16–34 (2002)
8. Karp, R.M.: Reducibility among combinatorial problems. In: Miller, R.E., Thatcher, J.W., Bohlinger, J.D. (eds.) Complexity of Computer Computations. The IBM Research Symposia Series, pp. 85–103. Springer, New York (1972). https://doi.org/10.1007/978-1-4684-2001-2_9
9. Edmonds, J.: Submodular Functions, Matroids, and Certain Polyhedra. Combinatorial Structures and their Applications, pp. 68–87. Gordon and Breach, New York (1970)
10. Gabow, H.N., Stallmann, M.: Efficient algorithms for graphic matroid intersection and parity. In: Brauer, W. (ed.) ICALP 1985. LNCS, vol. 194, pp. 210–220. Springer, Heidelberg (1985). https://doi.org/10.1007/BFb0015746

Variations of Cops and Robbers Game on Grids

Sandip Das and Harmender Gahlawat[(✉)]

Indian Statistical Institute, Kolkata, India
harmendergahlawat@gmail.com

Abstract. Cops and robber is a two player turn based game played on a graph where the cops try to catch the robber. The *cop number* of a graph is the minimum number of cops required to catch the robber. We consider two variants of this game, namely *cops and attacking robber*, and *lazy cops and robber*. In cops and attacking robber, the robber can attack a cop and remove him from the game, whereas in lazy cops and robber only one cop is allowed to move during the cop's turn. We prove that the cop number for both these variants in finite square grids is two. We show that the cop number for cops and attacking robber in n-dimensional hypercube is at most n, and the cop number in the same version of a 3-dimensional grid is three.

1 Introduction

Cops and robber is a two player game played on finite connected graphs. One player controls the k cops and the other controls the robber. First the k cops and then the robber occupy some vertices of the graph. More than one cop can occupy a vertex. Then cops and robber make alternating moves. On cop's turn, any number of the k cops can move to their neighboring vertices, or can pass by staying on the same vertex. On robber's turn, he does the same. Each player can see all the moves. If at least one of the cops succeed in occupying the same vertex as the robber, we call it a *capture*. The cops win if they capture the robber in finite time, otherwise the robber wins. The cop number of a graph G, denoted by $c(G)$, is the minimum number of cops required to capture the robber. The cop number is well defined as the number of vertices of a graph is clearly an upper bound.

The cops and robber game (for $k = 1$) was introduced by Nowakowski and Winkler [7], and independently by Quillot [8]. Aigner and Fromme [1] generalized this game to k cops, as presented here, and introduced cop number. Goldstien and Reingold [9] proved that determining if k cops can catch a robber is EXPTIME-complete if the initial positions are given. Later Fomin et al. [5] proved that finding the minimal number of cops needed to catch the robber is NP-hard. For further results, we refer the readers to the book by Bonato and Nowakowski [10]. Here we consider finite square grids only, hence we simply call them grids. Bhattacharya et al. [12], and recently Luccio and Pagli [6], proved that cop number of a finite grid is 2.

© Springer International Publishing AG 2018
B. S. Panda and P. P. Goswami (Eds.): CALDAM 2018, LNCS 10743, pp. 249–259, 2018.
https://doi.org/10.1007/978-3-319-74180-2_21

Many variants of this game has been studied. Some versions allow cops and robbers to have different speeds, some allow players to be on edges, some allow cops and robber to move simultaneously etc. A brief survey can be found in [10, see Chap. 8]. In this article we consider two variants of cops and robbers: *cops and attacking robber*, and *lazy cops and robber*.

The game of Cops and attacking Robber was introduced by Bonato et al. [11]. In this variant, robber is able to strike back against the cops. If on a robber's turn, there is a cop in its neighborhood, then robber can attack the cop and remove it from the game. However if more than one cops occupy the neighboring vertex and the robber attacks them, then only one of the cops get eliminated, and then the robber is captured. The cop number for capturing an attacking robber on a graph G is represented as $cc(G)$. Bonato et al. [11] proved that $c(G) \leq cc(G) \leq 2c(G)$. This can be easily verified as $cc(G)$ cops can catch the robber in the classical version, and if we play the classical version with $2c(G)$ cops (two at each initial $c(G)$ cop positions) we can catch the attacking robber. Our result concerning the cop number for attacking robber on finite grids $G_{m,n}$ is as follows.

Theorem 1. *For $m \geq 2$ and $n \geq 2$, $cc(G_{m,n}) = 2$.*

The game of Lazy cops and robbers was introduced by Offner and Ojakian [2] and they gave bounds for hypercubes. Later their results were improved by Bal et al. [3]. Recently Sim et al. [4] gave bounds for the generalised hypercubes. In this variant only one cop can move during the cop's turn. This restricts the ability of the cops with respect to the classical version. The cop number for lazy cops to capture a robber in a graph G is denoted by $lc(G)$. Clearly $c(G) \leq lc(G)$, as $lc(G)$ cops can capture a robber in the classical version. Our result concerning the lazy cop number of finite grids $G_{m,n}$ is as follows.

Theorem 2. *For $m, n \geq 2$, $lc(G_{m,n}) = 2$.*

We revisit the game of cops and attacking robbers. We consider this game played on n-dimension hypercubes Q_n and give a strategy for n cops to capture the attacking robber.

Theorem 3. *For $n > 0$, $cc(Q_n) \leq n$.*

We also consider the 3-dimensional grid $G_{m,n,p}$ and use the strategy given in proof of Theorem 2 to prove the following.

Theorem 4. *For $m, n, p \geq 2$, $cc(G_{m,n,p}) = 3$.*

In the next section we present some notations and definitions required in this article. Sections 3, 4, 5 and 6 contains the proofs of Theorems 1, 2, 3 and 4 respectively.

2 Definitions

We follow the standard notation of West [13]. An $m \times n$ grid, $G_{m,n}$ is the Cartesian product of two paths, P_m and P_n. Similarly an $m \times n \times p$ grid, $G_{m,n,p}$ is the Cartesian product of three paths, P_m, P_n and P_p. We use the following alternate definition of a grid. An $m \times n$ *grid* is a graph whose vertex set is $\{(i,j) \mid 0 \le i \le m-1,\ 0 \le j \le n-1\}$ and vertex (i,j) is adjacent to $(i-1,j),(i+1,j),(i,j-1),(i,j+1)$, whenever these coordinates stay inside the closed interval $[0, m-1]$ and $[0, n-1]$ respectively. So the vertices of a grid are arranged in m rows and n columns. We fix an orientation such that $(0,0)$ is the bottommost left vertex of the $m \times n$ grid. A vertex is either a *corner vertex*, *boundary vertex* or *internal vertex* if its degree is 2, 3 or 4 respectively. A grid has the following 4 corner vertices $(0,0),(0,n-1),(m-1,n-1)$ and $(m-1,0)$, which are arranged in clockwise order. The boundary vertices are *top*, *bottom*, *right* or *left* if they belong to $\{(i,n-1) \mid 0 \le i \le m-1\}$, $\{(i,0) \mid 0 \le i \le m-1\}$, $\{(m-1,j) \mid 0 \le j \le m-1\}$ or $\{(0,j) \mid 0 \le j \le m-1\}$ respectively. In Fig. 1, c_1 is in a corner vertex, c_2 is in a boundary vertex and c_3 is in an internal vertex. We have similar definitions for the 3-dimensional grid $G_{m,n,p}$. In particular, in a 3-dimensional grid, degree 3 vertices are corner vertices, degree 6 vertices are internal vertices and rest are boundary vertices.

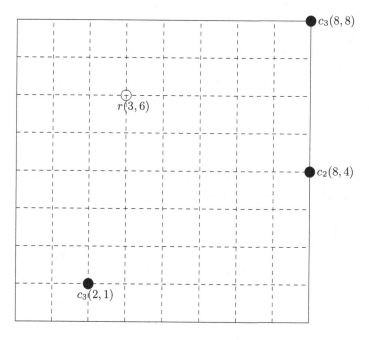

Fig. 1. Cops and robber on a 8×8 grid. Here $h = 1$ and $v = 2$.

Now consider the grid $G_{m,n}$. Let cop c_i be at vertex (x_i, y_i) and robber r be at (x_r, y_r). Let $h_{c_i} = \mid x_i - x_r \mid$ and $v_{c_i} = \mid y_i - y_r \mid$. Let $h = \min_{c_i \in C} h(c_i)$ and $v = \min_{c_i \in C} v(c_j)$ (see Fig. 1). We say c_i *moves towards* r if either h_{c_i} or v_{c_i} decreases. Similarly r *moves away from* c_i if either h_{c_i} or v_{c_i} increases. We also define c_i moves vertically (horizontally) towards r if h_{c_i} (or v_{c_i}) decreases, and r moves vertically (horizontally) away from c_i if h_{c_i} (or v_{c_i}) increases. Cop c_i at (x_i, y_i) moves towards a vertex (p, q) if $\mid x_i - p \mid$ or $\mid y_i - q \mid$ decreases. We have similar definitions for the 3-dimensional grid $G_{m,n,p}$.

A *round* consists of a robber's turn followed by the cop's turn. In a round, cop c_i *mirrors* the move of the robber r if $h(c_i)$ and $v(c_i)$ do not change after that round i.e. in this round c_i moves exactly in the same direction as r.

In $G_{m,n}$, we say the robber r is in cop c_i's *guard* if $h_{c_i} = 1$ and $v_{c_i} = 1$. Suppose c_i guards r. The two vertices that are common neighbors of both c_i and r are called as *coguard positions*. If we draw the co-ordinate axes with origin at c_i it divides the grid in four quadrants each containing one corner point. Let C denote the corner point of the grid that is in the same quadrant as r.

In $G_{m,n}$, the robber r is *trapped* by a cop c_i when r is in a corner vertex and c_i guards r. When r is trapped it cannot make any move without getting captured.

Each of our strategies are described in terms of rounds i.e. they say how the cops will move depending on the robber's move. So in order to avoid confusion, unless mentioned otherwise, we fix the following. After the cops and robbers are placed, the cops pass their first turn. So the first round begins with the first turn of the robber.

The n-dimensional hypercube Q_n is defined recursively in terms of cartesian products of two graphs as follows: $Q_1 = K_2$ and $Q_n = Q_{n-1} \times K_2$, where K_2 is an edge and the symbol $\times$ refers to the cartesian product of two graphs (see West [13]).

3   Proof of Theorem 1

In the classical cops and robbers game, the cop number of finite grids is 2 as proved in [12]. Since $c(G) \leq cc(G)$, in the cops and attacking robber version, the cop number is at least 2. Now it suffices to give a strategy to catch the robber using two cops.

We begin with an outline of such a strategy. Two cops c_1 and c_2 are initially placed at $(0, 0)$ and $(1, 0)$. They remain on adjacent vertices for the rest of the game, so that if robber attacks one of the cops, he gets caught in next move. In further rounds, cops move such that neither h nor v ever increases. Then we ensure that after a finite number of rounds either h or v or both decreases. So after a finite number of rounds both h and v reduces to 0, which implies capture. Now we give the detailed strategy.

Cops c_1 and c_2 are initially placed at $(0, 0)$ and $(1, 0)$. In each round cops move trying to decrease either h or v or both, while ensuring that after their move neither h nor v increases in that round. The following two cases can arise.

1. $h > 0$ & $v > 0$: If the robber r moves such that h or v increases, the cops c_1 and c_2 mirror r's move, hence restoring the values of h and v at the beginning of this round. Upon any other move of the r, c_1 moves towards r and c_2 mirrors c_1's move resulting in a decrease in the value of either h or v or both. After each round either $h > 0$ & $v > 0$ or at least one of them is 0. Accordingly the game continues in this case or enters case 2 respectively.

2. $h = 0$ or $v = 0$: Without loss of generality assume that $h = 0$. This case is further divided into two sub-cases 2(a) and (b):

   (a) $h(c_1) = h(c_2) = 0$: In this sub-case c_1, c_2 and r lie in one vertical line. Without loss of generality assume c_1 is closer to r. If r moves such that v increases, then both c_1 and c_2 mirror r's move. If r moves such that h increases, then c_2 moves to position of c_1 and c_1 mirrors r's move, resulting in case 2(b). Upon any other move of r, both c_1 and c_2 move towards r resulting in decrease of value of v. After any round at least one of $h(c_1)$ or $h(c_2)$ is zero. Hence when the game enters this sub-case, it either remains in this sub-case i.e. 2(a) or enters the next sub-case i.e. 2(b).

   (b) $h(c_1) = 0$ & $h(c_2) = 1$: In this sub-case c_1 and r lie in one vertical line. If r moves such that v or h increases, then both c_1 and c_2 mirror r's move. Upon any other move of r, both c_1 and c_2 move vertically towards r. So in this sub-case after any round $h(c_1) = 0$ & $h(c_2) = 1$. Hence once the game enters this sub-case it remains in this sub-case until the robber is caught.

In our strategy if initially $h > 0$ & $v > 0$, the cops move till one of them is 0 (i.e. case 2), then they move till the other is 0 (i.e. capture). It remains to prove that in each of the above mentioned cases there cannot be infinite rounds where h and v remains unchanged.

In each round in case 1, the value of h or v remains unchanged only if r's move results in an increase in h or v. In order to do so r has to move away vertically or horizontally from any of c_1 or c_2 (since c_1 and c_2 are adjacent and both h and v are positive, moving away from c_1 is same as moving away from c_2). Once r reaches a boundary vertex, one of the values of h or v cannot be further increased by r's move. Upon reaching a corner vertex neither h or v can be increased by r's move. Hence, in case 1, the sequence of rounds where both h and v remains unchanged is finite.

In any round in case 2(a) (without loss of generality assume $h = 0$), if r moves to increase value of h, then r's move results in case 2(b). So in each round in case 2(a), the value of v remains unchanged only if r's move results in an increase in v. In order to do so r has to move away vertically from c_1. Once it reaches a top/bottom boundary vertex or a corner vertex, v cannot be further increased by r's move. Hence, in case 2(a), the sequence of rounds where both h and v remains unchanged is finite.

The analysis in case 2(b) is similar to case 1 and it follows here also that the sequence of rounds where both h and v remains unchanged is finite. This completes the proof of Theorem 1.

4   Proof of Theorem 2

Since $lc(G)$ number of cops can catch the robber in the classical cops and robbers game, $c(G) \leq lc(G)$. And since the cop number on finite grid is 2 as proved in [12], $lc(G) \geq 2$. So it suffices to give a strategy to catch the robber using two lazy cops.

We begin with an outline of such a strategy. Let c_1 and c_2 be the two lazy cops. If one traps the robber r, the other lazy cop can come and capture it in finite time. Let c_1 and c_2 be initially placed at $(0,0)$ and $(1,0)$. Initially c_1 follows r till it takes him in its guard. In further rounds, c_1 and c_2 will force r to go in a trap position such that r is trapped by c_1. Then c_2 moves towards r till it captures r. The whole strategy relies on the following observation.

Observation 1. *If robber r gets into trap position of a cop c_1, r cannot move without being captured.*

Proof. This is because the two neighbors of r in the corner vertex are also neighbors of c_1.   □

Now we give the detailed strategy. Initially c_1 moves trying to first guard the robber r i.e. try to make $h(c_1) = 1$ and $v(c_1) = 1$. Depending on whether c_1 guards r or not at the beginning of each round, we have the following cases.

1. $h(c_1) \neq 1$ or $v(c_1) \neq 1$: In this case c_1 is not guarding r. But c_1 moves trying to guard r; once it reaches a guard position we move to the next case. Depending on values of $h(c_1)$ and $v(c_1)$ at the beginning of each round, we have the following two sub-cases.
   (a) $h(c_1) > 0$ & $v(c_1) > 0$: In this sub-case, if r moves away from c_1, then c_1 mirrors r's move, hence restoring the values of $h(c_1)$ and $v(c_1)$ at the beginning of this round. After r's move if it is guarded by c_1, then c_1 passes his turn. Upon any other move of the robber, c_1 moves towards the robber such that $\max\{h(c_1), v(c_1)\}$ decreases. After each round, if c_1 guards r we enter the next case, else either $h(c_1) > 0$ & $v(c_1) > 0$ or one them is 0. Accordingly the game continues in this sub-case or moves to the next sub-case 1(b) respectively.
   (b) $h(c_1) = 0$ or $v(c_1) = 0$: If $h(c_1) = 0$, c_1 and r lie in one vertical line. If r moves such that $v(c_1)$ increases, then c_1 mirrors r's move, hence restoring the values of $h(c_1)$ and $v(c_1)$ at the beginning of the round. If r moves such that $h(c_1)$ increases and c_1 guards r, then c_1 passes his move. If r moves such that $h(c_1)$ increases and c_1 doesn't guard r, then c_1 moves vertically towards r, hence decreasing $v(c_1)$ and making $h(c_1) = 1$. Upon any other move of r, c_1 moves towards r, decreasing the value of $v(c_1)$. Similarly if $v(c_1) = 0$, c_1 and r lie in one horizontal line. If r moves such that $h(c_1)$ increases, then c_1 mirrors r's move, hence restoring the values of $h(c_1)$ and $v(c_1)$ at the beginning of the round. If r moves such that $v(c_1)$ increases and c_1 guards r, then c_1 passes his turn. If r moves such that $v(c_1)$ increases and c_1 doesn't guard r, then c_1 moves horizontally

towards r, hence decreasing $h(c_1)$ and making $v(c_1) = 1$. Upon any other move of r, c_1 moves towards r, decreasing the value of $h(c_1)$.

In this sub-case, if $h(c_1) = 0$, then c_1 always moves vertically, and if $v(c_1) = 0$, then c_1 always moves horizontally. After each round if c_1 guards r we enter case 2, else either $h(c_1) > 0$ & $v(c_1) > 0$ or one them is 0. Accordingly the game moves to the previous sub-case i.e. $1(a)$ or continues in this sub-case respectively.

Notice that the game can move from case $1(a)$ to case $1(b)$ and vice versa. When the game changes from case $1(a)$ to case $1(b)$ the value of $h(c_1) + v(c_1)$ decreases. However when the game changes from case $1(b)$ to case $1(a)$ the value of $h(c_1) + v(c_1)$ remains same.

2. $h(c_1) = 1$ & $v(c_1) = 1$: In this case c_1 is guarding r. Once r is guarded but not trapped, it can move to only two or one of its neighbors depending on whether r is in an internal or boundary vertex, since its other neighbors are adjacent to the cop.

Recall that once r moves, it gets closer to the corner point C which lies in the same quadrant as r when origin is fixed at c_1.

If r moves, c_1 mirrors its move. So c_1 always guards r. If r does not move, then c_2 moves towards the closest coguard position. Once c_2 reaches the coguard position, r is forced to move, and hence gets closer to C. Once r reaches C, c_1 traps it. Then c_2 moves towards C and captures r.

Once c_1 guards r, it always guards r irrespective of whether r moves or not. Hence once the game enters this case it remains in this case till the end.

In our strategy if initially none of the cops guard the robber r, then c_1 moves till it guards r (enters case 2). Once c_1 guards r, it traps r with the help of c_2 and then c_2 captures r. It remains to prove that the game does not continue indefinitely.

Recall that the game can move from case $1(a)$ to case $1(b)$ and vice versa. When the game changes from case $1(a)$ to case $1(b)$ the value of $h(c_1) + v(c_1)$ decreases. However when the game changes from case $1(b)$ to case $1(a)$ the value of $h(c_1) + v(c_1)$ remains same.

In each round in case $1(a)$, the value of $h(c_1) + v(c_1)$ remains unchanged only if r's move increases $h(c_1)$ or $v(c_1)$. In order to do so r has to move away horizontally or vertically from c_1. It can keep on moving away from c_1 till it reaches a corner point which is in the same quadrant as r when origin is placed at c_1. Once r reaches this corner point $h(c_1) + v(c_1)$ decreases. Also if it enters case $1(b)$, then $h(c_1) + v(c_1)$ decreases.

In each round in case $1(b)$, if $h(c_1) = 0$, the value of $h(c_1) + v(c_1)$ remains unchanged only if r's move increases $h(c_1)$ or $v(c_1)$. If r's move increases $h(c_1)$, then $h(c_1) + v(c_1)$ remains unchanged but the game enters case $1(a)$, where we saw that $h(c_1) + v(c_1)$ decreases after finite rounds. If r's move increases $v(c_1)$, then $h(c_1) + v(c_1)$ remains unchanged till it reaches a top or bottom boundary vertex or a corner after which r's move cannot increase $v(c_1)$ anymore.

Similarly in each round in case $1(b)$, if $v(c_1) = 0$, the value of $h(c_1) + v(c_1)$ remains unchanged only if r's move increases $h(c_1)$ or $v(c_1)$. If r's move increases

$v(c_1)$, then $h(c_1)+v(c_1)$ remains unchanged but the game enters case $1(a)$, where we saw that $h(c_1)+v(c_1)$ decreases after finite rounds. If r's move increases $h(c_1)$, then $h(c_1)+v(c_1)$ remains unchanged till it reaches a left or right boundary vertex or a corner after which r's move cannot increase $h(c_1)$ anymore.

Hence after a finite number of rounds in case 1, value of $h(c_1) + v(c_1)$ decreases. So after a finite number of rounds $h(c_1) + v(c_1)$ becomes 2 i.e. $h(c_1) = 1$, $v(c_1) = 1$, or $h(c_1) = 0$, $v(c_1) = 2$ or $h(c_1) = 2$, $v(c_1) = 0$. Notice that $h(c_1) = 1$, $v(c_1) = 1$ is the desired guard position. If $h(c_1) = 0$, $v(c_1) = 2$, r moves such that either $h(c_1)$ increases in which case c_1 moves vertically towards r to a guard position; or $v(c_1)$ increases in which case it cannot increase when it reaches a top or bottom boundary vertex or a corner vertex, after which it is forced to move such that $h(c_1)$ increases and then c_1 moves to a guard position. Hence after a finite number of rounds c_1 guards r.

Once r is guarded, each time it moves it gets closer to corner C. If it does not move c_2 reaches a coguard position after which it has to move. Hence in finite number of rounds, r is trapped by c_1, after which c_2 captures r in a finite number of rounds. Hence after a finite number of rounds r is captured. This completes the proof of Theorem 2.

5   Proof of Theorem 3

In this section the cop does not skip his first move, as assumed in Sect. 2 i.e. after the robber is placed the cops move, else the robber might attack a suitable cop.

We use induction to prove this theorem. But first we need the following definitions. The hypercube Q_{n+1} is cartesian product of Q_n and K_2. So it is union of two Q_n's with edges between the corresponding vertices in each Q_n. Let they be represented as A_n and B_n. If the robber is in B_n, then the vertex in A_n adjacent to the vertex containing robber is called image of robber r and is denoted by I_r. If a robber is in B and a cop c_i is neighbour of I_r in A, then robber cannot enter A and we say that c_i is *protecting* A.

Since Q_1 has only 2 vertices, and the cop occupies one of them in first move, robber can enter only at other vertex and hence will be captured in next move of cop. For Q_2 also, if we place two cops at any two distinct positions, they will capture the robber in next round as the other two vertices are in the neighborhood of the cops. In Q_3 all three of the cops start at one of the Q_2's, say A_2. Now, if robber enters A, he will be captured in next move of the cops. If robber enters at a vertex on B, two of the cops will use their strategy to capture I_r. Once, one of cops, say c_i is at a neighboring vertex of I_r, that cop will move such that it always remains adjacent to I_r. Now remaining cops will readjust their positions such that every cop is adjacent to another cop, or two cops are in the same vertex (if they are not already in such positions). Then they all move to B, while the cop c_i protecting A remains in A and will move only in such a way that it keeps protecting A. So robber can't enter A. The 2 cops in B will capture the robber as $cc(Q_2) = 2$.

Now assume $cc(Q_n) \leq n$.

For Q_{n+1} we show that $n+1$ cops are sufficient. The hypercube Q_{n+1} is union of two Q_n: A_n and B_n with their corresponding vertices connected. Initially we place all our cops in A_n where n cops are placed as per our strategy for Q_n and remaining one cop is placed with any of the cop and will move with that cop as long as it is in A_n. If the robber is in A_n and never goes to B_n, then these n cops will capture it (by our induction hypothesis). If robber is in B_n, or enters B_n in some round from A_n, these n cops will move to capture I_r. Again by our induction hypothesis they will be able to capture I_r in a finite number of steps. When one of the cops, say c_i, is at a neighbouring vertex of I_r, then it protects A_n. In the further rounds it always stays in the neighborhood of I_r, and hence protects A_n till the end of the game. So robber cannot enter A_n without being captured.

The rest n cops will readjust their positions such that for every cop there is at least one cop in its neighborhood or in its position (if they are not already in such positions). Then they all move to B. Since $cc(Q_n) \leq n$, the cops in B_n capture the robber.

So in Q_{n+1}, we have a strategy to capture the attacking robber by $n + 1$ cops. This completes the proof of Theorem 3.

6   Outline of Proof of Theorem 4

This proof is similar to proof of Theorem 2. So we omit some of the details evident from the proof of Theorem 2. In the classical cops and robber game, the cop number of 3-dimensional finite grids is 3 as proved in [12]. Since $c(G) \leq cc(G)$, in the cops and attacking robber version, the cop number is at least 3. Now it suffices to give a strategy to catch the robber using three cops.

We begin with a few definitions. In a *guard configuration*, the three cops c_1, c_2 and c_3 are at (x, y, z), $(x + 1, y, z)$ and $(x, y + 1, z)$ and the robber is at $(x+1, y+1, z+1)$. In a *post-guard configuration*, the three cops c_1, c_2 and c_3 are at $(x, y+1, z)$, $(x+1, y, z)$ and $(x+1, y+1, z)$ and the robber is at $(x+1, y+1, z+1)$. In all these configurations, the cops are connected; so if the robber attacks one of the cops, then he gets captured in the next move. Also in one round, the cops can reach the post-guard configuration from the guard configuration. Let C be the corner vertex $(m - 1, n - 1, p - 1)$.

We define $d_1 = |x_{c_1} - x_r|$, $d_2 = |y_{c_1} - y_r|$ and $d_3 = |z_{c_1} - z_r|$. We also extend the definitions of *moves away from* and *moves towards* given in Sect. 2, to 3-dimensions.

We first give an outline of the strategy. Let c_1, c_2 and c_3 be the three cops positioned at $(0, 0, 0)$, $(1, 0, 0)$ and $(0, 1, 0)$ respectively. First the cops move till the guard configuration is reached.

Once the guard configuration is attained, if the robber is not in the corner vertex C, the cops attain the post-guard configuration and then the robber is forced to move. Once the post-guard configuration is attained, the cops just mirror the robber's move. So after every round they always attain the post-guard configuration and force the robber to move further. Whenever the robber moves (after a guard configuration is attained), it gets closer to the corner vertex C. After some rounds the robber is forced to the corner vertex C and then captured by the cops.

If the robber is in the corner vertex C, when the guard configuration is reached, then the cops attain the post-guard configuration and capture the robber.

We have the following trivial observations.

1. In the guard configuration, all the cops and robber lie in one cube i.e. Q_3. Every other vertex of this cube, except the position of robber is either a cop's position or is adjacent to some cop.
2. If the robber moves after a guard position is attained, then it moves towards the corner vertex C. This is because if we fix origin at position of c_1, then the robber is in the same octant at the corner vertex C. Every time the cops move, this octant this gets smaller. In the beginning of each round, the robber and the cop c_1 lie in the same cube with $d_1 = 1$, $d_2 = 1$ and $d_3 = 1$. As the octant gets smaller the robber moves closer to the corner vertex C.
3. In the post-guard configuration, the robber is forced to move, and it always moves towards the corner vertex C (same reason as above).
4. Once the robber is in the corner vertex in a guard configuration, it cannot move without being captured. Once the robber is in the corner vertex in a post-guard configuration, it gets captured in the next cops turn.

Now we give the detailed strategy. Let c_1, c_2 and c_3 be the three cops positioned at $(0,0,0)$, $(1,0,0)$ and $(0,1,0)$ respectively. In this strategy, cop c_1 is placed such that whenever the robber moves away from the cop c_1, it moves towards the corner vertex C (similar to Observation 2 above). We have the following cases.

1. *Robber is not in guard configuration*: In a round, if the robber moves towards the corner C, then the cops mirror its move. So $d_1 + d_2 + d_3$ is same as at the beginning of this round. For any other move of the robber, the cop c_1 moves towards the robber such that $\max\{d_1, d_2, d_3\}$ decreases. The cops c_2 and c_3 mirror the move of c_1.

   Extending the arguments given in the proof of Theorem 2, the value of $d_1 + d_2 + d_3$ eventually decreases after finitely many rounds. As $d_1 + d_2 + d_3$ decreases and reaches 3, the cops attain the guard configuration if $d_1 = d_2 = d_3 = 1$. For all other cases of $d_1 + d_2 + d_3 = 3$, by case analysis (using the argument in proof of Theorem 2), the robber will be forced to be in a guard position. We discuss one such case below.

   If the robber remains in the XY-plane (assuming the standard X, Y, Z orientations) and does not leave the XY plane till it reaches the corner vertex $(m - 1, n - 1, 0)$. When $d_1 + d_2 + d_3 = 3$, here either $d_1 = 2, d_2 = 1, d_3 = 0$ or $d_1 = 1, d_2 = 2, d_3 = 0$. Then if c_1, c_2 and c_3 reach $(m - 2, n - 2, 0)$, $(m - 1, n - 2, 0)$ and $(m - 2, n - 1, 0)$, then $d_1 + d_2 + d_3 = 2$. Hence the robber is forced to move to $(m - 1, n - 1, 1)$ there by attaining the guard position.

2. *Robber is in guard configuration*: If the robber is in the corner vertex C, the cops attain the post-guard configuration and then capture the robber. If the robber is not in the corner vertex C, and it moves (towards the corner vertex C), then the cops mirror its move. If the robber is not in the corner vertex C, and it does not move, then the cops attain the post-guard configuration and force the robber to move towards the corner C. After this the cops mirror robber's move, so they always stay in the post-guard configuration. This continues till the robber reaches C, and then gets captured.

The finiteness of the game can be proved just by following the arguments in the proof of Theorem 2. This completes the outline of proof of Theorem 4.

7   Conclusion

We intend to calculate the cop number of n-dimensional grid in the cops and attacking robber version. Notice that a n-dimensional hypercube, whose cop number we have bounded, is a cell of a n-dimensional grid. We are working on a strategy to force the robber to the corner and then capture it in that corner hypercube.

References

1. Aigner, M., Fromme, M.: A game of cops and robbers. Discrete Appl. Math. **8**, 1–12 (1984)
2. Offner, D., Okajian, K.: Variations of cops and robber on the hypercube. Aust. J. Comb. **59**(2), 229–250 (2014)
3. Bal, D., Bonato, A., Kinnersley, W.B., Pralat, P.: Lazy cops and robbers on hypercubes. Comb. Probab. Comput. **24**(6), 829–837 (2015)
4. Sim, K.A., Tan, T.S., Wong, K.B.: Lazy cops and robbers on generalized hypercubes. Discrete Math. **340**(7), 1693–1704 (2017)
5. Fomin, F., Golovach, P., Kratochvil, J., Nisse, N., Suchan, K.: Pursuing a fast robber on a graph. Theor. Comput. Sci. **411**, 1167–1181 (2010)
6. Luccio, F., Pagli, L.: Cops and robber on grids and tori. CoRR, abs/1708.08255 (2017)
7. Nowakowski, R., Winkler, P.: Vertex-to-vertex pursuit in a graph. Discrete Math. **43**, 253–259 (1983)
8. Quillot, P.: A short note about pursuit games payed on a graph of given genus. J. Comb. Theory Ser. B **38**, 89–92 (1985)
9. Goldstein, A.S., Reingold, E.M.: The complexity of pursuit on a graph. Theor. Comput. Sci. **143**, 93–112 (1995)
10. Bonato, A., Nowakowski, R.: The Game of Cops and Robbers on Graphs. American Mathematical Society, Providence (2011)
11. Bonato, A., et al.: The robber strikes back. In: Krishnan, G.S.S., Anitha, R., Lekshmi, R.S., Kumar, M.S., Bonato, A., Graña, M. (eds.) Computational Intelligence, Cyber Security and Computational Models. AISC, vol. 246, pp. 3–12. Springer, New Delhi (2014). https://doi.org/10.1007/978-81-322-1680-3_1
12. Bhattacharya, S., Paul, G., Sanyal, S.: A cops and robber game in multidimensional grids. Discrete Appl. Math. **58**, 1745–1751 (2010)
13. West, D.B.: Introduction to Graph Theory, 2nd edn. Pearson Education, Essex (2002)

Alternation, Sparsity and Sensitivity: Combinatorial Bounds and Exponential Gaps

Krishnamoorthy Dinesh[(✉)] and Jayalal Sarma

Department of Computer Science and Engineering,
Indian Institute of Technology Madras, Chennai, India
{kdinesh,jayalal}@cse.iitm.ac.in

Abstract. The well-known SENSITIVITY CONJECTURE regarding combinatorial complexity measures on Boolean functions states that for any Boolean function $f : \{0,1\}^n \to \{0,1\}$, block sensitivity of f is polynomially related to sensitivity of f (denoted by $\mathsf{s}(f)$). From the complexity theory side, the XOR LOG-RANK CONJECTURE states that for any Boolean function, $f : \{0,1\}^n \to \{0,1\}$ the communication complexity of a related function $f^{\oplus} : \{0,1\}^n \times \{0,1\}^n \to \{0,1\}$, (defined as $f^{\oplus}(x,y) = f(x \oplus y)$) is bounded by polynomial in logarithm of the sparsity of f (the number of non-zero Fourier coefficients for f, denoted by $\mathsf{sparsity}(f)$). Both the conjectures play a central role in the domains in which they are studied.

A recent result of Lin and Zhang (2017) implies that to confirm the above two conjectures it suffices to upper bound alternation of f (denoted $\mathsf{alt}(f)$) for all Boolean functions f by polynomial in $\mathsf{s}(f)$ and logarithm of $\mathsf{sparsity}(f)$, respectively. In this context, we show the following results:

- We show that there exists a family of Boolean functions for which $\mathsf{alt}(f)$ is at least *exponential* in $\mathsf{s}(f)$ and $\mathsf{alt}(f)$ is at least *exponential* in $\log \mathsf{sparsity}(f)$. Enroute to the proof, we also show an exponential gap between $\mathsf{alt}(f)$ and the decision tree complexity of f, which might be of independent interest.
- As our main result, we show that, despite the above exponential gap between $\mathsf{alt}(f)$ and $\log \mathsf{sparsity}(f)$, the XOR LOG-RANK CONJECTURE is true for functions with the alternation upper bounded by $\mathsf{poly}(\log n)$. It is easy to observe that the SENSITIVITY CONJECTURE is also true for this class of functions.
- The starting point for the above result is the observation (derived from Lin and Zhang (2017)) that for any Boolean function f, $\deg(f) \le \mathsf{alt}(f)\deg_{\mathbb{F}_2}(f)^2$ where $\deg(f)$ and $\deg_{\mathbb{F}_2}(f)$ are the degrees of f over $\mathbb{R}$ and $\mathbb{F}_2$. We give two further applications of this bound: (1) We show that Boolean functions with bounded alternation have high sparsity ($\Omega(\sqrt{\deg(f)})$), thus partially answering a question of Kulkarni and Santha (2013). (2) We observe that the above relation improves the upper bound for influence to $\deg_{\mathbb{F}_2}(f)^2 \cdot \mathsf{alt}(f)$ improving Guo and Komargodski (2017).

© Springer International Publishing AG 2018
B. S. Panda and P. P. Goswami (Eds.): CALDAM 2018, LNCS 10743, pp. 260–273, 2018.
https://doi.org/10.1007/978-3-319-74180-2_22

1 Introduction

A central theme of research in Boolean function complexity is relating the complexity measures of Boolean functions (see [5] for a survey). For a Boolean function $f : \{0,1\}^n \to \{0,1\}$, *sensitivity* of f on $x \in \{0,1\}^n$, is the maximum number of indices $i \in [n]$, such that $f(x \oplus e_i) \neq f(x)$ where $e_i \in \{0,1\}^n$ with exactly the ith bit as 1. The sensitivity of f (denoted by $\mathsf{s}(f)$) is the maximum sensitivity of f over all inputs. A related parameter is the *block sensitivity* of f (denoted by $\mathsf{bs}(f)$), where we allow disjoint blocks of indices to be flipped instead of a single bit. Another parameter is the *degree* (denoted by $\deg(f)$) of a multilinear polynomial over reals that agrees with f on Boolean inputs. If the polynomial is over $\mathbb{F}_2$, then the degree of the polynomial is denoted by $\deg_{\mathbb{F}_2}(f)$.

Nisan and Szegedy conjectured that for an arbitrary function $f : \{0,1\}^n \to \{0,1\}$, $\mathsf{bs}(f) \leq \mathsf{poly}(\mathsf{s}(f))$ and this is popularly known as the SENSITIVITY CONJECTURE [21]. Though the measures $\mathsf{bs}(f)$ and $\mathsf{s}(f)$ were introduced to understand the CREW-PRAM model of computation [8,20], subsequent works [5,24] showed connection to other Boolean function parameters, in particular, $\sqrt{\mathsf{bs}(f)} \leq \deg(f) \leq \mathsf{bs}(f)^3$. Hence the SENSITIVITY CONJECTURE can equivalently be stated as: for any Boolean function f, $\deg(f) \leq \mathsf{poly}(\mathsf{s}(f))$. This question has been extensively studied in [1,14,23] (see [12] for a survey) and for various restricted classes of Boolean functions in [2,6,7,20,26]. There are also recent approaches to settle the conjecture via a formulation in terms of a communication game [9] and via a formulation in terms of distributions on the Fourier spectrum of Boolean functions [10].

For an $f : \{0,1\}^n \to \{-1,1\}$, define $f^\oplus(x,y) = f(x \oplus y)$. From the complexity theory side, the XOR LOG-RANK CONJECTURE (proposed in [28]) says that the deterministic communication complexity of $f^\oplus$ (denoted by $\mathsf{CC}_\oplus(f)$) is polynomially upper bounded by the logarithm of sparsity of f (denoted by $\mathsf{sparsity}(f)$). The best known bound for any Boolean function f is due to Tsang et al., [25] who showed that the $\mathsf{CC}_\oplus(f)$ is $O(\sqrt{\mathsf{sparsity}(f)} \log \mathsf{sparsity}(f))$. The conjecture was proved for restricted classes of Boolean functions like monotone functions [19], symmetric functions [27], functions computable by constant depth polynomial size circuits [15] and functions of small spectral norm [25].

Recently, Lin and Zhang [16] studied both the above stated conjectures in connection to alternation, a measure of non-monotonicity of f (denoted by $\mathsf{alt}(f)$, see Sect. 2 for definition), by proving that for any Boolean function f, $\mathsf{bs}(f) = O(\mathsf{s}(f)\mathsf{alt}(f)^2)$ and $\mathsf{CC}_\oplus(f) = O(\log \mathsf{sparsity}(f)\mathsf{alt}(f)^2)$. These results shows that to settle the SENSITIVITY CONJECTURE, it suffices to show that for any Boolean function f, $\mathsf{alt}(f) \leq \mathsf{poly}(\mathsf{s}(f))$ and to settle the XOR LOG-RANK CONJECTURE, it suffices to show that for any Boolean function f, $\mathsf{alt}(f) \leq \mathsf{poly}(\log \mathsf{sparsity}(f))$.

Our Results: As a first step, we ask is it indeed true that for all Boolean functions f, $\mathsf{alt}(f) = O(\mathsf{poly}(\mathsf{s}(f)))$ and for all Boolean functions f, $\mathsf{alt}(f) \leq \mathsf{poly}(\log \mathsf{sparsity}(f))$. We answer both of these questions in the negative by exhibiting a family of Boolean functions $\mathcal{F} = \{f_k \mid k \in \mathbb{N}\}$ (Definition 1) for which $\mathsf{alt}(f_k)$ is at least *exponential* in $\mathsf{s}(f_k)$ and $\mathsf{alt}(f_k)$ is at least *exponential* in $\log\mathsf{sparsity}(f_k)$.

Theorem 1. *There exists a family of Boolean functions* $\mathcal{F} = \{f_k : \{0,1\}^{n_k} \to \{0,1\} \mid k \in \mathbb{N}\}$ *such that* $\mathsf{alt}(f_k) \geq 2^{\mathsf{s}(f_k)} - 1$ *and* $\mathsf{alt}(f_k) \geq 2^{(\log\mathsf{sparsity}(f_k))/2} - 1$.

The main property of $f_k \in \mathcal{F}$ which we exploit to prove Theorem 1 is that $\mathsf{alt}(f_k) = 2^{\mathsf{DT}(f_k)} - 1$ (Theorem 5) where $\mathsf{DT}(f_k)$ is the depth of the optimal decision tree computing f_k (see Sect. 2 for definition). We also show an asymptotically matching upper bound for alternation of any Boolean function. More precisely, for any $f : \{0,1\}^n \to \{0,1\}$, with $\mathsf{DT}(f)$ as the depth of an optimal decision tree computing f, we show that $\mathsf{alt}(f) \leq 2^{\mathsf{DT}(f)+1} - 1$ (Theorem 7).

Though the function family $\mathcal{F}$ rules out settling the SENSITIVITY CONJECTURE (XOR LOG-RANK CONJECTURE resp.) via upper bounding alternation by a polynomial in sensitivity (polynomial in logarithm of sparsity resp.) for all Boolean functions, it is partly unsatisfactory since it can be shown that both the conjectures are true for all $f_k \in \mathcal{F}$.

In fact, any $f : \{0,1\}^n \to \{0,1\}$ for which $\mathsf{alt}(f) = 2^{\Omega(\mathsf{DT}(f))}$, must satisfy $\mathsf{DT}(f) = O(\log n)$. In addition, if f depends on all the input variables, the SENSITIVITY CONJECTURE is true for f. Notice that, for all $f_k \in \mathcal{F}$, $\mathsf{DT}(f_k) = \log n_k$ and f_k depends on all the n_k variables. Hence a natural question is, does there exist another family of functions f where $\mathsf{alt}(f)$ is at least super-polynomial in $\mathsf{s}(f)$, but $\mathsf{DT}(f)$ is not logarithmic in n. To this end, we exhibit a family of Boolean functions $\mathcal{G}$, such that for all $g \in \mathcal{G}$, $\mathsf{alt}(g)$ is *super-linear* in $\mathsf{s}(g)$ and $\mathsf{DT}(g)$ is $\omega(\log n)$ where n is the number of variables in g.

Theorem 2. *There exists a family of Boolean functions* $\{g_k : \{0,1\}^{n_k} \to \{0,1\} \mid k \in \mathbb{N}\}$ *such that* $\mathsf{alt}(g_k) \geq \mathsf{s}(g_k)^{\log_3 5}$ *while* $\mathsf{DT}(g_k)$ *is* $\Omega(n_k^{\log_6 3})$.

The main tool used in proving Theorem 2 is a bound on the alternation of composed Boolean functions (Lemma 1).

As mentioned before, Lin and Zhang [16] showed that XOR LOG-RANK CONJECTURE is true for all Boolean functions satisfying $\mathsf{alt}(f) \leq \mathsf{poly}(\log\mathsf{sparsity}(f))$. As our main result, we further strengthen this when $\mathsf{sparsity}(f) < n$.

Theorem 3 (Main). *For large enough* n, *the* XOR LOG-RANK CONJECTURE *is true for all* $f : \{0,1\}^n \to \{0,1\}$, *such that* $\mathsf{alt}(f) \leq \mathsf{poly}(\log n)$ *where* f *depends on all its variables.*

Our starting point in proving the above result is a relation connecting $\deg$, $\deg_{\mathbb{F}_2}$ and alt. For all Boolean functions f,

$$\deg(f) \leq \mathsf{alt}(f) \cdot \deg_{\mathbb{F}_2}(f)^2 \tag{1}$$

We remark that for special cases, Eq. 1 is known to be true. For instance, if f is a monotone, it can be shown[1] that $\deg(f) \leq \deg_{\mathbb{F}_2}(f)^2$. However, there are

---

[1] When f is monotone, it is known that $\mathsf{DT}(f) \leq \mathsf{s}(f)^2$ and $\mathsf{s}(f) \leq \deg_{\mathbb{F}_2}(f)$ (Corollary 5 and Proposition 4 of [5]). Proposition 4 of [5] though states that $\mathsf{s}(f) \leq \deg(f)$ for any monotone f, the argument is valid for $\deg_{\mathbb{F}_2}(f)$ also. Since, $\deg(f) \leq \mathsf{DT}(f)$ (cf. [5]), $\deg(f) \leq \deg_{\mathbb{F}_2}(f)^2$.

functions of large alternation where $\deg_{\mathbb{F}_2}(f)$ is constant while $\deg(f)$ is n (for instance, parity on n bits). Hence we cannot upper bound degree by $\mathbb{F}_2$-degree in general but Eq. 1 says that we can indeed upper bound $\deg(f)$ by $\deg_{\mathbb{F}_2}(f)$ using $\mathsf{alt}(f)$. The result is implicit in [16].

We now give two further applications of Eq. 1. As our first application, we show that Boolean functions with bounded alternation have high sparsity. Kulkarni and Santha [15] had studied the relation between $\log \mathsf{sparsity}(f)$ and $\deg(f)$ in the case of a restricted families of monotone functions and asked if they are linearly related in the case of monotone functions (see Sect. 5.1). In this direction, we show the following lower bound for $\log_2 \mathsf{sparsity}(f)$ in terms of $\deg(f)$.

Theorem 4. *For Boolean functions f with $\mathsf{alt}(f) = O(1)$, $\log \mathsf{sparsity}(f) = \Omega(\sqrt{\deg(f)})$.*

As a second application, we observe that Eq. 1 implies an improved upper bound for influence (denoted by $\mathbf{I}[f]$, see Sect. 2 for a definition) to $\deg_{\mathbb{F}_2}(f)^2 \cdot \mathsf{alt}(f)$. This improves the result of Guo and Komargodski [11] who showed that $\mathbf{I}[f] = O(\mathsf{alt}(f)\sqrt{n})$, thus giving faster learning algorithms for functions of bounded alternation in the PAC learning model (see Sect. 5.2).

2   Preliminaries

We introduce the notations and definitions used in this paper. All logarithms are to the base 2 unless otherwise stated. Let $[n] \stackrel{\text{def}}{=} \{1, 2, \ldots, n\}$. For $i \in [n]$, define e_i to be an n bit Boolean string with one in i^{th} location and zero elsewhere.

A Boolean function $f : \{0,1\}^n \to \{0,1\}$ is *monotone* if $\forall x, y \in \{0,1\}^n$, $x \prec y \implies f(x) \le f(y)$ where, $x \prec y$ iff $\forall i \in [n], x_i \le y_i$. The *alternation* of a Boolean function is a measure of non-monotonicity of the Boolean function. More precisely, if we define a collection of distinct inputs $x_0, x_1, x_2 \ldots, x_n \in \{0,1\}^n$ satisfying $0^n = x_0 \prec x_1 \prec x_2 \prec \cdots \prec x_n = 1^n$ as a *chain* in the Boolean hypercube $\mathcal{B}_n$ then, alternation of f (denoted by $\mathsf{alt}(f)$) is defined as $\max \{\mathsf{alt}(f, \mathcal{C}) \mid \mathcal{C} \text{ is a chain in } \mathcal{B}_n\}$ where $\mathsf{alt}(f, \mathcal{C})$ is $|\{i \mid f(x_{i-1}) \ne f(x_i), x_i \in \mathcal{C}, i \in [n]\}|$. Indeed for a monotone f, $\mathsf{alt}(f) = 1$.

Any chain $\mathcal{C}$ of a Boolean hypercube over $\{0,1\}^n$ is uniquely determined by a permutation $\sigma \in S_n$ and vice versa. An $x \in \{0,1\}^n$ belongs to a chain $\mathcal{C}$ defined by $\sigma \in S_n$ iff $x = 0^n$ or $x = \bigvee_{i=1}^{wt(x)} e_{\sigma(i)}$ where the OR is taken coordinate wise and $wt(x)$ is the number of ones in x. If a chain is defined using a permutation σ, we use σ to denote the chain $\mathcal{C}$.

For a Boolean function f on m variables and g on n variables, we denote $f \circ g$ as a function on mn variables $\{x_{11}, \ldots, x_{mn}\}$ defined as $f(g(x_{11}, \ldots, x_{1n}), g(x_{21}, \ldots, x_{2n}), \ldots, g(x_{m1}, \ldots, x_{mn}))$. We define $g^{\circ k}$ as the Boolean function on n^k variables as $g^{\circ(k-1)} \circ g$ for $k > 1$ and g for $k = 1$.

Given a Boolean function, there always exists a unique n variable multilinear polynomial over $\mathbb{R}[x_1, \ldots, x_n]$ such that the evaluation agrees with the function on $\{0,1\}^n$. The *degree* of function f is the degree of such a polynomial (denoted by $\deg(f)$). If we consider the polynomial over $\mathbb{F}_2[x_1, \ldots, x_n]$ instead, we get the $\mathbb{F}_2$-*degree* of f (denoted by $\deg_{\mathbb{F}_2}(f)$).

A deterministic Boolean *decision tree* is a rooted tree where the leaves are labeled 0 or 1 and non-leaf nodes labeled by a variable having two outgoing edges (corresponding to the value taken by the variable). A decision tree is said to compute a Boolean function f, if for all inputs x, the path from root to the leaf determined by x is labeled $f(x)$. Define $\mathsf{DT}(f)$ as the depth of the smallest depth decision tree computing f.

For an $x, y \in \{0,1\}^n$, we denote by $x \oplus y$, the input obtained by taking bitwise parity of x and y. For $B \subseteq [n]$, e_B denotes the characteristic vector of B. For $f : \{0,1\}^n \to \{0,1\}$ and $x \in \{0,1\}^n$, define the sensitivity of f on x (denoted by $\mathsf{s}(f,x)$) as $|\{i \mid f(x \oplus e_i) \neq f(x), i \in [n]\}|$. We define the block sensitivity of f on x (denote by $\mathsf{bs}(f,x)$) as the size of maximal collection of disjoint non-empty sets $\{B_i\}$ where each $B_i \subseteq [n]$ in the collection satisfy $f(x \oplus e_{B_i}) \neq f(x)$. The *sensitivity* of f (denoted by $\mathsf{s}(f)$) is defined as $\max_{x \in \{0,1\}^n} \mathsf{s}(f,x)$. The *influence* of a Boolean function f (denoted by $\mathbf{I}[f]$) is defined as $\mathbf{E}_{x \in \{0,1\}^n}[\mathsf{s}(f,x)]$. The *block sensitivity* of f (denoted by $\mathsf{bs}(f)$) is $\max_{x \in \{0,1\}^n} \mathsf{bs}(f,x)$. Note that $\mathbf{I}[f] \leq \mathsf{s}(f) \leq \mathsf{bs}(f)$. It is also known that $\mathbf{I}[f] \leq \deg(f) \leq \mathsf{DT}(f)$ (cf. [5]).

For $x \in \{0,1\}^n$ and $S \subseteq [n]$, define $\chi_S(x) = (-1)^{\sum_{i \in S} x_i}$. Any $f : \{0,1\}^n \to \{-1,1\}$ can be uniquely expressed as $\sum_{S \subseteq [n]} \widehat{f}(S) \chi_S(x)$ where $\widehat{f}(S) \in \mathbb{R}$, indexed by $S \subseteq [n]$, denotes the *Fourier coefficients* of f which is $\frac{1}{2^n} \sum_x f(x) \chi_S(x)$ (see [22] for more details). The *sparsity* of a Boolean function f (denoted by $\mathsf{sparsity}(f)$) is the number of non-zero Fourier coefficients of f. For Boolean functions f whose range is $\{0,1\}$, we define sparsity of f to be the sparsity of the function $1 - 2f$ in this paper.

3   Alternation Vs Sensitivity and Alternation Vs Logarithm of Sparsity

In this section, we show that there exists a family of function $\mathcal{F} = \{f_k \mid k \in \mathbb{N}\}$ with $\mathsf{alt}(f_k)$ is at least exponential in $\mathsf{s}(f_k)$, $\mathsf{DT}(f_k)$ and $\log \mathsf{sparsity}(f_k)$ respectively (Sect. 3.1). Complementing this, we show that for any Boolean function f, $\mathsf{alt}(f)$ can be at most exponential in $\mathsf{DT}(f)$ (Sect. 3.2). We prove a bound on the alternation of composed Boolean functions and use it to obtain a family of functions with super-linear gap between alternation and sensitivity with large decision tree depth unlike functions in $\mathcal{F}$ (Sect. 3.3).

3.1   Exponential Gaps : Alternation Vs Decision Tree Depth

We prove Theorem 1 in this section. We first show that there exists a family of function $\mathcal{F} = \{f_k \mid k \in \mathbb{N}\}$ with $\mathsf{alt}(f_k)$ equals $2^{\mathsf{DT}(f_k)} - 1$ (Theorem 5). Since for any Boolean function f, $\mathsf{s}(f) \leq \mathsf{DT}(f)$ (cf. [5]), we have, $\mathsf{alt}(f_k) = 2^{\mathsf{DT}(f_k)} - 1 \geq 2^{\mathsf{s}(f_k)} - 1$ and since for any Boolean function f, $\log \mathsf{sparsity}(f) \leq 2\deg(f) \leq$

$2\mathsf{DT}(f)$ (cf. [22]), we get that for f_k, $\mathsf{alt}(f_k) = 2^{\mathsf{DT}(f_k)} - 1 \geq 2^{0.5\log\mathsf{sparsity}(f_k)} - 1$ thereby proving Theorem 1.

Hence, one cannot hope to show that for all Boolean functions f, alternation is upper bounded polynomially by sensitivity of f or polynomially by logarithm of sparsity of f. We now define our family $\mathcal{F}$ of Boolean functions.

Definition 1. *Let* $\mathcal{F} = \{f_k \mid k \in \mathbb{N}\}$ *be a family of Boolean functions where for every* $k \in \mathbb{N}$, $f_k : \{0,1\}^{2^k - 1} \to \{0,1\}$ *is defined by the decision tree which is a full binary tree of depth* k *with each of the* $2^k - 1$ *internal node querying a distinct variable and each of the nodes at level* k *have left leaf child labeled* 0 *and right leaf child labeled* 1.

A Boolean function $f_3 \in \mathcal{F}$ is described using a decision tree as in Fig. 1a.

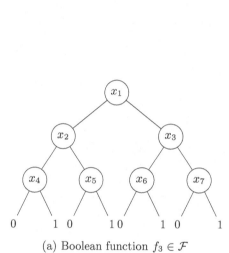

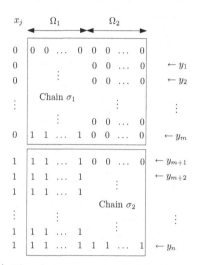

(a) Boolean function $f_3 \in \mathcal{F}$

(b) The chain σ constructed in the proof of Theorem 5. Note that Ω_1 and Ω_2 need not be contiguous.

Fig. 1. (a) Boolean function $f_3 \in \mathcal{F}$. (b) The chain σ constructed in the proof of Theorem 5. Note that Ω_1 and Ω_2 need not be contiguous.

Theorem 5. *For every* $k \geq 1$, $f_k \in \mathcal{F}$, $\mathsf{alt}(f_k) = 2^{DT(f_k)} - 1$.

Proof. By definition, f_k is computed by a decision tree of depth k. We show that for $k \geq 1$, $\mathsf{alt}(f_k) \geq 2^k - 1$. Since f_k is defined on $2^k - 1$ variables, we get that $\mathsf{alt}(f_k) = 2^k - 1$ thereby completing the proof.

Since we need to work with functions whose variable set is not necessarily numbered from 1 to n, we associate bijections (instead of permutations) with chains. For any set Ω, let B_Ω be defined as $\{\sigma : [|\Omega|] \to \Omega \mid \sigma \text{ is a bijection}\}$. For a Boolean function f defined on the variables $\{x_{i_1}, x_{i_2}, \ldots, x_{i_n}\}$, $var(f)$ be defined as $\{i_1, i_2, \ldots, i_n\}$.

We now show that $\mathsf{alt}(f_k) \geq 2^k - 1$ by induction on k. For $k = 1$, since f depends only on 1 variable the result holds.

Suppose that the result holds for $f_k \in \mathcal{F}$. For $f_{k+1} \in \mathcal{F}$ on $n = 2^{k+1} - 1$ variables $\{x_1, x_2, \ldots, x_n\}$, let T be the decision tree (as in Definition 1) computing f_{k+1} of depth $k + 1$ with x_j being the root variable for some $j \in [n]$. Let T_1 and T_2 be the left and right subtree of the root node x_j. Consider the Boolean function f_1 (resp. f_2) computed by the decision tree T_1 (resp. T_2). Note that since T_1 and T_2 are obtained from T in this way, by Definition 1, f_1 and f_2 belongs to $\mathcal{F}$ and computes the same function upto variable renaming. Also note that both f_1 and f_2 are on $m = 2^k - 1$ variables. Since both T_1 and T_2 are of depth k, by inductive hypothesis, $\mathsf{alt}(f_1) \geq m$ and $\mathsf{alt}(f_2) \geq m$. Using this, we now construct a chain for f_{k+1} of alternation $2^{k+1} - 1$.

Let $\Omega_1 = var(f_1)$, $\Omega_2 = var(f_2)$. Let $\sigma_1 \in B_{\Omega_1}$ and $\sigma_2 \in B_{\Omega_2}$ be such that $\mathsf{alt}(f_1, \sigma_1) = 2^k - 1$ and $\mathsf{alt}(f_2, \sigma_2) = 2^k - 1$. We now define a $\sigma \in B_{var(f)} = B_{\Omega_1 \cup \Omega_2 \cup \{j\}}$ as $\sigma(i) = \sigma_1(i)$ if $i \in \{1, 2, \ldots, m\}$, $\sigma(m+1) = j$ and $\sigma(m+1+i) = \sigma_2(i)$ if $i \in \{1, 2, \ldots, m\}$. By definition, σ is indeed a bijection. The σ obtained is pictorially represented in Fig. 1b for clarity of exposition.

We now claim that $\mathsf{alt}(f, \sigma) = 2^{k+1} - 1$. To show this, consider the chain corresponding to σ given by $0^n \prec y_1 \prec \cdots y_m \prec y_{m+1} \prec \cdots \prec y_n = 1^n$ all belonging to $\{0, 1\}^n$. By definition of σ, for $i \in [m]$ since j^{th} bit of y_i is 0, $f(y_i) = f_1(y_i|_{\Omega_1})$ and for $i \in \{0\} \cup [m]$, since j^{th} bit of y_{m+1+i} is 1, $f(y_{m+1+i}) = f_2(y_{m+1+i}|_{\Omega_2})$. Again by definition of σ, for $i \in [m]$, 0^m along with the elements $y_i|_{\Omega_1}$ for $i \in [m]$ (in that order) is a chain witnessing f_1 alternating m times and $y_{m+1+i}|_{\Omega_2}$ for $i \in \{0\} \cup [m]$ (in that order) is a chain witnessing f_2 alternating m times. Now observe that $f(y_m) \neq f(y_{m+1})$. This is because $f(y_m)$ is f_1 evaluated on $x_i = 1$ for all $i \in \Omega_1$ is the rightmost child of T_1 which is 1 and $f(y_{m+1})$ is f_2 evaluated on $x_i = 0$ for all $i \in \Omega_2$ is the leftmost child of T_2 which is 0. Hence $\mathsf{alt}(f, \sigma) = \mathsf{alt}(f_1, \sigma_1) + \mathsf{alt}(f_2, \sigma_2) + 1 = 2m + 1 = 2^{k+1} - 1$ completing the induction.   □

In the next section, we show that for any Boolean function f, we can indeed upper bound alternation of f by an exponential in decision tree depth of f.

3.2   Alternation is at most Exponential in Decision Tree Depth

In this section, we show that for all Boolean functions f, $\mathsf{alt}(f) \leq 2^{\mathsf{DT}(f)+1} - 1$. Markov [18] studied a parameter closely related to $\mathsf{alt}(f)$ defined as *decrease* (denoted by $\mathsf{dc}(f)$) where the definition is same as alternation except that the flips in the chain from 1 to 0 (corresponding to a decrease in the function value) alone are counted. Hence for any Boolean function f, $\mathsf{alt}(f) \in \{2\mathsf{dc}(f) - 1, 2\mathsf{dc}(f), 2\mathsf{dc}(f) + 1\}$. Markov [18] showed the following tight connection between negations needed to compute a Boolean function and its decrease.

Theorem 6 (Markov [18]). *Let* $\mathsf{negs}(f)$ *be the minimum number of negations needed in any circuit computing* f. *Then, for an* $f : \{0, 1\}^n \to \{0, 1\}$, $\mathsf{negs}(f) = \lceil \log(1 + \mathsf{dc}(f)) \rceil$.

We also use the notion of a *connector* for two Boolean functions which is a crucial idea used by Markov in proving the above theorem. A connector of two Boolean functions f_0 and f_1 is a function $g(b_0, b_1, x)$ on $n + 2$ bits such that $g(0, 1, x) = f_0(x)$ and $g(1, 0, x) = f_1(x)$. Markov showed the following remarkable bound that $\mathsf{negs}(g) \leq \max\{\mathsf{negs}(f_0), \mathsf{negs}(f_1)\}$ (for a proof, see Jukna [13]). We now prove our result which follows from an inductive application of Theorem 6.

Theorem 7. *For any $f : \{0,1\}^n \to \{0,1\}$, $\mathsf{alt}(f) \leq 2^{DT(f)+1} - 1$.*

Proof. Since $\mathsf{alt}(f) \leq 2\mathsf{dc}(f) + 1$, it suffices to show that $\mathsf{dc}(f) \leq 2^{DT(f)} - 1$. Proof is by induction on $DT(f)$. For $DT(f) = 1$, the function depends on at most 1 variable, giving $\mathsf{dc}(f) \leq 1 = 2^{DT(f)} - 1$. For any f with $DT(f) \geq 2$ computed by a decision tree T of depth k, let x_i be the variable queried at the root of T for some $i \in [n]$. Define f_0 as f restricted to $x_i = 0$ and f_1 as f restricted to $x_i = 1$. Removing the node x_i from T gives two decision trees which computes f_0 and f_1 (respectively) giving $DT(f_0) \leq DT(f) - 1$ and $DT(f_1) \leq DT(f) - 1$. Hence by induction, $\mathsf{dc}(f_0) \leq 2^{DT(f_0)} - 1 \leq 2^{DT(f)-1} - 1$ and similarly, $\mathsf{dc}(f_1) \leq 2^{DT(f)-1} - 1$.

Applying Theorem 6, we get, $\mathsf{negs}(f_0) = \lceil \log(\mathsf{dc}(f_0) + 1) \rceil$ (and similarly for f_1). Let g be the connector for f_0 and f_1. Since $f(x) = x_i \wedge f_1(x) \vee \neg x_i \wedge f_0(x)$, $f(x) = g(\neg x_i, x_i, x)$. Applying Markov's result on the number of negations needed in computing g, we get $\mathsf{negs}(f) \leq \mathsf{negs}(g) + 1 \leq 1 + \max\{\mathsf{negs}(f_0), \mathsf{negs}(f_1)\}$. Since, $\max\{\mathsf{negs}(f_0), \mathsf{negs}(f_1)\} \leq \lceil \log(2^{DT(f)-1} - 1 + 1) \rceil$ which is $DT(f) - 1$, $\mathsf{negs}(f) \leq DT(f)$. Applying Theorem 6, on f completes the induction.   □

Note that for the family of functions $\mathcal{F}$, Theorem 5 shows that the above result is asymptotically tight.

3.3 Super Linear Gaps Between Alternation and Sensitivity

In this section, we ask the question whether there exists a family of functions f where $\mathsf{alt}(f)$ grows faster than $\mathsf{s}(f)$ but $DT(f)$ is not very small. We exhibit a family of Boolean functions $\{g_k : \{0,1\}^{n_k} \to \{0,1\} \mid k \in \mathbb{N}\}$ with $\mathsf{alt}(g_k) = \omega(\mathsf{s}(g_k))$ and $DT(g_k)$ is $\Omega(n_k^{\log_6 3})$.

Before proceeding, we show a lower bound on the alternation of composition of two Boolean functions in terms of its alternation.

Lemma 1. *For any $g : \{0,1\}^n \to \{0,1\}$, with $g(0^n) \neq g(1^n)$ and any $f : \{0,1\}^m \to \{0,1\}$, $\mathsf{alt}(f \circ g) \geq \mathsf{alt}(f) \cdot \mathsf{alt}(g)$.*

Proof. Without loss of generality, assume $g(0^n) = 0$ and $g(1^n) = 1$ (otherwise work with $\neg g$ as $\mathsf{alt}(g) = \mathsf{alt}(\neg g)$). Let $A = (0^n = z_0 \prec z_1 \prec \cdots \prec z_n = 1^n)$ be a chain on $\{0,1\}^n$ such that the alternation of g is maximum. Consider any maximum alternation chain of f and let σ be the permutation associated with the chain. We exhibit a chain $B = (y_0, y_1, \ldots, y_{nm})$ on $\{0,1\}^{nm}$ with $\mathsf{alt}(f) \cdot \mathsf{alt}(g)$ many alternations.

We divide the inputs in the chain B into m blocks of size n each. We say that for a $k \in [nm]$, the input $y_k \in \{0,1\}^{nm}$ belongs to the *block* b if $b = \lceil \frac{k}{n} \rceil$.

We define the *position* of y_k in its block, $pos(k)$, as n if $n \mid k$ and $(k \mod n)$ otherwise. Let $y_k = (x_1^k, x_2^k, \ldots, x_m^k)$ where $x_i^k \in \{0,1\}^n$. For $k = 0$, define $x_i^k = 0^n$ for all $i \in [m]$ and for $k = nm$, define $x_i^k = 1^n$ for all $i \in [m]$. For the remaining values of k, x_i^k for $i \in [m]$ is defined as

$$
x_i^k = \begin{cases}
z_n = 1^n & \text{if } i \in \{\sigma(1), \sigma(2), \ldots, \sigma(b-1)\} \text{ and } b \geq 2 \\
z_{pos(k)} & \text{if } i = \sigma(b) \\
z_0 = 0^n & \text{otherwise}
\end{cases}
$$

We can see that $y_0 = 0^{nm}$ and $y_{nm} = 1^{nm}$ and for $k \in [nm]$, from the above definition, $y_{k-1} \prec y_k$ as $\forall i \in [m]$ $x_i^{k-1} \prec x_i^k$. We now argue that $f \circ g$ alternates at least $\mathsf{alt}(f) \cdot \mathsf{alt}(g)$ times in the chain B. Consider the input $(g(x_1^k), g(x_2^k), \ldots, g(x_m^k))$ to the function f and let y_k belong to the block b.

Consider the case when $b = 1$. In this case, all of x_i^k except $i = \sigma(b)$ is 0^n. As long as y_k stays within the block b, the input the function f changes its value only at $x_{\sigma(b)}^k = x_{\sigma(1)}^k$. Since $g(x_{\sigma(b)}^k)$ changes its value $\mathsf{alt}(g)$ times, $f \circ g$ will also alternate $\mathsf{alt}(g)$ times if value of f changes on flipping location $\sigma(b)$ in its input.

For k such that $b > 1$, by definition of y_k, $x_{\sigma(1)}^k, \ldots, x_{\sigma(b-1)}^k$ is 1^n and $x_{\sigma(b+1)}^k \cdots, x_{\sigma(m)}^k$ is 0^n. Since $g(0^n) = 0$ and $g(1^n) = 1$, the input to f will be either $r_1 = \vee_{i=1}^b e_{\sigma(i)}$ or $r_0 = \vee_{i=1}^{b-1} e_{\sigma(i)}$. Since g alternates $\mathsf{alt}(g)$ times thereby changing the input to f between r_0 and r_1, $f \circ g$ will also alternate $\mathsf{alt}(g)$ times if value of f changes on flipping location $\sigma(b)$.

Thus in both cases, if f alternates once, $f \circ g$ alternates $\mathsf{alt}(g)$ in the chain B. Since f alternates $\mathsf{alt}(f)$ times on σ, $f \circ g$ alternates $\mathsf{alt}(f) \cdot \mathsf{alt}(g)$ times in the chain B. □

For $f = \vee_n$ and g being a parity on m bits for any odd integer m, $\mathsf{alt}(f \circ g) \geq mn$ by Lemma 1 while $\mathsf{alt}(f) \cdot \mathsf{alt}(g) = m$. Thus, in general, it is not true that $\mathsf{alt}(f \circ g) \leq \mathsf{alt}(f) \cdot \mathsf{alt}(g)$ and hence Lemma 1 is not tight. Using Lemma 1, we can prove the following Corollary.

Corollary 1. *For any* $h : \{0,1\}^n \to \{0,1\}$, *with* $h(0^n) \neq h(1^n)$, *for any* $k \geq 2$, $\mathsf{alt}(h^{\circ k}) \geq \mathsf{alt}(h)^k$.

Super-linear Gap between Alternation and Sensitivity: We use Corollary 1 to exhibit a family of Boolean functions for which alternation is super-linear in sensitivity.

Theorem 2. *There exists a family of Boolean functions* $\{g_k : \{0,1\}^{n_k} \to \{0,1\} \mid k \in \mathbb{N}\}$ *such that* $\mathsf{alt}(g_k) \geq \mathsf{s}(g_k)^{\log_3 5}$ *while* $\mathsf{DT}(g_k)$ *is* $\Omega(n_k^{\log_6 3})$.

Proof. Consider the address function $\mathsf{ADDR}_t : \{0,1\}^{t+2^t} \to \{0,1\}$ defined as $\mathsf{ADDR}_t(x_1, x_2, \ldots, x_t, y_0 y_1, y_2, \ldots, y_{2^t-1}) = y_{int(x_1 x_2 \ldots x_t)}$ where $int(x)$ is the integer corresponding to the binary string x. Consider the chain $(000000, 001000, 101000, 101010, 111010, 111011, 111111)$. Since, ADDR_2 changes value 5 times along this chain, $\mathsf{alt}(\mathsf{ADDR}_2) \geq 5$ while $\mathsf{s}(\mathsf{ADDR}_2) = 3$. We consider the family of functions $\{g_k \mid k \in N\}$ obtained by composing ADDR_2 k times. Since sensitivity of composed function is at most the product of their sensitivity [24],

$s(g_k) \leq s(\mathsf{ADDR}_2)^k = 3^k$. Since $g_1 = \mathsf{ADDR}_2$ is 0 on all zero input and 1 on all ones input, applying Corollary 1, $\mathsf{alt}(g_k) \geq 5^k \geq s(g_k)^{\log_3 5}$. Note that $\mathsf{DT}(g_k) = \mathsf{DT}(\mathsf{ADDR}_2)^k$ (as decision tree depth multiplies under composition [24]). Hence $\mathsf{DT}(g_k) = 3^k$ which is $n_k^{\log_6 3}$ where n_k is the number of variables of g_k and hence does not grows logarithmic in n_k. □

4 XOR Log-Rank Conjecture for Bounded Alternation Boolean Functions

In this section, we prove the XOR LOG-RANK CONJECTURE for f when $\mathsf{alt}(f)$ is at most $\mathsf{poly}(\log n)$. Before proceeding, we give a short proof for the fact that for all Boolean functions f, $\deg(f) \leq \mathsf{alt}(f)\deg_{\mathbb{F}_2}(f)^2$ (Eq. 1 from Introduction).

Proof (of Eq. 1). The statement directly follows from Theorem 14 of [16] where it is shown that for a Boolean function f, $\mathsf{DT}(f) \leq \mathsf{alt}(f)\deg_{\mathbb{F}_2}(f)^2$. Observing that $\deg(f) \leq \mathsf{DT}(f)$ (cf. [5]) completes the argument. □

As a first step towards showing $\mathsf{alt}(f) \leq \mathsf{poly}(\log n)$ implies that the XOR LOG-RANK CONJECTURE holds for f, we prove the following bound on the weighted average of the Fourier coefficients, weighted by the number of elements.

Proposition 1. *For an $f : \{0,1\}^n \to \{-1,1\}$ that depends on all its inputs, $\sum_S |\widehat{f}(S)||S| \geq n$.*

Proof. It suffices to show that for every $i \in [n]$, $\sum_{S:i\in S} |\widehat{f}(S)| \geq 1$. Fix an $i \in [n]$. Since $f(x) = \sum_S \widehat{f}(S)\chi_S(x) = \sum_{S \subseteq [n]\setminus\{i\}} (\widehat{f}(S) + \widehat{f}(S\cup\{i\})(-1)^{x_i}) \prod_{j\in S}(-1)^{x_j}$, for any $S \subseteq [n] \setminus \{i\}$ and $b \in \{0,1\}$, $\widehat{f|_{x_i=b}}(S) = \widehat{f}(S) + (-1)^b \cdot \widehat{f}(S \cup \{i\})$. Hence we conclude that for any x, $f_{x_i=0}(x) - f_{x_i=1}(x) = \sum_{S\subseteq[n]\setminus\{i\}} 2\widehat{f}(S \cup \{i\}) \prod_{j\in S}(-1)^{x_j}$. Now taking absolute values on both sides and applying triangle inequality,

$$\left| \frac{f_{x_i=0} - f_{x_i=1}}{2} \right| = \left| \sum_{S\subseteq[n]\setminus\{i\}} \widehat{f}(S \cup \{i\}) \prod_{j\in S}(-1)^{x_j} \right| \leq \sum_{S\subseteq[n]\setminus\{i\}} \left| \widehat{f}(S\cup\{i\}) \right|$$

Since f is sensitive at i on some input $a \in \{0,1\}^n$, for the input a' obtained by removing i^{th} bit from a, $|f_{x_i=0}(a') - f_{x_i=1}(a')| = 2$ implying $\sum_{S:i\in S} |\widehat{f}(S)| \geq 1$ by the above equation which completes the proof. □

We show that if $\mathsf{alt}(f) \leq \mathsf{poly}(\log n)$, then $\deg(f) \leq \mathsf{poly}(\log \mathsf{sparsity}(f))$. This implies that the XOR LOG-RANK CONJECTURE holds[2] for f. As a first step, using Proposition 1, we show that if $\deg(f) \leq \mathsf{poly}(\log n)$, then $\deg(f) \leq \mathsf{poly}(\log \mathsf{sparsity}(f))$ (Lemma 2). We then argue using Eq. 1 that $\mathsf{alt}(f) \leq \mathsf{poly}(\log n)$ implies that $\deg(f) \leq \mathsf{poly}(\log n)$ proving Theorem 3.

---

[2] Using the facts that $\mathsf{CC}_\oplus(f) \leq 2\mathsf{DT}(f)$ [19] and $\mathsf{DT}(f) \leq \deg(f)^4$ [5], $\mathsf{CC}_\oplus(f) = O(\deg(f)^4)$ implying $\mathsf{CC}_\oplus(f) = O(\mathsf{poly}(\log \mathsf{sparsity}(f)))$.

Lemma 2. *For an $f : \{0,1\}^n \to \{0,1\}$ which depends on all its inputs and for large enough n, if $\deg(f) \le (\log n)^c$ for some $c > 0$, then $\deg(f) \le (\log \operatorname{sparsity}(f))^c$.*

Proof. Since the f depends on all the inputs, applying Proposition 1 to $g(x) = 1 - 2f(x)$, $n \le \sum_S |\hat{g}(S)||S| \le \deg(f) \sum_S |\hat{g}(S)| \le \deg(f) \sqrt{\operatorname{sparsity}(f)}$. In concluding this, we used the fact that maximum sized index $|S|$ for which $\hat{g}(S) \ne 0$ is $\deg(f)$ and $\sum_S |\hat{g}(S)| \le \sqrt{\operatorname{sparsity}(f)}$ [22]. Thus, $\sqrt{\operatorname{sparsity}(f)} \cdot \deg(f) \ge n$. Since $\deg(f) \le (\log n)^c$, we have $\sqrt{\operatorname{sparsity}(f)} \ge \frac{n}{(\log n)^c} \ge \sqrt{n}$ for large enough n. Hence $\deg(f) \le (\log n)^c \le (\log \operatorname{sparsity}(f))^c$. $\square$

Theorem 3. *For large enough n, the XOR LOG-RANK CONJECTURE is true for all $f : \{0,1\}^n \to \{0,1\}$, such that $\operatorname{alt}(f) \le \operatorname{poly}(\log n)$ where f depends on all its variables.*

Proof. If $\deg_{\mathbb{F}_2}(f) = 1$, then f is a parity function and the XOR LOG-RANK CONJECTURE holds for f. Hence we can assume that $\deg_{\mathbb{F}_2}(f) > 1$. If $\operatorname{alt}(f) \le \deg_{\mathbb{F}_2}(f)$, then by Eq. 1, we have $\deg(f) \le \deg_{\mathbb{F}_2}(f)^3 \le \log \operatorname{sparsity}(f)^3$ (since $\deg_{\mathbb{F}_2}(f) > 1$, $\deg_{\mathbb{F}_2}(f) \le \log \operatorname{sparsity}(f)$ [3]). Hence the XOR LOG-RANK CONJECTURE holds for f. If $\operatorname{alt}(f) > \deg_{\mathbb{F}_2}(f)$, then by Eq. 1, $\deg(f) < \operatorname{alt}(f)^3$. Since $\operatorname{alt}(f) \le \operatorname{poly}(\log n)$, we have $\deg(f) \le \operatorname{poly}(\log n)$. Applying Lemma 2, we get that $\deg(f) \le \operatorname{poly}(\log \operatorname{sparsity}(f))$. $\square$

Remark 1. It should be noted that for f satisfying conditions of Theorem 3, the SENSITIVITY CONJECTURE is true. This is because for f that depends on all its inputs, $\mathsf{s}(f) = \Omega(\log n)$ [23] implying that $\operatorname{alt}(f) \le \operatorname{poly}(\log n) \le \operatorname{poly}(\mathsf{s}(f))$. Hence the SENSITIVITY CONJECTURE is true for f by the result of Lin and Zhang [16].

5 Two Further Applications of the deg Vs $\deg_{\mathbb{F}_2}$ Relation

We showed that for all Boolean functions f, $\deg(f) \le \operatorname{alt}(f) \deg_{\mathbb{F}_2}(f)^2$ (Eq. 1) in Sect. 4. We now give two applications of this result. Firstly, we partially answer a question raised by Kulkarni and Santha [15] on the sparsity of monotone Boolean functions by show a variant of their statement. Secondly, we observe that Eq. 1 improves a bound on $\mathbf{I}[f]$ due to Guo and Komargodski [11].

5.1 Dense Fourier Spectrum for Bounded Alternation Functions

Kulkarni and Santha [15] studied certain special Boolean functions which are indicator functions $f_{\mathcal{M}}$ of a bridgeless matroids $\mathcal{M}$ on ground set $[n]$. While it is known that for any f, $\log \operatorname{sparsity}(f) \le 2 \cdot \deg(f)$ [22], Kulkarni and Santha showed that this upper bound is asymptotically tight for $f = f_{\mathcal{M}}$. They observed that $f_{\mathcal{M}}$ is a monotone function (by virtue of the underlying support set being a matroid) and asked if a similar statement holds for the general class of monotone Boolean functions. More precisely, they asked whether $\log \operatorname{sparsity}(f) = \Omega(\deg(f))$ for every monotone Boolean function f.

We show that for functions of constant alternation (which includes monotone functions), $\log \text{sparsity}(f)$ is relatively large. This can be seen as a variant of the question posed by Kulkarni and Santha.

Theorem 4. *For Boolean functions f with* $\text{alt}(f) = O(1)$, $\log \text{sparsity}(f) = \Omega(\sqrt{\deg(f)})$.

Proof. Observe that for $\deg_{\mathbb{F}_2}(f) = 1$, f is parity of constant number of variables or its negation as $\text{alt}(f) = O(1)$. Hence the result holds for this case. For $\deg_{\mathbb{F}_2}(f) > 1$, by Eq. 1 and the result that $\deg_{\mathbb{F}_2}(f) \leq \log \text{sparsity}(f)$ when $\deg_{\mathbb{F}_2}(f) > 1$ [3], we get that $\deg(f) = O(\deg_{\mathbb{F}_2}(f)^2) = O(\log \text{sparsity}(f)^2)$.   □

Note that Theorem 4 does show that logarithm of sparsity of monotone functions is nearly close to the upper bound possible but does not completely settles the question of Kulkarni and Santha.

5.2   Improved Upper Bound for I[f]

For an n bit Boolean function, the best known upper bound of $\mathbf{I}[f]$ in terms of $\text{alt}(f)$ is $\mathbf{I}[f] \leq O(\text{alt}(f)\sqrt{n})$ due to Guo and Komargodski [11] using a probabilistic argument. Since $\mathbf{I}[f] \leq \deg(f)$ [22], Eq. 1 gives an improvement over the known bound on $\mathbf{I}[f]$ when $\deg_{\mathbb{F}_2}(f) < \sqrt[4]{n}$.

Proposition 2. *For any $f : \{0,1\}^n \rightarrow \{0,1\}$, $\mathbf{I}[f] \leq \text{alt}(f) \cdot \deg_{\mathbb{F}_2}(f)^2$.*

This immediately gives improved learning algorithms for functions of bounded alternation in the PAC learning model.

Blais et al. [4] gave a uniform learning algorithm for the class of functions C_t computable by circuits with at most t negations that can learn an $f \in C_t$ from random examples with error $\epsilon > 0$ in time $n^{O(2^t \sqrt{n})/\epsilon}$ where $t \leq O(\log n)$. In terms of alternation, the runtime is $n^{O(\text{alt}(f)\sqrt{n})/\epsilon}$. The main tool used in this area is the low degree learning algorithm due to Linial et al. [17] using which the following result is derived in [22].

Lemma 3 (Corollary 3.22 and Theorem 3.36 [22]). *For $t \geq 1$, let $\mathcal{A}_t = \{f \mid f : \{-1,1\}^n \rightarrow \{-1,1\}, \mathbf{I}[f] \leq t\}$ and $\mathcal{B}_t = \{f \mid f : \{-1,1\}^n \rightarrow \{-1,1\}, \deg(f) \leq t\}$ Then $\mathcal{A}_t$ can be learned from random examples with error ϵ in time $n^{O(t/\epsilon)}$ for any $\epsilon \in (0,1)$ and $\mathcal{B}_t$ can be exactly learned from random examples in time $n^t \text{poly}(n, 2^t)$.*

The claimed result follows from Proposition 2. Applying Lemma 3, we obtain

- an exact learning algorithm from random examples with a runtime of $n^{O(\text{alt}(f)\deg_{\mathbb{F}_2}(f)^2)} \text{poly}(n, 2^{(\text{alt}(f)\deg_{\mathbb{F}_2}(f)^2)})$ and
- an ϵ error learning algorithm from random examples with a runtime $n^{O(\text{alt}(f)\deg_{\mathbb{F}_2}(f)^2/\epsilon)}$

thereby removing the dependence on the parameter n in the exponent and improving the runtime for those f such that $\deg_{\mathbb{F}_2}(f) < \sqrt[4]{n}$.

Acknowledgments. The authors would like to thank the anonymous reviewers for constructive comments which improved the presentation of the paper.

References

1. Ambainis, A., Prūsis, K., Vihrovs, J.: Sensitivity versus certificate complexity of Boolean functions. In: Kulikov, A.S., Woeginger, G.J. (eds.) CSR 2016. LNCS, vol. 9691, pp. 16–28. Springer, Cham (2016). https://doi.org/10.1007/978-3-319-34171-2_2
2. Bafna, M., Lokam, S.V., Tavenas, S., Velingker, A.: On the sensitivity conjecture for read-k formulas. In: 41st International Symposium on Mathematical Foundations of Computer Science, MFCS 2016 - Kraków, Poland, pp. 16:1–16:14 (2016)
3. Bernasconi, A., Codenotti, B.: Spectral analysis of Boolean functions as a graph eigenvalue problem. IEEE Trans. Comput. **48**(3), 345–351 (1999)
4. Blais, E., Canonne, C.L., Oliveira, I.C., Servedio, R.A., Tan, L.-Y.: Learning circuits with few negations. In: Approximation, Randomization, and Combinatorial Optimization. Algorithms and Techniques, APPROX/RANDOM 2015, 24–26 August 2015, Princeton, NJ, USA, pp. 512–527 (2015)
5. Buhrman, H., de Wolf, R.: Complexity measures and decision tree complexity: a survey. Theor. Comput. Sci. **288**(1), 21–43 (2002)
6. Karthik, C.S., Tavenas, S.: On the sensitivity conjecture for disjunctive normal forms. In: 36th IARCS Annual Conference on Foundations of Software Technology and Theoretical Computer Science, FSTTCS 2016, 13–15 December 2016, Chennai, India, pp. 15:1–15:15 (2016)
7. Chakraborty, S.: On the sensitivity of cyclically-invariant Boolean functions. Discrete Math. Theor. Comput. Sci. **13**(4), 51–60 (2011)
8. Cook, S., Dwork, C., Reischuk, R.: Upper and lower time bounds for parallel random access machines without simultaneous writes. SIAM J. Comput. **15**(1), 87–97 (1986)
9. Gilmer, J., Koucký, M., Saks, M.E.: A new approach to the sensitivity conjecture. In: Proceedings of the 2015 Conference on Innovations in Theoretical Computer Science, ITCS 2015, pp. 247–254. ACM, New York (2015)
10. Gopalan, P., Servedio, R.A., Wigderson, A.: Degree and sensitivity: tails of two distributions. In: 31st Conference on Computational Complexity, CCC 2016, Tokyo, Japan, pp. 13:1–13:23 (2016)
11. Guo, S., Komargodski, I.: Negation-limited formulas. Theor. Comput. Sci. **660**, 75–85 (2017)
12. Hatami, P., Kulkarni, R., Pankratov, D.: Variations on the sensitivity conjecture. Theor. Comput. Libr. Grad. Surv. **4**, 1–27 (2011)
13. Jukna, S.: Boolean Function Complexity: Advances and Frontiers. Algorithms and Combinatorics. Springer, Heidelberg (2012)
14. Kenyon, C., Kutin, S.: Sensitivity, block sensitivity, and ℓ-block sensitivity of Boolean functions. Inf. Comput. **189**(1), 43–53 (2004)
15. Kulkarni, R., Santha, M.: Query complexity of matroids. In: Spirakis, P.G., Serna, M. (eds.) CIAC 2013. LNCS, vol. 7878, pp. 300–311. Springer, Heidelberg (2013). https://doi.org/10.1007/978-3-642-38233-8_25
16. Lin, C., Zhang, S.: Sensitivity conjecture and log-rank conjecture for functions with small alternating numbers. In: 44th International Colloquium on Automata, Languages, and Programming (ICALP 2017), vol. 80, pp. 51:1–51:13, Dagstuhl, Germany (2017)

17. Linial, N., Mansour, Y., Nisan, N.: Constant depth circuits, fourier transform, and learnability. J. ACM **40**(3), 607–620 (1993)
18. Markov, A.A.: On the inversion complexity of a system of functions. J. ACM **5**(4), 331–334 (1958)
19. Montanaro, A., Osborne, T.: On the communication complexity of XOR functions. CoRR abs/0909.3392 (2009)
20. Nisan, N.: CREW PRAMs and decision trees. SIAM J. Comput. **20**(6), 999–1007 (1991)
21. Nisan, N., Szegedy, M.: On the degree of Boolean functions as real polynomials. In: Proceedings of the 24th Annual ACM Symposium on Theory of Computing, STOC 1992, pp. 462–467. ACM, New York (1992)
22. O'Donnell, R.: Analysis of Boolean Functions. Cambridge University Press, New York (2014)
23. Simon, H.-U.: A tight ω(loglog n)-bound on the time for parallel Ram's to compute nondegenerated boolean functions. In: Karpinski, M. (ed.) FCT 1983. LNCS, vol. 158, pp. 439–444. Springer, Heidelberg (1983). https://doi.org/10.1007/3-540-12689-9_124
24. Tal, A.: Properties and applications of Boolean function composition. In: Proceedings of the 4th Conference on Innovations in Theoretical Computer Science, ITCS 2013, pp. 441–454. ACM, New York (2013)
25. Tsang, H.Y., Wong, C.H., Xie, N., Zhang, S.: Fourier sparsity, spectral norm, and the log-rank conjecture. In: 54th Annual IEEE Symposium on Foundations of Computer Science, FOCS 2013, 26–29 October 2013, Berkeley, CA, USA, pp. 658–667 (2013)
26. Turán, G.: The critical complexity of graph properties. Inf. Process. Lett. **18**(3), 151–153 (1984)
27. Zhang, Z., Shi, Y.: Communication complexities of symmetric XOR functions. Quantum Inf. Comput. **9**(3), 255–263 (2009)
28. Zhang, Z., Shi, Y.: On the parity complexity measures of Boolean functions. Theor. Comput. Sci. **411**(26–28), 2612–2618 (2010)

On Oriented $L(p, 1)$-labeling

Sandip Das[1], Soumen Nandi[1(✉)], and Sagnik Sen[2]

[1] Indian Statistical Institute, Kolkata, India
sandipdas@isical.ac.in, soumen2004@gmail.com
[2] Ramakrishna Mission Vivekananda University, Kolkata, India
sen007isi@gmail.com

Abstract. An oriented graph is a directed graph without any directed cycle of length at most 2. In this article, we characterize the oriented $L(p, 1)$-labeling span $\lambda^o_{p,1}(\overrightarrow{G})$ of an oriented graph $\overrightarrow{G}$ using graph homomorphisms. Using this characterization and probabilistic techniques we prove the upper bound of $\lambda^o_{p,1}(\mathcal{G}_\Delta) \leq 2 . \Delta^2 . 2^\Delta + (p\Delta)$, where $\mathcal{G}_\Delta$ is the family of graphs with maximum degree at most Δ. Moreover, by proving a lower bound exponential in Δ for the same graph family we conclude that the upper bound is not too far from being optimal. We also settle an open problem given by Sen (DMGT 2014) for the family of outerplanar graphs $\mathcal{O}$ by showing $\lambda^o_{2,1}(\mathcal{O}) = 10$.

Keywords: Oriented graph · Graph homomorphism
Oriented $l(p, 1)$-labeling · Outerplanar graph · Maximum degree

1 Introduction

An oriented graph is a directed graph $\overrightarrow{G}$ having no directed cycle of length 1 or 2 with set of vertices $V(\overrightarrow{G})$ and set of arcs $A(\overrightarrow{G})$. The underlying undirected graph of $\overrightarrow{G}$ is denoted by G and $\overrightarrow{G}$ is an *orientation* of G. The *directed distance* $\overrightarrow{d}(u, v)$ between two vertices u and v is the *length* (number of edges) of a shortest directed path connecting u and v (can be either from u to v, or from v to u). The set of vertices and edges of an undirected graph G is denoted by $V(G)$ and $E(G)$, respectively. The *distance* $d(u, v)$ between two vertices u and v of a graph G is the length of a shortest path connecting u and v.

Griggs and Yeh [8] introduced the $L(2, 1)$-labeling problem as a variation of the channel assignment problem given by Hale [9]. Since then, the generalization, $L(p, q)$-labeling problem has been studied by several researchers (see the survey by Calamoneri [3] for details), with particular focus on $L(p, 1)$-labeling [6] and $L(2, 1)$-labeling [4,11]. Aardal et al. [1] pointed out that the same problem, modeled on directed or oriented graphs can potentially be better. After that two different oriented versions of $L(p, q)$-labeling have been introduced and studied.

Chang et al. [5] introduced the 2-dipath $L(p, q)$ labeling of oriented graphs. A *2-dipath k-L(p, q)-labeling* of an oriented graph $\overrightarrow{G}$ is a function $l : V(\overrightarrow{G}) \rightarrow$

© Springer International Publishing AG 2018
B. S. Panda and P. P. Goswami (Eds.): CALDAM 2018, LNCS 10743, pp. 274–282, 2018.
https://doi.org/10.1007/978-3-319-74180-2_23

$\{0, 1, \ldots, k\}$ such that $|l(u) - l(v)| \geq p$ for $\overrightarrow{d}(u, v) = 1$ and $|l(u) - l(v)| \geq q$ for $\overrightarrow{d}(u, v) = 2$. The 2-dipath span $\overrightarrow{\lambda}_{p,q}(\overrightarrow{G})$ of an oriented graph $\overrightarrow{G}$ is the minimum k such that $\overrightarrow{G}$ admits a 2-dipath k-$L(p, q)$-labeling.

Gonçalves et al. [7] introduced the oriented $L(p, q)$-labeling. An *oriented $k-$* $L(p,q)$-*labeling* of an oriented graph $\overrightarrow{G}$ is a 2-dipath k-$L(p, q)$ labeling l of $\overrightarrow{G}$ such that any two arcs $wx, yz \in A(\overrightarrow{G})$ with $l(w) = l(z)$ implies $l(x) \neq l(y)$. The oriented $L(p, q)$-labeling span $\lambda^o_{p,q}(\overrightarrow{G})$ of an oriented graph $\overrightarrow{G}$ is the minimum k such that $\overrightarrow{G}$ admits an oriented k-$L(p, q)$-labeling.

For an undirected graph G its 2-dipath $L(p, q)$-labeling span

$$\overrightarrow{\lambda}_{p,q}(G) = \max\{\overrightarrow{\lambda}_{p,q}(\overrightarrow{G}) | \overrightarrow{G} \text{ is an orientation of } G\}$$

and oriented $L(p, q)$-labeling span

$$\lambda^o_{p,q}(G) = \max\{\lambda^o_{p,q}(\overrightarrow{G}) | \overrightarrow{G} \text{ is an orientation of } G\}.$$

Also for a family $\mathcal{F}$ of graphs its 2-dipath (resp. oriented) $L(p, q)$-labeling span

$$\overrightarrow{\lambda}_{p,q}(\mathcal{F}) = \max\{\overrightarrow{\lambda}_{p,q}(G) | G \in \mathcal{F}\} \text{ (resp. } \lambda^o_{p,q}(\mathcal{F}) = \max\{\lambda^o_{p,q}(G) | G \in \mathcal{F}\}).$$

Several works has been done on the parameters $\overrightarrow{\lambda}_{p,q}(\cdot)$ and $\lambda^o_{p,q}(\cdot)$, with a particular focus on $(p, q) = (2, 1)$. Chang et al. [5] studied $\overrightarrow{\lambda}_{p,q}(\cdot)$ for bipartite and non-bipartite graphs. Calamoneri and Sinaimeri [4] studied $\overrightarrow{\lambda}_{2,1}(\cdot)$ for the families of prisms, halin graphs and cactus, whereas Calamoneri [2] studied the same parameter for square, triangular and hexagonal grid graphs. Gonçalves et al. [7] computed upper bounds of $\lambda^o_{p,q}(\overrightarrow{G})$ on trees, bipartite graphs and planar graphs as well as provided an upper bound of $\overrightarrow{\lambda}_{p,1}(\overrightarrow{G})$ in terms of the maximum degree of $\overrightarrow{G}$. Sen [11] studied both the parameters $\overrightarrow{\lambda}_{p,q}(\cdot)$ and $\lambda^o_{p,q}(\cdot)$ for complete multipartite graphs and provided close/tight bounds of $\overrightarrow{\lambda}_{2,1}(\cdot)$ and $\lambda^o_{2,1}(\cdot)$ for families of cactus, 2-trees, outerplanar, and planar graphs with girth restrictions.

Our main focus is to study the oriented $L(p, 1)$-labeling of oriented graphs with some of the results concerning the particular case $p = 2$. Gonçalves et al. [7] proved a quadratic upper bound $\overrightarrow{\lambda}_{p,1}(\mathcal{G}_\Delta) \leq \lfloor \frac{\Delta^2}{2} \rfloor + p\Delta$ where $\mathcal{G}_\Delta$ is the family of graphs with maximum degree at most Δ. For $\lambda^o_{p,1}(\mathcal{G}_\Delta)$ no such bound has been proved till date. However, using the relation $\lambda^o_{p,q}(\overrightarrow{G}) \leq \max\{p, q\} \cdot \chi_o(\overrightarrow{G}) - 1$ [4] and the result $\chi^o(\mathcal{G}_\Delta) \leq 2.\Delta^2.2^\Delta$ [10] one can prove the trivial upper bound $\lambda^o_{p,1}(\mathcal{G}_\Delta) \leq 2p.\Delta^2.2^\Delta - 1$.

In this article we formulate an equivalent definition of oriented $L(p, 1)$-labeling and its span using the notion of graph homomorphism. Then, using that alternative definition and probabilistic techniques we improve this upper bound to $\lambda^o_{p,1}(\mathcal{G}_\Delta) \leq 2.\Delta^2.2^\Delta + (p\Delta)$. Moreover, by proving a lower bound exponential in Δ we conclude that the upper bound is not too far from being optimal. To be precise, we prove the following result.

Theorem 1. *For the family of graphs $\mathcal{G}_\Delta$ with maximum degree $\Delta \geq 2$, we have*
$2^{\frac{\Delta}{2}} - 1 \leq \lambda_{p,1}^o(\mathcal{G}_\Delta) \leq 2\Delta^2 2^\Delta + 2p\Delta.$

Sen [11] proved $9 \leq \lambda_{2,1}^o(\mathcal{O}) \leq 10$, where $\mathcal{O}$ is the family of outerplanar graphs. Determining the exact value of $\lambda_{2,1}^o(\mathcal{O})$ was left as an open problem [11]. In this article, we settle the open problem for outerplanar graphs by showing the following result.

Theorem 2. *For the family $\mathcal{O}$ of outerplanar graphs we have $\lambda_{2,1}^o(\mathcal{O}) = 10$.*

2   Homomorphism and Oriented $L(p, 1)$-labeling

The following two results have been used to prove lower and upper bounds of $\overrightarrow{\lambda}_{p,q}(\cdot)$ and $\lambda_{p,q}^o(\cdot)$ in the previous works. The first one is regarding the obvious relation between the two parameters.

Lemma 1 ([7]). *For an oriented graph $\overrightarrow{G}$ we have $\overrightarrow{\lambda}_{p,q}(\overrightarrow{G}) \leq \lambda_{p,q}^o(\overrightarrow{G})$.*

The second result shows that both the parameters respects homomorphism in the following sense.

Lemma 2 ([7]). *If $\overrightarrow{G} \to \overrightarrow{H}$, then $\overrightarrow{\lambda}_{p,q}(\overrightarrow{G}) \leq \overrightarrow{\lambda}_{p,q}(\overrightarrow{H})$ and $\lambda_{p,q}^o(\overrightarrow{G}) \leq \lambda_{p,q}^o(\overrightarrow{H})$.*

Here we go one step further by providing an equivalent formulation of oriented $L(2, 1)$-labeling span using the notion of homomorphisms.

The graph $L_{p,n}$ is a graph with set of vertices $V(L_{p,n}) = \{0, 1, ..., n\}$ and set of edges $E(L_{p,n}) = \{uv \text{ such that } |u - v| \geq p\}$. Let $\overrightarrow{\mathcal{L}}_{p,n}$ denote the family of all orientations of $L_{p,n}$.

Theorem 3. *For an oriented graph $\overrightarrow{G}$ we have $\lambda_{p,1}^o(\overrightarrow{G}) \leq n$ if and only if there exists an $\overrightarrow{L}_{p,n} \in \overrightarrow{\mathcal{L}}_{p,n}$ such that $\overrightarrow{G} \to \overrightarrow{L}_{p,n}$.*

Proof. First assume that $\lambda_{p,1}^o(\overrightarrow{G}) \leq n$. Let l be an oriented n-$L(p, 1)$-labeling of $\overrightarrow{G}$.

Now we will construct an oriented graph $\overrightarrow{L}$ with set of vertices $V(\overrightarrow{L}) = \{0, 1, ..., n\}$. If there exists an arc $xy \in A(\overrightarrow{G})$ with $l(x) = u$ and $l(y) = v$, then include the arc $uv \in A(\overrightarrow{L})$. Now add an arc between each pair $\{u, v\}$ of non-adjacent vertices such that $|u - v| \geq p$. This will finish the construction of $\overrightarrow{L}$. Note that $\overrightarrow{G} \to \overrightarrow{L}$ and $\overrightarrow{L}$ is an orientation of $L_{p,n}$.

For the converse, assume that there exists an orientation $\overrightarrow{L}_{p,n}$ of $L_{p,n}$ such that $l : \overrightarrow{G} \to \overrightarrow{L}_{p,n}$ is a homomorphism. Therefore, note that $l : V(\overrightarrow{G}) \to V(\overrightarrow{L}_{p,n}) = \{0, 1, ..., n\}$, as a function, is an oriented n-$L(p, 1)$-labeling of $\overrightarrow{G}$.

Given two (oriented) graphs G, H their *disjoint union graph* $G \cup H$ is a graph with set of vertices $V(G \cup H) = V(G) \cup V(H)$ and set of edges $E(G \cup H) = E(G) \cup E(H)$. A family $\mathcal{F}$ of graphs is *complete* if $G, H \in \mathcal{F}$ implies that the disjoint union graph $G \cup H \in \mathcal{F}$.

Corollary 1. *For a complete family $\mathcal{F}$ of graphs $\lambda^o_{p,1}(\mathcal{F}) \leq n$ if and only if there exists an $\overrightarrow{L}_{p,n} \in \overrightarrow{\mathcal{L}}_{p,n}$ such that $\overrightarrow{G} \to \overrightarrow{L}_{p,n}$ for all $\overrightarrow{G} \in \mathcal{F}$.*

Proof. First we will prove the converse. Let there exist an $\overrightarrow{L}_{p,n} \in \overrightarrow{\mathcal{L}}_{p,n}$ such that $\overrightarrow{G} \to \overrightarrow{L}_{p,n}$ for all $\overrightarrow{G} \in \mathcal{F}$. Then $\lambda^o_{p,1}(\overrightarrow{G}) \leq n$ for all $\overrightarrow{G} \in \mathcal{F}$ by Theorem 3.

Now we will prove the 'if' part by contradiction. Thus assume that, for each $\overrightarrow{L}_{p,n} \in \overrightarrow{\mathcal{L}}_{p,n}$ there exists a $\overrightarrow{G}_{\overrightarrow{L}_{p,n}} \in \mathcal{F}$ such that $\overrightarrow{G}_{\overrightarrow{L}_{p,n}}$ does not admit a homomorphism to $\overrightarrow{L}_{p,n}$. Let $\overrightarrow{X}$ be the disjoint union graph of $\overrightarrow{G}_{\overrightarrow{L}_{p,n}}$'s for all $\overrightarrow{L}_{p,n} \in \overrightarrow{\mathcal{L}}_{p,n}$. Thus $\overrightarrow{X}$ does not admit a homomorphism to any $\overrightarrow{L}_{p,n} \in \mathcal{F}$. This implies $\lambda^o_{p,1}(\mathcal{F}) > n$ by Theorem 3, a contradiction.

In the following two sections we will deal with three complete family of graphs, namely, the family of graphs with maximum degree Δ, the family of planar graphs and the family of outerplanar graphs. We will use the above two results to prove our bounds.

3   Proof of Theorem 1

Let $\overrightarrow{G}$ be an oriented graph having the arc uv. Then v is a $+$-*neighbor* of u and u is a $-$-*neighbor* of v. The set of all $+$-neighbors and $-$-neighbors of v is denoted by $N^+(v)$ and $N^-(v)$, respectively. Let $\boldsymbol{a} = (a_1, a_2, ..., a_j)$ be a j-*vector* such that $a_i \in \{+, -\}$ where $i \in \{1, 2, ..., j\}$. Let $J = (v_1, v_2, ..., v_j)$ be a j-*tuple* (without repetition) of vertices from $\overrightarrow{G}$. Then we define the set $N^{\boldsymbol{a}}(J) = \{v \in V | v \in N^{a_i}(v_i) \, for \, all \, 1 \leq i \leq j\}$. Finally, we say that $\overrightarrow{G}$ has property $Q^{t,j}_{g(j)}$ if for each j-vector $\boldsymbol{a}$ and each j-tuple J we have $|N^{\boldsymbol{a}}(J)| \geq g(j)$ where $j \in \{0, 1, ..., t\}$ and $g : \{0, 1, ..., t\} \to \{0, 1, ...\infty\}$ is an integral function.

While proving $\chi_o(\mathcal{G}_\Delta) \leq 2.\Delta^2.2^\Delta$ Kostochka et al. [10] showed the following (see the proof of Theorem 5 in [10]):

Lemma 3 ([10]). *If $\overrightarrow{L}$ has property $Q^{\Delta,j}_{1+(\Delta-j)(\Delta-1)}$, then $\overrightarrow{G} \to \overrightarrow{L}$ for all $\overrightarrow{G} \in \mathcal{G}_\Delta$.*

Therefore, to prove the upper bound of Theorem 1 it is enough to prove the following lemma.

Lemma 4. *There exists an $\overrightarrow{L}_{p,n} \in \overrightarrow{\mathcal{L}}_{p,n}$ having property $Q^{t,j}_{1+(t-j)(t-1)}$, where $n = 2t^2.2^t + 2pt$ and $t \geq 2$.*

Proof. Consider the graph $L_{p,n}$ and let $C_J = \cap_{i=1}^j N(v_i)$ for any j-tuple $J = (v_1, v_2, ..., v_j)$. For a fixed v_i, as $|l(v_i) - l(u)| \geq p$ for $\overrightarrow{d}(v_i, u) = 1$, at most $(2p - 1)$ vertices labeled with $l(v_i), l(v_i) \pm 1, l(v_i) \pm 2, ..., l(v_i) \pm (p - 1)$ cannot be the neighbors of v_i.

So, number of non-neighbors of a vertex v_i is at most $(2p - 1)$, which implies that

$$\bigcup_{i=1}^j (V(L_{p,n}) - N(v_i)) \leq (2p - 1)j.$$

That means, $|C_J| \geq 2t^2.2^t + 2pt - 2pj + j \geq 2t^2.2^t - j$. Now, fix a set $C'_J \subseteq C_J$ for each j-tuple J in such a way that $|C'_J| = 2t^2.2^t - j = m - j$, where $m = 2t^2.2^t$.

We want to show that there exists an orientation $\overrightarrow{L}_{p,n} \in \overrightarrow{\mathcal{L}}_{p,n}$ such that for each j-vector $\boldsymbol{a}$ and each j-tuple J we have $|N^{\boldsymbol{a}}(J)| \geq 1 + (t - j)(t - 1)$. This will imply that $\overrightarrow{L}_{p,n}$ has property $Q_{1+(t-j)(t-1)}^{t,j}$.

Thus let us take a random orientation of $L_{p,n}$ where probability of the existence of an arc is $\frac{1}{2}$. We will show that the probability of $\overrightarrow{L}_{p,n}$ not having property $Q_{1+(t-j)(t-1)}^{t,j}$ is strictly less than 1 when $|\overrightarrow{L}_{p,n}| = n = 2t^2.2^t + 2pt$. Let $P(J, \boldsymbol{a})$ denote the probability of the event $|N^{\boldsymbol{a}}(J)| < 1 + (t - j)(t - 1)$ where J is a j-tuple of $\overrightarrow{L}_{p,n}$ and $\boldsymbol{a}$ is a j-vector for some $j \in \{0, 1, ..., t\}$. Call such an event a *bad event*. Thus,

$$
\begin{aligned}
P(J, \boldsymbol{a}) &\leq \sum_{i=0}^{(t-j)(t-1)} \binom{|C'_J|}{i} 2^{-ij}(1 - 2^{-j})^{|C'_J|-i} \\
&= \sum_{i=0}^{(t-j)(t-1)} \binom{m-j}{i} 2^{-ij}(1 - 2^{-j})^{m-i-j}.
\end{aligned}
\tag{1}
$$

So, from Eq. 1 we can write by [10] that $P(J, \boldsymbol{a}) < e^{-m2^{-j}} m^{(t-j)(t-1)+1}$.

Let $P(B)$ denote the probability of the occurrence of at least one bad event. To prove this lemma it is enough to show that $P(B) < 1$. Let T^j denote the set of all j-tuples and W^j denote the set of all j-vectors. Then by [10] we can conclude that

$$
\begin{aligned}
P(B) = \sum_{j=0}^{t} \sum_{J \in T^j} \sum_{\boldsymbol{a} \in W^j} P(J, \boldsymbol{a}) &< \sum_{j=0}^{t} \binom{m}{j} 2^j e^{-m2^{-j}} m^{(t-j)(t-1)+1} \\
&< 1.
\end{aligned}
\tag{2}
$$

Hence, the result follows.

Finally, we are ready to prove Theorem 1.

Proof of Theorem 1. We know that given any oriented graph $\overrightarrow{G}$ the following inequality $\lambda_{p,1}^o(\overrightarrow{G}) \geq \chi_o(\overrightarrow{G}) - 1$ due to Gonçalves et al. [7] and $\chi_o(\overrightarrow{G}) \geq 2^{\frac{A}{2}}$ due to Kostochka et al. [10]. These two inequalities together implies the lower bound.

The upper bound follows from Lemmas 3 and 4.   □

4   Proof of Theorem 2

Before proving the theorem we will introduce some notations, definitions and constructions to aid the proof. After that we will prove some lemmas, combining which we will finally prove Theorem 2.

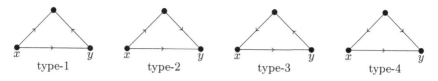

Fig. 1. The spikes.

The spikes: The four graphs depicted in Fig. 1 are called spikes. The names of the spikes are as suggested in the pictures.

Merging a spike with an arc: Let G be a graph having an arc uv. Merging a type-k spike with uv is to place a copy of a type-k spike by identifying the vertices x and y with the vertices u and v, respectively.

A spiked oriented path: Let $X = v_0 v_1 \cdots v_{4n}$ be a path with an orientation $\overrightarrow{X}$. Let the arc in $\overrightarrow{X}$ corresponding to the edge $v_{i-1} v_i$ of X be a_i for all $i \in \{1, 2, \cdots 4n\}$. Merge

- a type-1 spike with each of the first n arcs $a_1, a_2, \cdots, a_n$ of $\overrightarrow{X}$,
- a type-2 spike with each of the next n arcs $a_{n+1}, a_{n+2}, \cdots, a_{2n}$ of $\overrightarrow{X}$,
- a type-3 spike with each of the next n arcs $a_{2n+1}, a_{2n+2}, \cdots, a_{3n}$ of $\overrightarrow{X}$,
- a type-4 spike with each of the last n arcs $a_{3n+1}, a_{3n+2}, \cdots, a_{4n}$ of $\overrightarrow{X}$.

The so-obtained graph is the *spiked oriented path* of $\overrightarrow{X}$. The operation by which we obtain the spiked oriented path of $\overrightarrow{X}$ is called *adding spikes* on $\overrightarrow{X}$. The new arcs are the *outer arcs* and the oriented path induced by the outer arcs is the *outer oriented path* of the spiked oriented path of $\overrightarrow{X}$.

An i-spiked oriented path: Let $X = v_0 v_1 \cdots v_{4^i n}$ be a path with an orientation $\overrightarrow{X}$. First obtain the spiked oriented path $\overrightarrow{X}_1$ of $\overrightarrow{X}$ and alternatively call it the 1-*spiked oriented path* of $\overrightarrow{X}$. After that consider the outer oriented path of $\overrightarrow{X}_1$ and add spikes on it. The so-obtained graph is the 2-*spiked oriented path* of $\overrightarrow{X}$ with the *outer oriented path* being the path induced by the newly added arcs. Finally we recursively define the i-spiked oriented path $\overrightarrow{X}_i$ of $\overrightarrow{X}$ by the following: the i-spiked oriented path $\overrightarrow{X}_i$ of $\overrightarrow{X}$ is the graph obtained by adding spikes to the outer oriented path $\overrightarrow{X}_{i-1}$ where the oriented path induced by the newly added arcs is the outer oriented path of $\overrightarrow{X}_i$. Also the outer oriented path added in the jth iteration is called the jth outer path for all $j \leq i$. An arc of a jth outer path for some $j < i$ is a *trendy arc*.

The n-fan: Take two directed paths $u_{-1} u_0 u_1 \cdots u_n u_{n+1}$ and $v_{-1} v_0 v_1 \cdots v_n v_{n+1}$ on n vertices each and make them adjacent to a vertex v with such orientations that $N^+(v) = \{u_{-1}, u_0, u_1, \cdots, u_{n+1}\}$ and $N^-(v) = \{v_{-1}, v_0, v_1, \cdots, v_{n+1}\}$.

The i-spiked fan: Let $\overrightarrow{F}$ be the 4^{i+1}-fan where $\overrightarrow{X} = u_0 u_1 \cdots u_n$ and $\overrightarrow{Y} = v_0 v_1 \cdots v_n$ denotes the induced directed paths. Now construct i-spiked oriented paths of $\overrightarrow{X}$ and $\overrightarrow{Y}$ to obtain the i-spiked fan.

Let l be an oriented n-labeling of an oriented graph $\overrightarrow{G}$. Consider the *label graph* $\overrightarrow{T}_{G,l}$ of $\overrightarrow{G}$ and l with set of vertices $V(\overrightarrow{T}_{G,l}) = \{0, 1, \cdots, n\}$ and set of arcs

$$A(\overrightarrow{T}_{G,l}) = \{ij| \text{ there exists } u, v \in A(\overrightarrow{G}) \text{ with } l(u) = i \text{ and } l(v) = j\}.$$

If two vertices i and j of $\overrightarrow{T}_{G,l}$ have $|N^\alpha \cap N^\beta| \geq 1$ for any $\alpha, \beta \in \{+, -\}$, then we say that i and j are *conjugates*.

Lemma 5. *If l is an oriented n-labeling of an oriented graph $\overrightarrow{G}$, then $\bar{l}$ given by $\bar{l}(i) = n - l(i)$ is also an oriented n-labeling of $\overrightarrow{G}$.*

The above lemma suggests a symmetry between the labels i and $n - i$ which we will use inside our proofs.

Now we will describe our example to prove the lower bound. Start with a 0-fan $\overrightarrow{F}_0$. After that we glue a copy of a 0-fan on each vertex x of $\overrightarrow{F}_0$ by identifying the vertex x with the vertex v of the corresponding 0-fan. Call this so-obtained graph as $\overrightarrow{F}_1$. Now glue a copy of a i-spiked fan on each vertex x of $\overrightarrow{F}_1$ by identifying the vertex x with the vertex v of the corresponding i-spiked fan. Call this so-obtained graph as $\overrightarrow{F}_2$. Note that $\overrightarrow{F}_2$ is an outerplanar graph. Thus we will be done if we can show that $\lambda^o_{2,1}(\overrightarrow{F}_2) \geq 10$.

We will assume the contrary, that is, suppose that $\lambda^o_{2,1}(\overrightarrow{F}_2) \leq 9$. Let l be an oriented n-labeling of $\overrightarrow{F}_2$. Now we will prove some structural properties of $\overrightarrow{T}_{\overrightarrow{F}_2,l}$ and the labeling l.

Lemma 6. *There exists a vertex x of $\overrightarrow{F}_1$ in $\overrightarrow{F}_2$ such that $l(x) \in \{4, 5\}$.*

Proof. We know that $\overrightarrow{F}_0$ has oriented chromatic number 7 due to Sopena [12]. Thus by pigeonhole principle there exists a vertex w of $\overrightarrow{F}_0$ in $\overrightarrow{F}_2$ such that $l(w) \in \{1, 4, 5, 8\}$.

If $l(w) \in \{4, 5\}$, then we are done. Otherwise $l(w) \in \{1, 8\}$. Note that $|N(w) \cap (V(\overrightarrow{F}_1) \setminus V(\overrightarrow{F}_0))| = 6$ and thus, we need to use 6 labels from $\{0, 1, \cdots, 9\} \setminus \{l(w) - 1, l(w), l(w) + 1\}$ on the vertices of $N(w) \cap (V(\overrightarrow{F}_1) \setminus V(\overrightarrow{F}_0))$. Therefore, we must need to use either 4 or 5 on a vertex of $\overrightarrow{F}_1$.

Due to the above result, we know that there exists a vertex x of $\overrightarrow{F}_1$ in $\overrightarrow{F}_2$ such that either $l(x) = 4$ or $\bar{l}(x) = 4$. Without loss of generality we may assume that there exists a vertex x of $\overrightarrow{F}_1$ in $\overrightarrow{F}_2$ such that $l(x) = 4$.

An arc $x_1 y_1 \in A(\overrightarrow{F}_2)$ is *repeat* of another arc $xy \in A(\overrightarrow{F}_2)$ if $l(x) = l(w)$ and $l(y) = l(z)$. Moreover, if xy has at least 4 repeats, say, $x_1 y_1, x_2, y_2, x_3 y_3$ and $x_4 y_4$, such that $x_k y_k$ has a type k-spike on it in $\overrightarrow{F}_2$ for each $k \in \{1, 2, 3, 4\}$, then the pair of labels $\{l(x), l(y)\}$ is called a *couple*. If this happens, then we also say that the couple $\{l(x), l(y)\}$ has *all types of repetitions*. If xy is a trendy arc, then $\{l(x), l(y)\}$ as above is a *trendy couple*.

Lemma 7. *If $\{i,j\}$ is a trendy couple with $1 \leq i,j \leq 8$ and $|i-j| \geq 3$, then $\{i,k\}$ and $\{j,k\}$ are also trendy couples for each $k \in \{0,1,\cdots,9\} \setminus \{i-1,i,i+1,j-1,j,j+1\}$.*

The following lemma directly follows from the above result.

Lemma 8. *If $\lambda^o_{2,1}(\mathcal{O}) \leq 9$, then there exists a label graph $T_{G,l}$ on 10 vertices in which each pair $i,j \in \{1,2,\cdots,8\}$ with $|i-j| \geq 3$ are conjugates.*

Now we are ready to prove our final lemma.

Lemma 9. *There does not exist any label graph $T_{G,l}$ on 10 vertices in which each pair $i,j \in \{1,2,\cdots,8\}$ with $|i-j| \geq 3$ are conjugates.*

Proof. Suppose the contrary. Without loss of generality assume that $6 \in N^\gamma(1)$ for some $\{\gamma,\bar{\gamma}\} = \{+,-\}$. As $1,4$ are conjugates having $N(1) \cap N(4) = \{6,7,8,9\}$, we must have

$$|N^\gamma(1) \cap \{6,7,8,9\}| = |N^{\bar\gamma}(1)\{6,7,8,9\}| = 2.$$

Notice that as $1,5$ are conjugates having $N(1) \cap N(5) = \{3,7,8,9\}$, we must have

$$|N^\gamma(1) \cap \{3,7,8,9\}| = |N^{\bar\gamma}(1)\{3,7,8,9\}| = 2.$$

This implies,

$$3 \in N^\gamma(1).$$

Furthermore, as $1,8$ are conjugates having $N(1) \cap N(8) = \{3,4,5,6\}$, we must have

$$|N^\gamma(1) \cap \{3,4,5,6\}| = |N^{\bar\gamma}(1)\{3,4,5,6\}| = 2.$$

This implies,

$$4,5 \in N^{\bar\gamma}(1) \text{ as } 3,6 \in N^\gamma(1).$$

Also as $1,7$ are conjugates having $N(1) \cap N(7) = \{3,4,5,9\}$, we must have

$$|N^\gamma(1) \cap \{3,4,5,9\}| = |N^{\bar\gamma}(1)\{3,4,5,9\}| = 2.$$

This implies,

$$9 \in N^\gamma(1) \text{ as } 4,5 \in N^{\bar\gamma}(1).$$

Moreover, as $1,4$ are conjugates having $N(1) \cap N(4) = \{6,7,8,9\}$, we must have

$$|N^\gamma(1) \cap \{6,7,8,9\}| = |N^{\bar\gamma}(1)\{6,7,8,9\}| = 2.$$

This implies,

$$7,8 \in N^{\bar\gamma}(1) \text{ as } 6,9 \in N^\gamma(1).$$

Therefore, we have

$$3,6,9 \in N^\gamma(1) \text{ and } 4,5,7,8 \in N^{\bar\gamma}(1).$$

Similarly, if we assume that $3 \in N^\mu(8)$ for some $\{\mu, \bar{\mu}\} = \{+, -\}$, then we can conclude that

$$0, 3, 6 \in N^\mu(1) \text{ and } 1, 2, 4, 5 \in N^{\bar{\mu}}(1).$$

This will contradict the fact that $1, 8$ are conjugates having common neighbors $N(1) \cap N(8) = \{3, 4, 5, 6\}$ as

$$N^\gamma(1) \cap N^\mu(8) = \{3, 6\} \text{ and } N^{\bar{\gamma}}(1) \cap N^{\bar{\mu}}(8) = \{4, 5\}.$$

This concludes the proof.

Finally, we are ready to prove Theorem 2.

Proof of Theorem 2. The proof directly follows from Lemmas 8 and 9.   □

5   Conclusions

In this article, we considered oriented $L(p, 1)$-labeling of graphs having bounded maximum degree and oriented $L(p, 1)$-labeling of outerplanar graphs. In particular we showed that $\Omega(2^{\frac{\Delta}{2}}) = \lambda_{p,1}^o(\mathcal{G}_\Delta) = O(\Delta^2 . 2^\Delta)$. We also settled an open question left by Sen [11] by proving $\lambda_{2,1}^o(\mathcal{O}) = 10$.

References

1. Aardal, K.I., Van Hoesel, S.P.M., Koster, A.M.C.A., Mannino, C., Sassano, A.: Models and solution techniques for frequency assignment problems. Ann. Oper. Res. **153**(1), 79–129 (2007)
2. Calamoneri, T.: The $L(2, 1)$-labeling problem on oriented regular grids. Comput. J. **54**(11), 1869–1875 (2011)
3. Calamoneri, T.: The $L(h, k)$-labelling problem: an updated survey and annotated bibliography. Comput. J. **54**(8), 1344–1371 (2011)
4. Calamoneri, T., Sinaimeri, B.: $L(2, 1)$-labeling of oriented planar graphs. Discrete Appl. Math. **161**(12), 1719–1725 (2013)
5. Chang, G.J., Chen, J.J., Kuo, D., Liaw, S.C.: Distance-two labelings of digraphs. Discrete Appl. Math. **155**(8), 1007–1013 (2007)
6. Gonçalves, D.: On the $L(p, 1)$-labelling of graphs. Discrete Math. **308**(8), 1405–1414 (2008)
7. Gonçalves, D., Raspaud, A., Shalu, M.A.: On oriented labelling parameters. Ser. Mach. Percept. Artif. Intell. **66**, 33–45 (2006)
8. Griggs, J.R., Yeh, R.K.: Labelling graphs with a condition at distance 2. SIAM J. Discrete Math. **5**(4), 586–595 (1992)
9. Hale, W.K.: Frequency assignment: theory and application. Proc. IEEE **68**, 1497–1516 (1980)
10. Kostochka, A.V., Sopena, É., Zhu, X.: Acyclic and oriented chromatic numbers of graphs. J. Graph Theor. **24**, 331–340 (1997)
11. Sen, S.: $L(2, 1)$-labelings of some families of oriented planar graphs. Discuss. Math. Graph Theor. **34**(1), 31–48 (2014)
12. Sopena, É.: The chromatic number of oriented graphs. J. Graph Theor. **25**, 191–205 (1997)

Radius, Diameter, Incenter, Circumcenter, Width and Minimum Enclosing Cylinder for Some Polyhedral Distance Functions

Sandip Das[1(✉)], Ayan Nandy[1], and Swami Sarvottamananda[2]

[1] Indian Statistical Institute, Kolkata, India
sandipdas@isical.ac.in, idaaning@gmail.com
[2] Ramakrishna Mission Vivekananda Educational and Research Institute,
Howrah, India
sarvottamananda@rkmvu.ac.in

Abstract. In this paper we present some efficient and a few optimal algorithms to compute the radius, diameter, incenter, circumcenter, width and k-dimensional enclosing cylinder for convex polyhedral and convex polyhedral offset distance functions in plane and in $\Re^d$. The radius, incenter and circumcenter of a convex polygon can be optimally computed in linear time, i.e. $O(n+m)$, in $\Re^2$ and in time $O(n+m+|R|)$, if the incenter or circumcenter is additionally constrained in a convex polygon R, where n is the size of input polygon and m is the size of polygon in convex polygonal or convex polygonal offset distance functions. In $\Re^d$, the radius, incenter and circumcenter of a convex polyhedron as well as of a set of convex polyhedra, unconstrained or constrained, can be computed in $O(nm)$ and $O(nm+|R|)$ time respectively, for convex constraint polyhedron R. The diameter of a convex polygon in plane can be computed in $O(n+m)$ time. The diameter of a polyhedron in $\Re^d$ can be computed in $O(nm)$ time by our methods. The width of a convex polyhedron can be computed in $O(n+m)$ time and $O(nm)$ time, for $\Re^2$ and $\Re^d$, respectively. The diameter and width of a set of convex polyhedra can be computed in $O(nm)$-time. We also show how the k-dimensional minimum enclosing cylinder can be computed in $O(n^{d-k+1}nm^2)$ in $\Re^d$ and in $O(n^d m(n+m))$ in $\Re^d$ for $k = 1$. The k-dimensional minimum enclosing cylinder for a set of convex polyhedra in $\Re^d$ can be computed in $O(n^{d-k+1}nm^2)$-time. The diameter and width problems can also be solved for a set of points in $\Re^d$, either in time $O(nm)$ or in time $O(mT(n))$, where $T(n)$ is the time complexity of the best convex hull computation algorithm for the given set of points. We also compute the minimum stabbing sphere for a set of convex polyhedra, unconstrained or constrained, for the above mentioned distance functions in $O(nm)$-time or $O(nm + |R|)$-time respectively for convex constraint polyhedron R.

Keywords: Geometric optimization
Polyhedral distance function · Radius · Diameter · Incenter
Circumcenter ·.Width · Minimum enclosing cylinder

© Springer International Publishing AG 2018
B. S. Panda and P. P. Goswami (Eds.): CALDAM 2018, LNCS 10743, pp. 283–300, 2018.
https://doi.org/10.1007/978-3-319-74180-2_24

1 Introduction

In this paper, we look at the familiar geometric problems of computing the diameter, incenter, radius (or equivalently circumcenter), width and minimum enclosing cylinders specifically for a convex polyhedra in the context of some convex polyhedral distance functions in $\Re^d$. In some cases we can generalize the problem to a set of points and to a set of convex polyhedra. The results of this paper are equally applicable to cases of other distance functions when (i) these distance functions are linear, (ii) the hyperspheres in $\Re^d$ for these distance functions are finite size polyhedron that are linear functions of distances and (iii) the lemmas, mentioned in this paper, of convex polyhedral distance functions are equally applicable to these distance functions. We can solve the related problems even if the distance functions are known to be non-metric.

The authors in a previous result [4] had showed how the circumcenter of a set of points in $\Re^d$, which is also known as 1-center, for convex polyhedral distance function, can be computed in linear time. In this paper we give an alternate algorithm for the 1-center of a set of points in $\Re^d$ for convex polyhedral distance function which not only improves the complexity in terms of m, the size of polyhedron in the distance function, but also extends the result for other type of convex polyhedral distance functions such as convex polyhedral offset distance functions.

We solve the problems of the radius, diameter, incenter, circumcenter, width, k-dimensional minimum enclosing cylinders and minimum stabbing sphere for three types of polyhedral distance functions: (i) *convex polyhedral distance function*, (ii) *real convex polyhedral offset distance function* and (iii) *normalized convex polyhedral offset distance function*. These functions are defined later. The convex polyhedral distance function was previously studied by Chew et al. in [3], and the normalized convex polyhedral offset distance function by Barequet et al. in [1]. We define real convex polyhedral offset function as the real offset from the polyhedra such that interior points are at a zero distance from origin of reference. In this paper, we effectively use the fact that the spheres in the case of convex polyhedral distance function and the two convex polyhedral offset distance functions are some convex polyhedra. This observation and that the concerned distance functions are convex leads us to some elegant formulations of the problems and some equally elegant solutions.

The distance functions mentioned in the following discussion is any one of the three distance functions mentioned above. Also, the size of input polyhedron is n and the size of convex polyhedron in distance function is m.

The *radius* of a polyhedron is defined as the radius of *circumcircle* or *circumsphere* which is the minimum radius circle or sphere that encloses or circumscribes the given polyhedron. We can define the circumcircle of a set of points and a set of polyhedra in a similar fashion. The *incircle* or *insphere* of a polyhedron is the maximum radius sphere that can be enclosed by or inscribed in the given polyhedron. We can define the incircle of a set of polyhedra as maximum radius sphere that is inscribed in all the polyhedra of the set. We solve both the problems in linear time, $O(n + m)$, in plane by suitably formulating the problems as LPP. Moreover, we can solve the problems in $O(nm)$ time in $\Re^d$ by using solutions of similar LPP (see [2,6–8]). We can also solve the constrained version of the problems in similar time complexities, where the incenter or circumcenter is constrained to lie inside a convex polyhedron. We also solve the problem of circumsphere where we need to circumscribe a set of points in $\Re^d$ and its generalization, where we compute minimum stabbing sphere that intersects a given set of convex polyhedra by the sphere's interior (which is actually a convex polyhedron for our distance functions). We also show, how we can solve the constrained version of these two problems too, where the circumcenter or center of minimum stabbing sphere lies in a constraint convex polyhedron. The minimum stabbing sphere problem can be solved in $O(nm)$ time and $O(nm+|R|)$-time for the constrained version, where the center of the stabbing sphere is constrained to lie in the polyhedron R.

The *diameter* of a set is defined as the distance between the farthest pair of points in the set for any given distance function. The computation of the diameter is usually a different problem than the computation of the radius for a non-Euclidean metric. The *width* of a polyhedron is the distance between nearest pair of parallel hyperplanes that encloses the given set for any given distance function. We compute the diameter and width of a convex polygon, by taking advantage of the polyhedral spheres, in $O(n + m)$ time in plane. We essentially traverse simultaneously the two polygons, the given polygon and the polygon of distance functions, keeping track of the distances between pair of vertices with some additional conditions discussed in the paper. We further show how we can compute the diameter and width of a convex polyhedron and a set of convex polyhedra in $\Re^d$ in $O(nm)$-time.

The concept of Euclidean cylinders can be generalized for any given distance function and for higher dimensions in $\Re^d$ such that we have axes as flats of some fixed dimension from 1 to $d - 1$. Then the cylinder is the set of points at a distance less than or equal to the fixed radius from the axis. We show how we can compute the k-dimensional minimum enclosing cylinders of a convex polyhedron and a set of convex polyhedra in $O(n^{d-k+1}nm^2)$-time if $k \neq 1$ and in $O(n^d m(n + m))$-time if $k = 1$, by the concepts that we develop to solve some problems mentioned above.

The paper is organized as follows. In Sect. 2 we present a few definitions, ideas and concepts related to the problems that we solve. All the sections deal with the three polyhedral distance functions mentioned earlier. In Sect. 3 we formulate the problem of computing the radius, the circumcenter and the incenter as LPP and give an algorithm to solve it optimally and efficiently. We also solve constrained version of the problems as well as the problem of computing the minimum stabbing sphere of a set of convex polyhedra. In Sect. 4 we give an optimal algorithm to compute the diameter of a polyhedron and of a set of convex polyhedra. In Sect. 5 we optimally compute the width of a polyhedron and of a set of convex polyhedra. In Sect. 7 we give an algorithm to compute the k-dimensional minimum enclosing cylinder. In Sect. 8 we develop the method of

traversing two convex polyhedra simultaneously and computing relevant tangent
hyperplanes used in the algorithms in the paper.

2   Ideas, Concepts and Definitions

Let P be a convex polyhedron of size n in $\Re^d$ for which we are computing radius,
diameter, incenter, circumcenter, width or k-dimensional minimum enclosing
cylinders. Let Q be a convex polyhedron of size m in $\Re^d$ used in the polyhedral
distance functions with the origin of reference in its interior. Let P_k and Q_k
denote the set of k-dimensional subfaces of P and Q respectively. For example
P_0 is set of vertices of P and P_{d-1} is set of hyperfaces of P. We use these and
distance functions defined in this section throughout the paper.

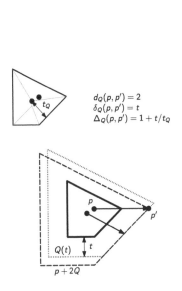

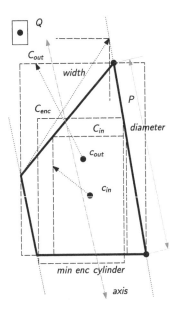

Fig. 1. Convex polyhedral distance functions $d_Q(p,p')$, $\delta_Q(p,p')$ and $\Delta_Q(p,p')$

Fig. 2. Radius, diameter, incenter c_{in}, circumcenter c_{out}, width and min 1-dimensional enclosing cylinder of polygon P for convex polygonal distance function d_Q for convex polygon Q in $\Re^2$

Let S and S' be two convex sets in $\Re^d$ (for example, points, lines, flats or
bounded convex regions) not necessarily of the same type.

The *convex polyhedral distance function* [3] from S to S', denoted by
$d_Q(S, S')$, is defined as $d_Q(S, S') = \inf_{p \in S, p' \in S'} \{t \geq 0 \mid p' \in (tQ + p)\}$, where
$tQ + p$ is the polyhedron Q translated to p and scaled by a factor t. See Fig. 1.

Let $Q(t)$ denote the *offset polyhedron* of Q offset by the Euclidean distance
t as in [3]. The *real convex polyhedral offset distance function* [3] from S to S',

denoted by $\delta(S, S')$, is defined as $\delta(S, S') = \inf_{p \in S, p' \in S'}\{t \geq 0 \mid p' \in p + Q(t)\}$, where $p + Q(t)$ is the polyhedron $Q(t)$ translated to p. See Fig. 1. The real convex polyhedral offset distance function from point p is zero in the interior of polyhedron $p + Q$.

Next we define *normalized convex polyhedral offset distance function* [3]. An additional condition for this definition is that the fixed origin of reference must be a medial center (a point on the medial axis that is farthest from the boundary of Q). The *normalized convex polyhedral offset distance function* from S to S', denoted by $\Delta(S, S')$, is defined as $\Delta(S, S') = \inf_{p \in S, p' \in S'}\{1+t/t_Q \mid \epsilon \geq 0, p' \in p + Q(t)\}$, where $p+Q(t)$ is polyhedron $Q(t)$ translated to p and t_Q is the (minimum) Euclidean distance from origin of reference to boundary of polyhedron Q. See Fig. 1. The normalized convex polyhedral offset distance of point p to itself is 0 and of any point in boundary of Q from origin is 1.

Below, we formalize the definitions of radius, diameter, incenter, circumcenter, width and k-dimensional enclosing cylinders in the context for convex polyhedral distance and convex polyhedral offset distance functions. See Fig. 2 for some examples. The distance function $d()$ in the following definitions stands for one of the distance functions $d_Q()$, $\delta_Q()$ or $\Delta_Q()$ depending on the context.

The radius, circumcenter and circumsphere are related to the minimum sphere that contains the given set of objects in $\Re^d$ for a distance metric. We use the following definition for the distance functions under consideration.

Definition 1 (Radius, Circumcenter and Circumsphere). *The circumcircle or circumsphere, C_{out}, of a convex polyhedron P is the minimum translated and scaled polygon Q in case of d_Q or translated and offset polygon Q in case of δ_Q and Δ_Q which encloses polyhedron P. The circumcenter, c_{out}, is the translated origin of C_{out}. The radius or alternatively circumradius is the radius of C_{out}. The radius of a set S of geometric objects is given by the expression:* $\text{Min}_{c \in \Re^d}\text{Max}_{s \in x, x \in S} d(c, s)$.

The incenter, incircle and insphere are related to the maximum sphere that is contained inside a continuous set in $\Re^d$. We give the definition for our special case.

Definition 2 (Incenter, Incircle and Insphere). *The incircle or insphere, C_{in}, of convex polyhedron P is the maximum translated and scaled polyhedron Q in case of d_Q or translated and offset polyhedron Q in case of δ_Q and Δ_Q which can be inscribed inside polyhedron P. The incenter, c_{in}, is the translated origin of C_{in}. The incenter of the polyhedron P is the optimal value of c in the optimization problem:* $\text{Max}_{c \in P}\text{Min}_{p \in boundary(P)}d(c, p)$. *The incircle or insphere of a set of convex polyhedra, S, is the maximum inscribed translated and scaled/offset circle/sphere drawn inside all the polyhedra in the set S.*

The diameter of a set is the distance between the farthest pair of points in the set. We define the diameter as follows.

Definition 3 (Diameter of set of objects S). *The diameter of set of geometrical objects, S, is the distance between farthest points in the set S. The farthest pair of points in polyhedron P is given by the expression:* $\text{Max}_{p, p' \in P} d(p, p')$

The width of a set is the minimum distance of all the pairs of parallel hyperplanes that bounds the set between them. The width of a set of points is defined below.

Definition 4 (Width). *The width of a set of geometrical objects, S, is defined as the minimum distance between parallel hyperplanes that enclose the set S. The width of the polyhedron P is given by the expression: $\mathrm{Min}_{h,h'}\{ d(h,h') \mid P \text{ is contained in between parallel hyperplanes } h \text{ and } h'\}$*

An *affine set* A, also called a *flat*, is a set of points such that any linear combination of a (finite) subset of points in A also belongs to A. For example, points, lines, hyperplanes are all affine sets. The dimension of an affine set A is the size of minimum set of independent points in A, called a *basis*, such that every point in A can be written as a linear combination of the basis. For example, points have affine dimension 0 and lines have affine dimension 1.

The concept of enclosing cylinders is related to the concept of width and is a generalization of concept of Euclidean cylinders in higher dimensions. An *affine cylinder* of radius r with axis f, which is a flat, is defined as set $\{p \in \Re^d \mid d(f,p) \leq r\}$. For comparison, a Euclidean cylinder (infinite, not truncated) of radius r in $\Re^3$ has axis as flat of dimension 1 and contains all the points at a distance less than or equal to r from the axis.

Definition 5 (Minimum enclosing cylinder of set S). *The k-dimensional minimum enclosing cylinder, C_{enc}, of set S is the smallest (minimum radius) cylinder that contains set S. The minimum enclosing cylinder of polyhedron P can be computed by solving the optimization problem: $\mathrm{Min}_{k\text{-flat } f}\{r \mid P \subseteq k\text{-dimensional affine cylinder of radius } r \text{ and axis } f)\}$.*

Below we mention some lemmas that allow us to compute the above mentioned entities efficiently.

Lemma 1. *Let the farthest distance for distance functions d_Q, δ_Q or Δ_Q in a given convex polyhedron P be d. Then there exists a pair of vertices of P such that they are at distance d.*

Proof. The farthest pair of points are in boundary of the convex polyhedron, say P. Because if one of them is not we can extend the line connecting points and take another point(s) which are at same or at a farther distance.

Now assume that one or both of the end points of farthest distance is on some edge or face of the convex polyhedron, and there exist no pair of vertices with the same distance. By convexity of the distance functions we can find another point on the boundary of face with a distance that is at least the farthest distance. This boundary face will have dimension one less than the dimension of the face containing the end point. By continuing we can get to a vertex with a distance that is at least the farthest distance, thus contradicting our assumption. Thus there always exists a pair of vertices in a convex polyhedron that are farthest. □

Lemma 1 ensures that there exists a pair of vertices in polyhedron P that determines the diameter. The next lemma (Lemma 2) characterizes the candidate pairs of vertices in polyhedron P for the diameter.

Lemma 2. *Let p, p' be the farthest vertices in convex polyhedron P for distance functions d_Q, δ_Q or Δ_Q. Let f be the corresponding face of relevant $p + tQ$ or $p + Q(\epsilon)$ for farthest distance from p to p'. Then tangential hyperplanes of P parallel to f touch P at p'.*

Proof. Since p and p' are farthest pair of points in convex polyhedron P, f will have to be tangent at p', otherwise p' will not be farthest.   □

For the following lemma we assume non-degeneracy. No non-point face of P is parallel to any non-point face of Q. This lemma characterizes C_{in}, C_{out} and C_{enc}.

Lemma 3. *Let C_{in} be the insphere polyhedron of polyhedron P. Let C_{out} be the circumsphere polyhedron of polyhedron P. Let C_{enc} be the minimum k-dimensional enclosing cylinder of P. Let there be no degeneracy. Then $(d + 1)$-faces of P will be tangent to C_{in}, $(d + 1)$-faces of C_{out} will be tangent to P and $(d - k + 1)$-faces of C_{enc} will be tangent to P. Moreover the sum of dimension of subfaces of P on which C_{enc} is tangent is k.*

Proof. Since there are no degeneracies, C_{in}, C_{out} and k-dimensional C_{enc} will be determined by $d + 1$, $d + 1$ and $d - k + 1$ points respectively, similar to the concepts of Euclidean spheres and cylinders.   □

Let $P(d)$ be the polyhedron consisting of points that are at a distance less than or equal to d from polyhedron P for polyhedral distance functions d_Q, δ_Q, or Δ_Q, then following lemma holds. This lemma characterizes the polyhedral space at a distance less than or equal to d from polyhedron P. This is used in computing minimum stabbing sphere.

Lemma 4. *Polyhedron $P(d)$ will have faces parallel to tangential hyperplanes on some subface $f \in P$ that are parallel to some subface $f' \in Q$.*

Proof. $P(d)$ is simply either the polyhedron $P + dQ$ or the polyhedron $P + Q(d)$ depending on the distance function. The faces will be parallel to faces of Q tangential on faces of P.   □

We denote the tangential hyperplane passing through f that is parallel to subfaces $f \in P$ and $f' \in Q$ of lemma 4, which is a flat, as $f \times f'$.

Let there be no degeneracies for the lemmas below. Consider the vertices of polyhedra Q and the $(d-1)$-faces of polyhedron P. Let us mark the $(d-1)$-faces of P with the vertices of Q, say the vertex p, such that the tangent hyperplanes parallel to the $(d - 1)$-face touches p and Q is on the same side. Let us call the resulting subdivision of subfaces of P as an f-*division* of P with Q and vice versa (f for front). We get a similar subdivision for tangent hyperplanes such that P

and Q are on the opposite side. We call this subdivision as the *b-division* of P with Q. The f-division and b-division of polyhedron P with Q is homeomorphic to a polyhedron. We have the following lemma for the k-dimensional face of these subdivisions.

Lemma 5. *Every k-face of the subdivisions, f-division and b-division of convex polyhedron P with convex polyhedron Q, is homeomorphic (can be continuously mapped) to a k-dimensional sphere.*

Proof. If two subfaces f and f' of P are part of same k-face of a subdivision, corresponding to subface q of Q then (i) we can continuously tilt the tangent hyperplane from f to f' suitably so that we get a sequence of k-faces all part of the same subdivision, and (ii) any two paths on a k-face of the subdivision can be continuously mapped from one to another so that every points remains inside that k-face of the subdivision. Point (i) ensures that we have a $k-1$ dimensional object at all times and point (ii) ensures homeomorphy.   □

By Lemma 4, Polyhedron $P(d)$ will have faces parallel to $f \times f'$ where $f \in P$, $f' \in Q$, subface f is in the f-division of P with respect to Q corresponding to subface f'.

Below, Lemma 6 characterizes the subdivisions in $f-division$ and $b-division$ of polyhedron P with Q. Lemma 7 characterizes the parallel hyperplanes for the width problem of polyhedron P.

Lemma 6. *If a k-face S of the subdivision in P is paired with a k'-face S' of the subdivision in Q, for either f-division or b-division, then for every k-face f in S and k'-face in S', there will be a tangent hyperplane $f \times f'$ on P and $f' \times f$ on Q.*

Lemma 7. *Let h and h' be the hyperplanes corresponding to the width of a polyhedron P. Let h and h' will be tangent on faces f and f' of P. If there is no degeneracy then sum of the dimensions of the subfaces f and f' on which h and h' are tangents is $d-1$. Moreover, either h or h' is parallel to either $f \times f_Q$ or $f' \times f_Q$ respectively where f_Q is the corresponding k-dimensional subface of either f or f' respectively in a k-face of the f-division of P and Q respectively.*

Proof. If the sum of dimensions of the subfaces f and f' is less than $d-1$ then there is a degree of freedom to rotate h and h' which will reduce the width. Since width is by definition minimum distance between h and h', it cannot be reduced further. Hence the sum of dimensions of the subfaces f and f' will be $d-1$. Furthermore, if the distance between h and h' is t then either h' is a face of $h+tQ$ or $h+Q(t)$, depending on the distance function, or h is a face of $h'+tQ$ or $h'+Q(t)$, depending on the distance function.   □

3   Radius, Incenter and Circumcenter of a Polyhedron

In the following sections the distance function $d()$ stands for one of the distance functions $d_Q()$, $\delta_Q()$ or $\Delta_Q()$ depending on the context, i.e., the expression is common for all three types of distance functions under consideration unless specified. We also assume non-degeneracy.

3.1   Circumcenter

In this section we compute circumcenter of a convex polyhedron P. We can write an LPP for radius and circumcenter as follows. Let us denote a face, f_j, of polyhedron Q as hyperplane $\hat{\boldsymbol{h}}_j \cdot \boldsymbol{x} = d_j$ ($\hat{\boldsymbol{h}}_j$ is the normal to hyperplane and d_j is the distance from origin) for each $f_j \in Q_{d-1}$. The interior of the polyhedron will be given by $\hat{\boldsymbol{h}}_j \cdot x \leq d_j$, $\forall f_j \in Q_{d-1}$. Q_{d-1} is the set of $(d-1)$-faces, i.e. hyperfaces, of Q, as defined earlier. Let the vertex v_i of polyhedron P be denoted by vector $\boldsymbol{p}_i$. We can write an LPP for distance $d_Q(\boldsymbol{c}, \boldsymbol{p})$ as follows.

$$\text{Min} \quad t$$
$$\text{subject to}$$
$$\hat{\boldsymbol{h}}_j \cdot (\boldsymbol{p} - \boldsymbol{c}) \leq t d_j \qquad \forall f_j \in Q_{d-1}$$
$$\text{where} \quad t \geq 0$$

The term $\hat{\boldsymbol{h}}_j \cdot (\boldsymbol{p} - \boldsymbol{c}) \leq t$ represents the relevant face of polyhedron $tQ + \boldsymbol{c}$. δ_Q and Δ_Q will have similar formulations. By the definition of radius, we can write an LPP for radius of the polyhedron P for polyhedral distance functions d_Q, δ_Q and Δ_Q as follows (P_0 is the set of 0-faces, i.e. vertices, of P as defined earlier):

$$\text{Min} \quad r$$
$$\text{subject to}$$
$$d(\boldsymbol{c}, \boldsymbol{p}_i) \leq r \qquad \forall v_i \in P_0$$
$$\text{where} \quad \boldsymbol{c} \in \Re^d, r \geq 0$$

We can rewrite above by substituting the LPP for distances as

$$\text{Min} \quad r$$
$$\text{subject to}$$
$$\hat{\boldsymbol{h}}_j \cdot (\boldsymbol{p}_i - \boldsymbol{c}) \leq r d_j \qquad \forall v_i \in P_0, \forall f_j \in Q_{d-1}$$
$$\text{where} \quad \boldsymbol{c} \in \Re^d, r \geq 0$$

The above LPP is of $d+1$ variables and $O(nm)$ constraints. This LPP can be further rewritten as

$$\text{Min} \quad r$$
$$\text{subject to}$$
$$\hat{\boldsymbol{h}}_j \cdot \boldsymbol{c} + r d_j \geq \text{Max}_{\forall v_i \in P_0} \{\boldsymbol{h}_j \cdot \boldsymbol{p}_i\} \qquad \forall f_j \in Q_{d-1}$$
$$\text{where} \quad \boldsymbol{c} \in \Re^d, r \geq 0$$

The constraints in above LPP essentially need computation of all the tangent hyperplanes parallel to faces of Q on vertices of P. We show in the Sect. 8 how we can efficiently compute such tangents in linear time in plane and $O(nm)$ in $\Re^d$. Once we compute all the tangents we only have m constraints in the above LPP which can then be solved using any efficient linear time algorithm for fixed dimensions [2, 6–8]. This significantly improves the complexity of the algorithm from $O(nm)$ to $O(n + m)$ in plane. The radius is the optimal solution of the LPP given above. The circumcenter is the corresponding value of $\boldsymbol{c}$ in the optimal solution.

If we have a set of convex polyhedra, S, instead of a single polyhedron we can use the last but one LPP for each polyhedron in the set S and then solve the LPP. Thus we have the following theorem.

Theorem 1. *The circumcenter of a convex polyhedron P for polyhedral distance functions d_Q, δ_Q and Δ_Q can be calculated in $O(|P| + |Q|)$-time in $\Re^2$ and $O(|P| \cdot |Q|)$ in $\Re^d$. The circumcenter of a set of convex polyhedra, S, for polyhedral distance functions d_Q, δ_Q and Δ_Q can be calculated in $O(|S| \cdot |Q|)$ in $\Re^d$.*

3.2 Incenter

In this section we compute incenter of a convex polyhedron P. We assume that the polyhedra P and Q are non-degenerate. We use following lemma for computing incenter.

Lemma 8. *If all the vertices of a convex polyhedron P' are inside convex polyhedra P then $P' \subseteq P$.*

Proof. Every point of convex polyhedra can be written as a convex combination of vertices. Since all vertices of P' are inside P, therefore any convex combination of the vertices is also inside P (due to convexity of P). □

Let $\hat{\boldsymbol{H}}_i \cdot \boldsymbol{x} = D_i$ ($\hat{\boldsymbol{H}}_i$ is normal to the hyperplane passing through the face and D_i is the distance from origin) represent the $(d-1)$-face f_i of P, that is, $f_i \in P_{d-1}$, such that P lies in $\hat{\boldsymbol{H}}_i \cdot \boldsymbol{x} \le D_i$. By Lemma 8 we only need to check if each vertex of translated and scaled polyhedron $\boldsymbol{c} + rQ$ or translated and offset polyhedron $c + Q(r)$ is in P. For incenter of convex polyhedral distance function d_Q we will solve the following LPP. For distance functions δ_Q and Δ_Q we can do similar simplifications.

$$\begin{aligned}
&\text{Max} \quad r \\
&\text{subject to} \\
&\hat{\boldsymbol{H}}_i \cdot (\boldsymbol{c} + r\boldsymbol{q}_j) \le D_i \qquad \forall f_i \in P_{d-1}, \forall v_j \in Q_0 \\
&\text{where} \quad c \in \Re^d, r \ge 0
\end{aligned}$$

The LPP given above can be solved in $O(nm)$ time. We can solve the incenter problem more efficiently if we apply our improvements discussed in the Sect. 3.1. The LPP above can be rewritten as

$$\begin{aligned}
&\text{Max} \quad r \\
&\text{subject to} \\
&D_i - \hat{\boldsymbol{H}}_i \cdot (\boldsymbol{c}) \le r \cdot \text{Max}_{\forall v_j \in Q_0}\{\hat{\boldsymbol{H}}_i \cdot \boldsymbol{q}_j\} \qquad \forall f_i \in P_{d-1} \\
&\text{where} \quad c \in \Re^d, r \ge 0
\end{aligned}$$

In the LPP above the term $\text{Max}_{\forall v_j \in Q_0}\{\hat{\boldsymbol{H}}_i \cdot \boldsymbol{q}_j\}$ for each $f_i \in P_{d-1}$ denotes the tangent hyperplane parallel to face $f_i \in P_{d-1}$ that touches the polyhedron Q. We compute all the tangent hyperplanes parallel to $(d-1)$-faces of P on Q and replace the term with the relevant computed value. Then we can solve the incenter problem in $O(n + m)$ time in plane and in $O(nm)$ time in $\Re^d$.

If we are given a set of convex polyhedra S, we can use the last but one LPP to compute the maximum sphere that is inscribed inside all the convex polyhedra in S. This leads us to the following theorem.

Theorem 2. *The incenter of a convex polyhedron P for polyhedral distance functions d_Q, δ_Q and Δ_Q can be calculated in $O(|P| + |Q|)$-time in plane and in $O(|P| \cdot |Q|)$ time in $\Re^d$. The incenter of a set of convex polyhedra, S, for polyhedral distance functions d_Q, δ_Q and Δ_Q can be calculated in $O(|S| \cdot |Q|)$ in $\Re^d$.*

3.3   Constrained Radius, Incenter and Circumcenter

In this section we solve the constrained versions of radius, incenter and circumcenter problems. Suppose we are given an additional constraint that center of incenter or circumcenter must lie inside a convex polyhedron and we need to compute the circumcenter. We can solve this problem using the same LPP formulation as in above section (see Sect. 3) along with additional linear constraints. The case for circumcenter is given below.

Let R be the constraint polyhedron. Let the faces of the polyhedron R be given as $\hat{h}_k^r \cdot \boldsymbol{x} \leq d_k^r$ for faces $f_k^r \in R_{d-1}$. For convex polyhedral distance function d_Q we modify the LPP as follows. For distance functions δ_Q and Δ_Q similar steps can be taken.

$$
\begin{aligned}
\text{Min} \quad & r \\
\text{subject to} \\
\hat{h}_j \cdot \boldsymbol{c} + r d_j \geq & \text{Max}_{\forall v_i \in P_0}(\boldsymbol{h}_j \cdot \boldsymbol{p}_i) \qquad \forall f_j \in Q_{d-1} \\
\hat{h}_k^r \cdot \boldsymbol{c} \leq & d_k^r \qquad\qquad\qquad\quad \forall f_k^r \in R_{d-1} \\
\text{where} \quad & \boldsymbol{c} \in \Re^d, r \geq 0
\end{aligned}
$$

Since there are $O(m + |R|)$ constraints in the LPP above we can solve the LPP in $O(nm + m + |R|)$ time where $O(nm)$-time is needed to compute the values of $\text{Max}_{\forall v_i \in P_0}(\boldsymbol{h}_j \cdot \boldsymbol{p}_i), \forall f_j \in Q_{d-1}$.

We can also apply our improvements and solve the above problem in time $O(n + m + |R|)$ in plane as follows. We observe that if incenter or circumcenter does not lie inside the constraint polyhedron R then the constraint incenter or circumcenter will lie on the boundary of constraint polyhedron R. The authors proved this for the *Euclidean 1-center* in an earlier result. See [5]. The concepts in [5] are applicable here too. Following earlier sections, we calculate all the tangent hyperplanes parallel to $(d-1)$-faces of Q on P. This reduces the number of constraints in the LPP to $O(nm+|R|)$ to only $O(n+|R|)$. Then we apply any linear time solution for LPP in fixed dimension to solve the LPP [2,6–8]. This reduces the time complexity of computing constrained incenter or circumcenter to $O(n + m + |R|)$ time in plane which is optimal. The worst case complexity in $\Re^d$ will be $O(nm + |R|)$. For a set of convex polyhedra we get the same results.

Theorem 3. *The constrained radius, circumcenter and incenter of a convex polyhedron P for polyhedral distance functions d_Q, δ_Q and Δ_Q can be calculated in $O(|P| + |Q| + |R|)$-time in $\Re^2$ and in $O(|P| \cdot |Q| + |R|)$-time in $\Re^d$, where R is the constraint convex polyhedron. The constrained radius, circumcenter and incenter of a set of convex polyhedra S for polyhedral distance functions d_Q, δ_Q and Δ_Q can be calculated in $O(|S| \cdot |Q| + |R|)$-time in $\Re^d$, where R is the constraint convex polyhedron.*

3.4   Minimum Stabbing Sphere of a Set of Convex Polyhedra

We again assume non-degeneracy of P and Q. We compute the minimum stabbing sphere of a set of convex polyhedra in this section. Let S be a set of convex polyhedra $P^i \in S$, $i \leq N$. We consider the problem of computing minimum stabbing sphere of this set for polyhedral distance functions d_Q, δ_Q and Δ_Q. The distance from a point c to a convex set C is given by the expression $d(c, C) = \text{Min}_{p \in C} d(c, p)$. Let the faces of polyhedron P^i be denoted by hyperplanes $\hat{\boldsymbol{h}}^i_j \cdot \boldsymbol{x} = d^i_j$, $\forall f_j \in P^i_{d-1}$. Again $\hat{\boldsymbol{h}}^i_j$ is the normal to hyperplane and d^i_j is the distance from origin. P^i should be on the side $\hat{\boldsymbol{h}}^i_j \cdot \boldsymbol{x} \leq d^i_j$. We can introduce this expression in the LPP in Sect. 3.1 as follows.

$$
\begin{aligned}
&\text{Min} \quad r \\
&\text{subject to} \\
&\quad d(\boldsymbol{c}, \boldsymbol{p}_i) \leq r &&\forall P^i \in S \\
&\quad \hat{\boldsymbol{h}}^i_j \cdot \boldsymbol{p}_i \leq d^i_j &&\forall f_j \in P^i_{d-1} \\
&\quad \text{where} \quad \boldsymbol{c} \in \Re^d, \boldsymbol{p}_i \in \Re^d, 1 \leq i \leq N, r \geq 0
\end{aligned}
$$

Unfortunately, the number of variables in the LPP above is $(N + 1)d + 1$. We cannot solve this LPP in deterministic linear time yet. So we rewrite the distance expression in a way that is amenable to efficiently solving of LPP. For the convex polyhedral distance function d_Q, we compute all the tangent hyperplanes corresponding to faces of P^i and Q (the f-divisions) for all $P^i \in S$ separately as in Sect. 8. The distance function as above is then $d(c, P^i) = \text{Min}_{f \in P^i}\{d(c, f)\}$ for any convex polyhedron P^i such that $f \in P^i$ are subfaces of P^i of any dimension. The distance in the interior of the polyhedron is 0, we include the interior of P_c as a d-face of P_c.

We compute the minimum stabbing sphere for distance function d_Q. For distance functions δ_Q and Δ_Q the computations are similar. We observe that for distance function d_Q, the minimum distance from c to a polyhedron P^i is the distance from c to some tangent hyperplane which is parallel to a face of P^i and a face of Q. So we can rewrite the expression for $d(c, P^i)$ as $d(c, P^i) = \text{Max}_{f \in P^i, f' \in Q, f \times f' \text{ is tangent to} P^i}\{\hat{\boldsymbol{h}}_{f,f'} \cdot (\boldsymbol{p}_f - \boldsymbol{c})\}$ where $\boldsymbol{p}_f$ is any point in the flat corresponding to subface f of P^i and $\hat{\boldsymbol{h}}_{f,f'}$ is the normal perpendicular to tangent hyperplane corresponding to face f of P^i and f' of Q.

Now we can rewrite the LPP as

$$\text{Min} \quad r$$
$$\text{subject to}$$
$$\text{Max}_{f \in P^i, f' \in Q, f \times f' \text{is tangent to} P^i} \{ \boldsymbol{h}_{f,f'} \cdot (\boldsymbol{p}_f - \boldsymbol{c}) \} \leq r \qquad \forall P^i \in S$$
$$\text{where } \boldsymbol{c} \in \Re^d, r \geq 0$$

This computation can be done more efficiently if we rewrite it as

$$\text{Min} \quad r$$
$$\text{subject to}$$
$$\boldsymbol{h}_{f,f'} \cdot \boldsymbol{p}_f \leq \boldsymbol{h}_{f,f'} \cdot \boldsymbol{c} + r \qquad \forall f' \in Q, \forall P^i \in S, \forall f \in P^i,$$
$$\text{s.t.} f \times f' \text{ is tangent to } P^i$$
$$\text{where} \quad \boldsymbol{c} \in \Re^d, r \geq 0$$

In Sect. 8 we show how we can calculate all the tangents in linear time of the size of any convex polyhedron P and convex polyhedron Q. Suppose the N polyhedra have size $s_1, s_2, \ldots, s_N$ respectively, then the time to compute all required tangent hyperplanes would be $\sum |P^i|m$. The number of constraints in the above LPP is $\sum |P^i|m$. Hence the time complexity to compute minimum stabbing sphere of a set of convex polyhedra in $O(nm)$, where $n = \sum |P^i|$, the size of input. We can also solve the constrained version where the center of minimum stabbing sphere lies inside a convex polyhedron R using the similar method as earlier by adding suitable constraints in time $O(nm + |R|)$-time.

Theorem 4. *The minimum stabbing sphere of a set S of convex polyhedra in $\Re^d$ for polyhedral distance functions d_Q, δ_Q and Δ_Q can be calculated in $O(|S| \cdot |Q|)$-time, and in $O(|S| \cdot |Q| + |R|)$-time for constrained version, where R is the constraint convex polyhedron.*

4   Diameter of a Polyhedron

We compute diameter of a convex polyhedron P in this section. The algorithm to compute diameter for all three distance functions is similar. Let us assume that there are no degeneracies, that is, no non-point subface of polyhedron P is parallel to any non-point subface of polyhedron Q. We show the computation for convex polyhedral distance function d_Q. For δ_Q and Δ_Q the computation is similar.

Initially, for every $d - 1$ dimensional face f in Q, we compute the tangent hyperplane on the both sides of P. Let these tangent hyperplanes touch P at p' such that P is on the same side as Q, and p such that P is on the other side. We check if p lies inside the relevant face corresponding to f in $p + tQ$ for d_Q and $p + Q(\epsilon)$ for δ_Q and Δ_Q (see Lemma 1). If it lies inside then it is a potential farthest distance pair, if not then we drop this pair. The diameter is the farthest pair of points among all the faces f. To find out all the tangents parallel to faces of Q on P we do a simultaneous traversal of both P and Q. The simultaneous traversal of P and Q is detailed in Sect. 8. If we are given a set of convex polyhedra then we have to do the computation for each convex polyhedron in S, but the algorithm essentially remains same.

Theorem 5. *The diameter of a convex polyhedron P for polyhedral distance functions d_Q, δ_Q and Δ_Q can be calculated in $O(|P| + |Q|)$-time in plane and in $O(|P| \cdot |Q|)$-time in $\Re^d$. The diameter of a set of convex polyhedra S for polyhedral distance functions d_Q, δ_Q and Δ_Q can be calculated in $O(|S| \cdot |Q|)$-time in $\Re^d$.*

5   Width of a Polyhedron

We present an algorithm to compute width of a convex polyhedron P for convex polyhedral distance function d_Q in this section. Again we assume non-degeneracy and only show the computation for distance function d_Q. The computation for distance functions δ_Q and Δ_Q are similar. The algorithm to compute width is similar to diameter except we do not apply the check for p lying inside the relevant face of $p + tQ$. We do the computation as follows.

Initially, we compute all the tangent hyperplanes of P parallel to faces of Q as detailed in Sect. 8. For every face of Q we shall have two tangents on opposite sides of P. We can compute the tangent hyperplane corresponding to f-division as in Sect. 8. However, we also simultaneously pivot at the opposite side of polyhedron Q. Since we do not allow degeneracies, on the opposite side the tangent hyperplane will only touch a vertex. While traversing we simultaneously calculate the relevant distances between the tangent hyperplanes. Since the distance functions are not symmetric, we take the smaller distance of the two. This can be done in $O(n + m)$ time in plane and in $O(nm)$ time in $\Re^d$. For a set of convex polyhedra S we can use the same technique as discussed in this section.

Theorem 6. *The width of a convex polyhedron P for polyhedral distance functions d_Q, δ_Q and Δ_Q can be calculated in $O(|P| + |Q|)$-time in plane and in $O(|P| \cdot |Q|)$-time in $\Re^d$. The width of a set of convex polyhedra S for polyhedral distance functions d_Q, δ_Q and Δ_Q can be calculated in $O(|S| \cdot |Q|)$-time in $\Re^d$.*

6   Diameter and Width of a Set of Points

The diameter and width of a set of points, S, in $\Re^d$ for distance functions d_Q, δ_Q and Δ_Q in this section. There are two ways to compute diameter and width of the set of points S. First method is by first computing the convex hull and then computing the diameter or width by the algorithms in Sects. 4 and 5. Second method is by noticing that the diameter is the farthest pair of points and the width is farthest pair of points from the center of the relevant sphere. Since we do not have a precomputed convex hull of the set of points we need to calculate the farthest pair that touch halfspaces parallel to faces of Q containing all the points of S. This we can do taking faces of Q one at a time. In the plane first method may be more efficient whereas in the higher dimensions second method may be more efficient. Thus we have the following theorem.

Theorem 7. *The diameter and width of a set of points, S, for polyhedral distance functions d_Q, δ_Q and Δ_Q can be calculated in $O(|Q|+T(|S|))$-time in plane and in $O(T(|S|) \cdot |Q|)$-time in $\Re^d$, where $T()$ is the complexity of best algorithm to compute convex hulls in $\Re^d$.*

The diameter and width of a set of points, S, for polyhedral distance functions d_Q, δ_Q and Δ_Q can also be calculated in $O(|S| \cdot |Q|)$-time in $\Re^d$.

7 Minimum Enclosing Cylinder of Dimension k in $\Re^d$

In this section we give a method of computing minimum k-dimensional enclosing cylinder. First we need to characterize $f+tQ$ or $f+Q(t)$ polyhedron for distance functions. We need a stronger non-degeneracy condition which can be ensured by perturbation techniques. We assume that after translation and scaling or offsetting no three points are collinear, no four points are coplanar, etc. Then the minimum enclosing cylinder, C_{enc}, will touch P at $d-k+1$ subfaces and sum of the dimensions of the subfaces of P on which faces of C_{enc} are tangent will be k (see Lemma 3) (Fig. 3).

We choose $d-k+1$ subfaces of P such that the sum of dimensions of the subfaces is k. Then we choose a subface of Q among the related subdivision in f-division of P and Q (related to the chosen subfaces of P, since the cylinder will be tangent at these subfaces). The subfaces of P whose dimensions sum to k determine the axis. We can project the polyhedra P and Q in the direction of axis and compute the circumsphere and the radius by previous algorithm in the projected space. If we inverse project the projected circumsphere we get a k-dimensional enclosing cylinder. We choose the minimum radius enclosing cylinder among all k-dimensional enclosing cylinder. This will be the required k-dimensional minimum enclosing cylinder. The complexity of this algorithm is $O(n^{d-k+1}m) \cdot O(nm)$, i.e. $O(n^{d-k+1}nm^2)$ if $k \neq 1$ and $O(n^d m(n+m))$ if $k = 1$. $\binom{n}{d-k+1}$ is the complexity of choosing $d-k+1$ subfaces of P of total dimension k, m is the maximum number of subfaces of Q, and the complexity of computing circumsphere in projected space is $O(nm)$, if $k \neq 1$, and $O(n+m)$, if $k = 1$. If we are given a set S of convex polyhedra then the complexity of the algorithm remains same, because the time complexity of the algorithm is dominated by choosing of $d-k+1$ faces (Fig. 4).

Theorem 8. *The k-dimensional minimum enclosing cylinder in $\Re^d$ of a convex polyhedron P for polyhedral distance functions d_Q, δ_Q and Δ_Q can be calculated in $O(n^{d-k+1}nm^2)$-time if $k \neq 1$ and in $O(n^d m(n+m))$-time if $k = 1$, where $n = |P|$, and $m = |Q|$. The k-dimensional minimum enclosing cylinder in $\Re^d$ of a set S of convex polyhedra for polyhedral distance functions d_Q, δ_Q and Δ_Q can be calculated in $O(|S|^{d-k+1}nm^2)$-time.*

8 Computing Tangent Hyperplanes Parallel to Faces of P and Q

In this section we show how we can compute tangent hyperplanes such that if subface $f \in P$ or subface $f' \in Q$ is given then we can efficiently determine tangent hyperplanes parallel to f on Q or to f' on P respectively. We are given two convex polyhedra P and Q of size n and m respectively. We assume that the polyhedra are represented by their incidence graphs with required information (specially the flats and later in the algorithm pointers to f-divisions and b-divisions) on each subface. We call a subface of dimension k of a polyhedron as a k-face of the polyhedron. We assume in this section that there are no degeneracies. The degeneracies in our case are when some non-vertex subface of P is parallel to another non-vertex subface of Q, not necessarily of same dimension. We say two subfaces, a k-face of P and k'-face of Q, are parallel if the affine combination of the two subfaces is of dimension less than $k + k'$. Vertices are always parallel to any other subface since they are of dimension 0.

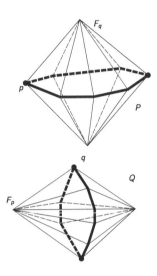

Fig. 3. A tentative subdivision of poly-hedra P

Fig. 4. A bad example for subdivisions

Note that tangent hyperplanes on P parallel to $(d-1)$-faces of Q are unique on either side of P. We, for the time being, consider the tangent hyperplanes such that P and Q are on the same side. The method remains same for the tangent hyperplanes for which P and Q are on the opposite sides. Let us consider tangent hyperplanes parallel to $(d-1)$-faces of Q on that are tangent on the same vertex of P. We mark such $(d-1)$-faces of Q to belong to same subdivision. We do the same for $d-1$-faces of P. Then we mark the boundaries of the subdivisions of dimensions $(d-2)$, $(d-1),\ldots,1$ in Q which will have tangent hyperplanes parallel to same subfaces of dimensions $1, 2,\ldots,(d-2)$ in P respectively. Then the polyhedra P and Q will be subdivided. The subdivisions will be a geometric structure of dimension d topologically equivalent to (homeomorphic to) a d-dimensional sphere. Our task is to compute this subdivision as efficiently as possible. The algorithm to subdivide polyhedra P and Q is as follows:

First we search for the hierarchy of subfaces of dimension $0, 1, \ldots, (d-1)$ in P and hierarchy of subfaces of dimension $(d-1), (d-2), \ldots, 0$ in Q respectively such that each face lies on the boundary of the subdivision of the same dimension and the each $k+1$-face is superface of k-face in the hierarchy. We can do this by starting with an arbitrary $(d-1)$-face of Q, finding the point of contact p on P of the tangent hyperplane parallel to the face, and then rotating the tangent hyperplanes and marking the faces, so that faces are not traversed twice, till we reach a $(d-1)$-face of Q which does not have tangent on p. Now we have got a $(d-2)$-face of Q on the $(d-2)$ dimensional boundary of the subdivision in Q. In P we get a 1-face on the 1-boundary of the subdivision of P. Let the line of contact of the corresponding tangent hyperplane be the 1-face l of P. We can traverse along the $(d-2)$ dimensional boundary on Q till we get a $(d-2)$-face that has the tangent hyperplane with the different line of contact. This will fetch us a $(d-3)$-face of Q on the $(d-3)$ dimensional boundary of the subdivision on Q. In P we will get a 2-face of P on the 2 dimensional boundary of the subdivision on P. We repeat till we get the whole hierarchy on each polyhedra P and Q. After we have initialized the sequence of faces of P and Q each on the boundary of subdivision of same division, we proceed to completely subdivide all of the polyhedra P and Q. The essential idea to compute all the tangents efficiently is to simultaneously compute the f-subdivision on P and its dual on Q. We individually complete k dimensional boundary of the subdivision of each dimension before taking up another k dimensional boundary. We can traverse the polyhedron using any kind of traversal that traverses the whole polyhedron. Thus, we can compute f-divisions and b-divisions of the polyhedra P and Q, with respect to each other, in $O(|P| + |Q|)$-time in $\Re^2$, and in $O(|P| \cdot |Q|)$-time in $\Re^d$.

References

1. Barequet, G., Dickerson, M.T., Goodrich, M.T.: Voronoi diagrams for polygon-offset distance functions. In: Dehne, F., Rau-Chaplin, A., Sack, J.-R., Tamassia, R. (eds.) WADS 1997. LNCS, vol. 1272, pp. 200–209. Springer, Heidelberg (1997). https://doi.org/10.1007/3-540-63307-3_60
2. Chazelle, B., Matousek, J.: On linear-time deterministic algorithms for optimization problems in fixed dimension. J. Algorithms **21**(3), 579–597 (1996)
3. Chew, L.P., Drysdale III., R.L.S.: Voronoi diagrams based on convex distance functions. In: O'Rourke, J. (ed.) Proceedings of the First Annual Symposium on Computational Geometry, pp. 235–244. ACM, June 1985
4. Das, S., Nandy, A., Sarvottamananda, S.: Linear time algorithm for 1-center in $\Re^d$ under convex polyhedral distance function. In: Zhu, D., Bereg, S. (eds.) FAW 2016. LNCS, pp. 41–52. Springer, Cham (2016). https://doi.org/10.1007/978-3-319-39817-4_5
5. Das, S., Nandy, A., Sarvottamananda, S.: Linear time algorithms for euclidean 1-center in $\Re^d$ with non-linear convex constraints. In: Govindarajan, S., Maheshwari, A. (eds.) CALDAM 2016. LNCS, vol. 9602, pp. 126–138. Springer, Cham (2016). https://doi.org/10.1007/978-3-319-29221-2_11

6. Dyer, M.E.: On a multidimensional search technique and its application to the Euclidean one-centre problem. SIAM J. Comput. **15**(3), 725–738 (1986)
7. Megiddo, N.: Linear-time algorithms for linear programming in $\Re^3$ and related problems. SIAM J. Comput. **12**(4), 759–776 (1983)
8. Megiddo, N.: Linear programming in linear time when the dimension is fixed. J. ACM **31**(1), 114–127 (1984)

Author Index

Printed in the United States
By Bookmasters